Materials Science and Technology of Optical Fabrication

Materials Science and Technology of Optical Fabrication

Tayyab I. Suratwala
Lawrence Livermore National Lab
7000 East Avenue
Livermore
CA, US, 94550

WILEY

The American Ceramic Society

This edition first published 2018

Registered Office
John Wiley & Sons, Inc., 111 River Street, Hoboken, NJ 07030, USA

Editorial Office
111 River Street, Hoboken, NJ 07030, USA

For details of our global editorial offices, customer services, and more information about Wiley products visit us at www.wiley.com.

Wiley also publishes its books in a variety of electronic formats and by print-on-demand. Some content that appears in standard print versions of this book may not be available in other formats.

Library of Congress Cataloging-in-Publication Data

Names: Suratwala, Tayyab, 1970- author.
Title: Materials science and technology of optical fabrication / by Tayyab I. Suratwala.
Description: First edition. | Hoboken, NJ : John Wiley & Sons, 2018. |
Includes bibliographical references and index. |
Identifiers: LCCN 2018009386 (print) | LCCN 2018021472 (ebook) | ISBN 9781119423744 (pdf) | ISBN 9781119423782 (epub) | ISBN 9781119423683 (cloth)
Subjects: LCSH: Optical materials–Design and construction. | Optical materials–Testing.
Classification: LCC TS514 (ebook) | LCC TS514 .S87 2018 (print) | DDC 620.1/1295–dc23
LC record available at https://lccn.loc.gov/2018009386

Cover Design: Wiley
Cover Description: (Foreground) Schematic illustration of some of the mechanical interactions between the workpiece, slurry particles, and lap, specifically describing the Ensemble Hertzian Multi-Gap (EHMG) model used to understand and predict surface roughness and material removal rate. (Background) Microscope image of a fixed abrasive-ground fused silica glass workpiece after removal of surface layer and a short chemical etch (1 μm) revealing the morphology of surface cracks (i.e. subsurface damage).

Set in 10/12pt WarnockPro by SPi Global, Chennai, India

10 9 8 7 6 5 4 3 2 1

Contents

Preface

The objective of this book is to re-examine the age-old field of optical fabrication from a materials science perspective. Optical fabrication is the manufacture of optical elements such as passive optics – for example, lenses, transmission flats, mirrors, and prisms – and active optics – for example, laser gain media, frequency converters, polarizers, and adaptive optics. These are crafted in a variety of shapes, sizes, and materials. The improved manufacturing of next-generation optics has been instrumental in boosting the performance of advanced lasers and imaging and spectroscopy systems. The interdisciplinary field of materials science and engineering explores how fabrication processes influence material structures and chemistry, and hence their properties and performance.

Optical fabrication dates back to Roman times, with the first use of metal convex mirrors and spherical glass optics [1]. Owing to the vast possibilities in workpiece materials, shapes, polishing materials, techniques, and process variables, the number of feasible fabrication recipes is essentially infinite. Historically, this field has been primarily an art in which a master optician developed a unique set of effective processing techniques through years of apprenticeship and trial and error. While there have been significant advances in the scientific understanding of optical fabrication over the past decades, this field has yet to reach the maturity required for full control, via a truly deterministic process, of the surface characteristics of workpieces made of various optical materials.

A wealth of optical fabrication books provide excellent guidance and references as to materials, grinding and polishing, processing methods, tools, optical-metrology techniques, and optical design [1–19]. Despite the seminal work of researchers in the past 40 years; however, notably Brown in the 1970s [10], Izumitani in the 1980s [16], Jacobs, Lambropoulos, and Cook in the 1990s [20–45], and Dornfeld and coworkers in the 2000s [46–50], a recent comprehensive review of optical fabrication has been lacking. The time is ripe for a single-source reference on this topic.

In the related field of chemical–mechanical planarization (CMP), which concerns the fabricating of integrated circuits, many reference works are available – some of which cover the materials science aspects of the optical-fabrication process [13–15, 17, 51–53]. However, because CMP has important differences from optical fabrication with respect to materials, amount of material removed, and specifications, the topic warrants a separate treatment.

Optical-fabrication processes, all of which involve grinding and polishing, share a basic set of interactions or phenomena among their three component parts, which are

- the workpiece – that is, the optic to be manufactured
- the lap or tool, which determines the time- and spatially dependent mechanical loading of the workpiece
- the slurry or lubricant, which may contain particles to remove material from the workpiece

These fabrication components interact as to their mechanics, tribology, physics, and chemistry to affect the four characteristics of optical fabrication (*surface figure, surface quality, surface roughness,* and *material removal rate*). We present these fundamental interactions in an organized, systematic, quantitative way, with illustrations of proposed models using experimental data. We also discuss the research and development in these basic elements that has led to novel characterization methods, fabrication techniques, and processing recipes not previously collected.

This book contains two parts. Part I: Fundamental Interactions – Materials Science begins with an introduction in Chapter 1. Chapters 2–5 discuss the mechanical and chemical phenomena in play during optical fabrication and their quantitative effects on *surface figure, surface quality, surface roughness,* and *material removal rate*. We begin with macroscopic interactions, such as those affecting *surface figure,* and move to the microscopic and molecular interactions that affect *surface quality, surface roughness,* and *material removal rate.* Numerous mechanisms and models have been proposed for the fundamental interactions, many of which are covered in this book. However, the aim of the book is to illustrate these concepts and to show details of some of the models (some of them from our own research group) in a standardized and organized fashion, as opposed to comprehensively describing the details of all the existing models.

Part II: Applications – Materials Technology introduces related topics that are tightly linked to the fundamentals in Part I. Chapter 6 discusses diagnosis and prevention of fracture-related failures, both catastrophic and those affecting *surface quality,* to improve optical fabrication yield. Chapter 7 describes novel supporting fabrication and characterization techniques to improve and better understand optical-fabrication processes. Novel optical fabrication processes are examined in Chapter 8. Chapter 9 covers how these techniques have been applied toward the fabrication of high-energy laser optics with high damage resistance.

This book is both an educational reference for students and scientists and a practical handbook for opticians, process engineers, and technicians in industry. Thus, Part I is in written textbook-style, describing basic physical principles and theoretical backgrounds. Many of these principles are discussed vis-à-vis brittle glass and ceramic workpieces. For fused silica and other glasses, quantitative examples are emphasized, as these materials have been the model systems in much of the research and development. Part II is a handbook for improving optical-fabrication processes. For example, the scratch forensics section in Chapter 6 is a useful tool set for the diagnosis of defects (i.e. scratches) and increasing yield.

Despite much recent progress in the ancient field of optical fabrication, it is still a young science. This book is offered as an overdue technological summary, reference, and compilation of advancements. It is hoped that the information in this volume will spur further scientific progress, pushing the envelope of optic performance and applications.

Lawrence Livermore National Laboratory, 2018 *Tayyab I. Suratwala*

References

1 Parks, R.E. (1981). Traditions of optical fabrication. *Proc. Soc. Photo-Opt. Instrum. Eng.* 315: 56–64.
2 Karow, H.H. (1992). *Fabrication Methods for Precision Optics* (ed. J.W. Goodman), 1–751. New York: Wiley.
3 Williamson, R. (2011). *Field Guide to Optical Fabrication*, xii, 121. Bellingham, WA: SPIE Press.
4 Bass, M. and Mahajan, V.N. (2010). *Optical Society of America. Handbook of Optics*, 3e, vol. 1, 4–5. New York: McGraw-Hill.
5 Schwiegerling, J. (2014). *Optical Specification, Fabrication, and Testing*, xi, 203. Bellingham, WA: SPIE Press.
6 Twyman, F. (1952). *Prism and Lens Making; A Textbook for Optical Glassworkers*, 2e, viii, 629. London: Hilger & Watts.
7 Fynn, G.W., Powell, W.J.A., and Fynn, G.W. (1988). *Cutting and Polishing Optical and Electronic Materials*, 2e, xxiii, 229. Bristol, PA: A. Hilger.
8 De Vany, A.S. (1981). *Master Optical Techniques*, viii, 600. New York: Wiley.
9 Malacara, D. (2007). *Optical Shop Testing*, 3e, xx, 862. Hoboken, NJ: Wiley-Interscience.
10 Brown, N. (1981). *A Short Course on Optical Fabrication Technology.* Lawrence Livermore National Laboratory.

11 Cook, L. (1990). Chemical processes in glass polishing. *J. Non-Cryst. Solids* 120 (1–3): 152–171.
12 Callister, W.D. and Rethwisch, D.G. (2010). *Materials Science and Engineering: An Introduction*, 8e, xxiii, 885, 82. Hoboken, NJ: Wiley.
13 Oliver, M.R. (2004). *Chemical–Mechanical Planarization of Semiconductor Materials*, x, 425. Berlin, New York: Springer-Verlag.
14 Liang, H. and Craven, D.R. (2005). *Tribology in Chemical–Mechanical Planarization*, 185. Boca Raton, FL: Taylor & Francis.
15 Babu, S.V. (2000). Chemical–Mechanical Polishing--Fundamentals and Challenges: Symposium Held, San Francisco, CA, USA (5–7 April, 1999), p. ix, 281. Warrendale, PA: Materials Research Society.
16 Izumitani, T. (1986). *Optical Glass*, x, 197. New York: American Institute of Physics.
17 Marinescu, I.D., Uhlmann, E., and Doi, T. (2006). *Handbook of Lapping and Polishing*. CRC Press.
18 Stavroudis, O., Boreman, G.D., Acosta-Ortiz, S.E. et al. (2001). *Handbook of Optical Engineering*. Marcel Dekker.
19 Jain, V.K. (2017). *Nanofinishing Science and Technology: Basic and Advanced Finishing and Polishing Processes*. CRC Press.
20 Cumbo, M., Fairhurst, D., Jacobs, S., and Puchebner, B. (1995). Slurry particle size evolution during the polishing of optical glass. *Appl. Opt.* 34 (19): 3743–3755.
21 Golini, D. and Jacobs, S.D. ed. (1990). Transition between brittle and ductile mode in loose abrasive grinding. San Dieg-DL Tentative. International Society for Optics and Photonics.
22 Lambropoulos, J.C., Fang, T., Funkenbusch, P.D. et al. (1996). Surface microroughness of optical glasses under deterministic microgrinding. *Appl. Opt.* 35 (22): 4448–4462.
23 Randi, J.A., Lambropoulos, J.C., and Jacobs, S.D. (2005). Subsurface damage in some single crystalline optical materials. *Appl. Opt.* 44 (12): 2241–2249.
24 Lambropoulos, J., Jacobs, S.D., Gillman, B. et al. (1997). Subsurface damage in microgrinding optical glasses. *Ceram. Trans.* 82: 469–474.
25 Shafrir, S.N., Lambropoulos, J.C., and Jacobs, S.D. (2007). Subsurface damage and microstructure development in precision microground hard ceramics using magnetorheological finishing spots. *Appl. Opt.* 46 (22): 5500–5515.
26 Cerqua, K.A., Lindquist, A., Jacobs, S.D., and Lambropoulos, J. (1987). Strengthened glass for high average power laser applications. SPIE: Conference on New Slab and Solid-State Laser Technologies and Applications, Volume 0736, pp. 13–21.
27 Guo, S., Arwin, H., Jacobsen, S.N. et al. (1995). A spectroscopic ellipsometry study of cerium dioxide thin films. *J. Appl. Phys.* 77 (10): 5369–5376.

28 Shoup, M.J., Jacobs, S.D., Kelly, J.H. et al. (1992). Specification of large aperture Nd:phosphate glass laser disks. Proceedings of SPIE, Volume 1627, pp. 192–201.

29 DeGroote, J.E., Marino, A.E., Wilson, J.P. et al. (2007). Removal rate model for magnetorheological finishing of glass. *Appl. Opt.* 46 (32): 7927–7941.

30 DeGroote, J.E., Jacobs, S.D., Gregg, L.L. et al. ed. (2001). Quantitative characterization of optical polishing pitch. International Symposium on Optical Science and Technology. International Society for Optics and Photonics.

31 Golini, D., Jacobs, S., Kordonski, W., and Dumas, P. (1997). Precision optics fabrication using magnetorheological finishing. SPIE, CR67–16, pp. 1–23.

32 Golini, D., Jacobs, S.D., Kordonski, V., and Dumas, P. ed. (1997). Precision optics fabrication using magnetorheological finishing. Proceedings Volume 10289, Advanced Materials for optics and precision structures: A Critical Review.

33 Shorey, A.B., Kwong, K.M., Johnson, K.M., and Jacobs, S.D. (2000). Nanoindentation hardness of particles used in magnetorheological finishing (MRF). *Appl. Opt.* 39 (28): 5194–5204.

34 Jacobs, S.D. (2007). Manipulating mechanics and chemistry in precision optics finishing. *Sci. Technol. Adv. Mat.* 8 (3): 153–157.

35 Jacobs, S.D., Golini, D., Hsu, Y. et al. ed. (1995). Magnetorheological finishing: a deterministic process for optics manufacturing. *International Conferences on* Optical Fabrication and Testing and Applications of Optical Holography. International Society for Optics and Photonics.

36 Kordonski, W.I. and Jacobs, S. (1996). Magnetorheological finishing. *Int. J. Mod. Phys. B* 10 (23–24): 2837–2848.

37 Lambropoulos, J.C., Miao, C., and Jacobs, S.D. (2010). Magnetic field effects on shear and normal stresses in magnetorheological finishing. *Opt. Express* 18 (19): 19713–19723.

38 Cerqua, K.A., Jacobs, S.D., and Lindquist, A. (1987). Ion-exchange strengthened phosphate laser glass. Development and applications. *J. Non-Cryst. Solids* 93: 361–376.

39 Jacobs, S.D. ed. (2004). International innovations in optical finishing. Optical Science and Technology, The SPIE 49th Annual Meeting. International Society for Optics and Photonics.

40 Jacobs, S. ed. (2003). Innovations in polishing of precision optics. International Progress on Advanced Optics and Sensors.

41 Shorey, A.B., Jacobs, S.D., Kordonski, W.I., and Gans, R.F. (2001). Experiments and observations regarding the mechanisms of glass removal in magnetorheological finishing. *Appl. Opt.* 40 (1): 20–33.

42 Abate, J.A., Brown, D.C., Cromer, C. et al. ed. (1977). Direct measurement of inversion density in silicate and phosphate laser glass. Laser Induced Damage in Optical Materials, Boulder, CO, USA (4–6 October 1977).

43 Jacobs, S.D., Kordonski, W., Prokhorov, I.V. et al. (2000). Deterministic magnetorheological finishing. Google Patents.
44 Arrasmith, S.R., Kozhinova, I.A., Gregg, L.L. et al. ed. (1999). Details of the polishing spot in magnetorheological finishing (MRF). SPIE's International Symposium on Optical Science, Engineering, and Instrumentation. International Society for Optics and Photonics.
45 Jacobs, R.R. and Weber, M.J. (1976). Dependence of the $^4F_{3/2}$-$^4I_{11/2}$ induced-emission cross section for Nd^{3+} on glass composition. *IEEE J. Quantum Electron.* QE-12 (2): 102–111.
46 Evans, C.J., Paul, E., Dornfeld, D. et al. (2003). Material removal mechanisms in lapping and polishing. *CIRP Ann.* 52 (2): 611–633.
47 Luo, J. and Dornfeld, D.A. (2001). Material removal mechanism in chemical mechanical polishing: theory and modeling. *IEEE Trans. Semicond. Manuf.* 14 (2): 112–133.
48 Luo, J. and Dornfeld, D.A. (2003). Effects of abrasive size distribution in chemical mechanical planarization: modeling and verification. *IEEE Trans. Semicond. Manuf.* 16 (3): 469–476.
49 Wang, C., Sherman, P., Chandra, A., and Dornfeld, D. (2005). Pad surface roughness and slurry particle size distribution effects on material removal rate in chemical mechanical planarization. *CIRP Ann. Manuf. Technol.* 54 (1): 309–312.
50 Moon, Y. (1999). Mechanical aspects of the material removal mechanism in chemical mechanical polishing (CMP). ProQuest Dissertations and Theses, thesis PhD. University of California, Berkeley, CA.
51 Li, Y. (2007). *Microelectronic Applications of Chemical Mechanical Planarization.* Wiley.
52 Babu, S. (2016). *Advances in Chemical Mechanical Planarization (CMP).* Woodhead Publishing.
53 Steigerwald, J.M., Murarka, S.P., and Gutmann, R.J. (2008). *Chemical Mechanical Planarization of Microelectronic Materials.* Wiley.

Acknowledgments

This work was performed under the auspices of the US Department of Energy by Lawrence Livermore National Laboratory (LLNL) under Contract DE-AC52-07NA27344. Special thanks to the LLNL midcareer award program for providing the funding and bandwidth needed to complete this book, and the Laboratory Directed Research and Development (LDRD) program for funding much of the research presented.

Thanks to Wiley and the American Ceramic Society for considering the important topic of optical fabrication for publishing and to the editors, graphic artists, and reviewers (especially Margaret Davis, Brian Chavez, and James Wickboldt).

Special thanks to the many coresearchers, collaborators and mentors, especially the LLNL coresearchers Rusty Steele, Mike Feit, Phil Miller, and Lana Wong, who joined this 15-year journey to unwind the mysteries and complexities of the fascinating field of optical fabrication. Their contributions greatly enriched this book. Finally, and most important, a debt of gratitude to my parents, who set the foundation of what I am today, and to my family: Maleka and my kids (Fatima, Maryam, and Aamina), who provided endless support and tolerated the endless hours.

Glossary of Symbols and Abbreviations

Symbols (with most common units and/or values in parenthesis).

Physical Constants

e_{el}	charge of electron (-1.602×10^{-19} C)
g	gravitational acceleration ($9.80\ \mathrm{m\ s^{-2}}$)
h	Planck's constant (6.626×10^{-34} J s)
k_b	Boltzmann's constant ($1.38 \times 10^{-23}\ \mathrm{J\ K^{-1}}$)
R	gas constant ($8.314\ \mathrm{J\ mol^{-1}\ K^{-1}}$)
ν_e	electron absorption frequency ($3.1 \times 10^{15}\ \mathrm{s^{-1}}$)
ε_o	permittivity in vacuum ($8.854 \times 10^{-12}\ \mathrm{F\ m^{-1}}$)

Variables

a	Hertzian contact radius (nm)
a_{ij}	Hertzian contact radius between two bodies (i and j) (1 for workpiece, 2 for lap, and 3 for particle) (nm)
a_j	Hertzian contact radius for particle j (nm)
a_0	vertical displacement of workpiece (μm)
a_2	workpiece curvature (μm)
a_c	crack flaw size (μm)
$\overline{a}_p$	average contact zone radius due to plastic removal (nm)
$\overline{a}_m$	average contact zone radius due to chemical removal (nm)
a_{KDP}	lattice parameter for KDP crystal (Å)
a_{DKDP}	lattice parameter for DKDP crystal (Å)
A	Auerbach's constant for Hertzian crack initiation ($\mathrm{N\ m^{-1}}$)
A_i	slurry island area; subscript i is a descriptor for each individual island (μm^2)
A_s	area of a single surface site (nm^2)
A_H	Hamaker constant (J)
A_o	pre-exponent constant (multiple units)

A_{p}	cross-sectional area of the plastic nanoscratch (nm^2)
A_{f}	area fraction of pitch buttons
A_{sf}	fracture surface area of a fractured workpiece (cm^2)
A_{pad}	contact area between workpiece and lap (mm^2)
A_{m}	fracture mirror marking constant for a given material ($\mathrm{MPa\,m}^{1/2}$)
b	empirical constant for single bond strength correction in MRF removal equation
B	friction constant for trailing indent cracks
B_i	index of brittleness of a material ($\mathrm{MPa\,m}^{-1/2}$)
B_{nd}	nanodiamond constant ($\mu\mathrm{m}^{4/3}$)
B_{CI}	carbonyl ion constant ($\mu\mathrm{m}^{4/3}$)
c	crack depth (μm)
c_i	crack depth ($i = h$ for Hertzian fracture, $i = \ell$ for lateral fracture, $i = r$ radial fracture, $i = t$ trailing indent fracture) (μm)
$\bar{c}$	mean crack depth (μm)
c_{90}	crack depth for 90% probability of removing (μm)
c_{max}	maximum crack depth or SSD depth (μm)
C_{L}	lap wear constant ($\mu\mathrm{m\,h}^{-1}$)
C	concentration of impurities (cm^{-3})
C_{o}	pre-exponential constant ($\mathrm{m\,s}^{-1}$)
C_K	K concentration ($\mathrm{atoms\,cm}^{-3}$)
$[\mathrm{Ce}]_{\mathrm{s}}$	steady state Ce concentration at the surface ($\mathrm{atoms\,cm}^{-3}$)
C_{p1}	heat capacity of workpiece ($\mathrm{J\,kg}^{-1}\,\mathrm{K}^{-1}$)
C_{p2}	heat capacity of lap ($\mathrm{J\,kg}^{-1}\,\mathrm{K}^{-1}$)
C_{pb}	linear rate of increase in deflection with area fraction for PBB (μm)
C_{nd}	nanodiamond concentration (cm^{-3})
C_{CI}	carbonyl-iron concentration (cm^{-3})
d	particle diameter or effective abrasive or polishing particle diameter (μm)
d_{a}	pre-exponential constant for workpiece roughness – slope of PSD relationship (μm)
$\bar{d}$	mean abrasive or polishing particle diameter (μm)
d_{b}	spacing between pitch buttons (mm)
d_{c}	critical depth of cut (μm)
d_{e}	effective distance from edge of workpiece (μm)
d_{g}	spacing between grooves on pad (mm)
d_{M}	moment arm distance from spindle pivot to workpiece–lap interface plane (mm)
d_{m}	chemical removal depth on workpiece by a polishing particle (nm)

$\overline{d}_{\mathrm{m}}$	average chemical removal depth on workpiece by all polishing particles (nm)
d_{p}	nanoplastic removal depth on workpiece by a polishing particle (nm)
$\overline{d}_{\mathrm{p}}$	average nanoplastic removal depth on workpiece by all polishing particles (nm)
d_{s}	stroke amplitude distance (cm)
d_{set}	settling distance (cm)
d_{t}	total depth of penetration of particle into the workpiece and lap (nm)
d_{ij}	depth of penetration between two bodies (i and j) (1 for workpiece, 2 for lap, and 3 for particle) (nm)
d_j	depth of removal of a given particle j
d_{PSD}	inverse exponential slope of the tail end in the slurry PSD (μm)
d_{rem}	depth of removal from a workpiece from slurry islands (nm)
D_{c}	diameter of etch cusp (μm)
D_{r}	ribbon penetration depth (μm)
D_{co}	initial cusp or crack size (μm)
D	impurity cation diffusivity ($cm^2\ s^{-1}$)
$D_{\mathrm{s}}(\mathrm{pH}_{\mathrm{MRF}})$	percent weight loss of the glass in the MR fluid supernatant as a function of MRF fluid pH
$E_{\mathrm{a}}, E_{\mathrm{b}}$	modulus components for viscoelastic lap (GPa)
E_{eff}	effective modulus of three-body system (GPa)
E_{ij}	composite modulus between two bodies (i and j) (1 for workpiece, 2 for lap, and 3 for particle) (GPa)
E_i	elastic modulus ($i = 1$ for workpiece, $i = 2$ for lap, $i = 3$ for particle) (GPa)
E_{rel}	stress relaxation function for the viscoelastic lap material (GPa)
E_{s}	elastic modulus of surface layer (GPa)
E_{sbs}	single bond strength ($kJ\ mol^{-1}$)
E_{r}	activation energy of the relative rates hydrolysis ($r_{\mathrm{Ce:Si}}$) ($kJ\ mol^{-1}$)
ΔE	activation energy for temperature dependence on removal rate ($kJ\ mol^{-1}$)
f	fill fraction of particles between workpiece and lap
f_{s}	fraction of the lap circumference loaded by the septum
f_{p}	fraction of particles contributing to nanoplastic removal
f_{m}	fraction of particles contributing to chemical removal
f_{r}	fraction of active particles at interface
f_{L}	fraction of applied load carried by particles (rather than by pad directly)

f_∞	fill fraction of particles
$f(r)$	fractional slurry PSD
f_A	fraction of pad area making contact with workpiece
f_{redep}	fraction of the glass products removed from the workpiece that deposits on a particular position of the lap
$f_{island}(h)$	slurry island height distribution
$f_{load}(c)$	fraction of particles being loaded for a given fracture depth
$f_{pad}(h)$	pad height distribution of the pad topography
$F_{pad}(h)$	cumulative height distribution of the pad topography
$f_o(c)$	instantaneous or fundamental crack depth distribution
$f_c(c)$	final incremental distribution of cracks
$F_c(c)$	final cumulative crack depth distribution
$f_o(r)$	fraction of the lap circumference loaded by the workpiece as a function of radial distance (r) from lap center
$f_s(r)$	fraction of the lap circumference loaded by the septum as a function of radial distance (r) from lap center
F_{ij}	force between two bodies (i and j) (1 for workpiece, 2 for lap, and 3 for particle) (N)
F_a	applied force during cleaning (N)
F_{vdw}	van der Waals force (N)
$F_{electrostatic}$	electrostatic image force (N)
F_{double}	electrical double layer force (N)
$F_{capillary}$	capillary force (N)
$F_{gravity}$	gravitation force (N)
F, F_x, F_y	friction force (N)
g_p, g_{pi}	interface gap or interface gap at pad asperity height i (μm)
$g_b(d)$	PSD of the base slurry abrasive
$g_r(d)$	PSD of the rogue particles
h_{ij}	heat transfer coefficient from various surfaces ($i = 1$ for workpiece, $i = 2$ for lap, $j =$ t for top, $j =$ b for bottom, and $j =$ s for side)
$h_L(x)$	relative height of the lap normalized to zero at the ends (μm)
h_{PV}	peak-to-valley height of the full lap (μm)
h_o	height of the workpiece or height of dome asperity (μm)
h_c	height at the center of the workpiece (μm)
h_f	pad compression or slurry island height upon workpiece loading (μm)
h_p	particle penetration into lap (μm)
h_m	maximum height of dome surface feature (nm)
h, h_i	height of pad asperity, slurry island or dome (μm)
h_t	heat transfer coefficient (W m^{-2} K^{-1})
$\Delta h_{oL}(x)$	workpiece–lap mismatch gap (μm)
$\Delta h_{oL}{}^*$	gap at the point of maximum pressure on the workpiece (μm)

$\bar{h}$	constant describing the rate at which pressure drops with increase in workpiece–lap mismatch (μm)
h_{vdw}	van der Waals constant (eV)
dh/dt	average thickness *material removal rate* on workpiece ($\mu m\,h^{-1}$)
dh_i/dt	instantaneous thickness *material removal rate* on workpiece ($\mu m\,h^{-1}$)
dh_{lap}/dt	thickness *material removal rate* on lap ($\mu m\,h^{-1}$)
ΔH	optic tilt height (μm)
H_i	hardness of workpiece (GPa) ($i = 1$ for workpiece, $i = 3$ for particle)
H_k	Knoop hardness of the workpiece (MPa)
$[H]_s$	H concentration at surface (atoms cm^{-3})
IEP_s	isoelectric point of slurry particle
J_z	flux of particles contacting surface ($m^{-2}\,s^{-1}$)
J_{gp}	glass reaction product flux into workpiece–lap interface ($mol\,m^{-2}\,s^{-1}$)
$J_{diffusion}$	flux of impurities into workpiece ($m^{-2}\,s^{-1}$)
$J_{material\ removal}$	flux of impurity removal during polishing ($m^{-2}\,s^{-1}$)
$J_s(r)$	evaporative flux of a drying fluid droplet ($mol\,s^{-1}\,cm^{-2}$)
$J(t)$	creep compliance function for viscoelastic lap material (GPa^{-1})
k, k'	constant relating material constants of two different materials (unitless or GPa^{-1})
$^{25}K_a$	equilibrium reaction constant for HF dissociation ($mol\,l^{-1}$)
^{25}K	equilibrium reaction constant for HF^{2-} formation ($mol\,l^{-1}$)
K_I	stress intensity ($MPa\,m^{1/2}$)
K_{Ic}	fracture toughness ($MPa\,m^{1/2}$)
K_{s1}, K_{s2}	equilibrium constants for change in surface hydroxyl surface charge ($mol\,l^{-1}$)
K_f	constant for fail-safe design of pressure differential windows ($MPa\,m^{1/2}$)
k_c	growth constant for etch cusps ($\mu m^2\,s^{-1}$)
k_i	thermal conductivity ($i = 1$ for workpiece, $i = 2$ for lap) ($W\,m^{-1}\,K^{-1}$)
k_p	Preston coefficient for workpiece ($m^2\,N^{-1}$)
k_{lap}	Preston coefficient for lap ($m^2\,N^{-1}$)
k_{olap}	Preston coefficient for lap at initial use of pad ($m^2\,N^{-1}$)
k_{max}	proportionality constant between *surface roughness* and crack depth
K_{sp}	spring constant of AFM tip
$[K]_s$	K concentration at surface (atoms cm^{-3})
ℓ_o	characteristic dimension (diameter or side length) of workpiece (m)

L	characteristic length for dome convergence during tumble finishing (μm)
L_{max}	maximum crack length (μm)
L_t	trailing indent fracture crack length (same as a scratch width) (μm)
$\overline{L}_t$	mean crack length (μm)
$\overline{L}_s$	mean length of scratch (μm)
LH	lapping hardness ($MPa^{-2}\ m^{-1/2}$)
m, n	nonnegative integers used for describing Zernike polynomials with $n \geq m$
m_s	constant in exponent describing influence of material properties to SSD
M, M_x, M_y	moment force (N m)
M_s	surface metal atom
M_{wp}, M_p	surface metal atom of workpiece (wp) or slurry particle (p)
M_{gp}	molecular concentration of glass reaction products at the workpiece–lap interface ($mol\ l^{-1}$)
MW_{gp}	molecular weight of glass reaction products ($gm\ mol^{-1}$)
n_s	areal number density of cracks at surface (cm^{-2})
n_{ps}	number of nanoscratching passes
n_o	counter ion concentration (cm^{-3})
n_i	refractive index ($i = 3$ for colloidal particle, $i = 4$ for liquid medium)
N	number of discrete, equally spaced measured points along a surface
N_b	number of pitch buttons on workpiece
N_c	power constant for empirical slow crack growth relationship
N_L	number of abrasive particles being loaded (i.e. active particles)
N_t	areal number density of particles at the interface (cm^{-2})
N_T	total number of particles between workpiece and lap
$O(c)$	SSD depth distribution represented as obscuration versus crack depth (c)
p	water vapor partial pressure (Pa)
p_o	saturation water vapor pressure (Pa)
p_{ad}	probability of an incident particle adsorbing to the surface
p_{si}	stoichiometry of atom i
P	applied load (N)
P_b	applied load for blunt impact (N)
P_s	applied load for sharp impact (N)
P_{ci}	critical load to initiate indentation fracture (i = h for Hertzian fracture, $i = \ell$ for lateral fracture, i = r for radial fracture, i = t for trailing indent fracture, and i = e for edge fracture) (N)
Pe	Peclet number

PV	peak-to-valley height (μm)
P_{tw}	Twyman stress ($N\,m^{-1}$)
q_d	surface charge density of particle ($C\,m^{-2}$)
$q(r)$	heat generation rate per unit area ($W\,m^{-2}$)
Q	frictional heat flux ($J\,m^{-2}\,s^{-1}$)
Q_I	activation energy for hydrolysis reaction ($kJ\,mol^{-1}$)
Q_{II}	activation energy for H_2O transport ($kJ\,mol^{-1}$)
Q_r	constant used in calculating SSD distribution with rogue particles
r	particle, island, or surface radius (μm) or radial distance (μm)
$\bar{r}$	average particle radius (nm)
r_i, r_j	radius of curvature of body (*i* and *j*) (1 for workpiece, 2 for lap, and 3 for particle)
r_A	pre-exponential constant
r_{arc}	arc radius; distance from lap center to a given leading edge point (x_L, y_L) on workpiece (mm)
r_b	bulk thickness etch rate on a flat surface ($\mu m\,h^{-1}$)
r_L	radius of lap (cm)
r_m	mirror marking radius on fracture surface (μm)
r_o	radius of workpiece (cm)
r_p	radius of pitch button (mm)
$r_{Ce:Si}$	hydrolysis reaction rate ratio between Ce–O–Ce and Si–O–Si
R_p	radius of a pinned droplet on a drying surface (mm)
R_L, $\vec{R}_L$	lap rotation rate (rpm)
R_o, $\vec{R}_o$	optic rotation rate (rpm)
R_s	stroke rotation rate (rpm)
R_{ts}	thermal shock FOM (K)
R_n^m	terms for describing Zernike polynomials
s, $\vec{S}$	separation distance from workpiece center and lap center (cm)
s_c	crack separation distance (μm)
S_p	areal density of Si atoms in fused-silica glass (nm^{-2})
t	time (min)
t_f	time-to-failure due to slow crack growth (h)
t_i	thickness ($i = 1$ for workpiece, $i = 2$ for lap) (mm)
t_s	thickness of surface layer (mm)
t_j	characteristic etch times (j = a, b, c) (min)
t_{Beilby}	effective thickness of the Beilby layer (nm)
t_{Lap}	removal rate decay constant for lap *material removal rate* (h)
$t_L(x,y)$	time of lap exposure at point x, y on the workpiece for the corresponding point on the lap (s)
T	temperature (K)
T_i	initial temperature (K)

T_f	final temperature (K)
T_{ave}	average temperature in the bulk (K)
T_{surf}	surface temperature (K)
ΔT	temperature difference, $T_{ave} - T_{surf}$ (K)
T_g	glass transition temperature (K)
T_e	edge toughness ($N\,mm^{-1}$)
U	work function difference between particle and workpiece (mV)
v_c	slow crack growth velocity ($m\,s^{-1}$)
v_I	slow crack growth velocity in Region I ($m\,s^{-1}$)
v_{II}	slow crack growth velocity in Region II ($m\,s^{-1}$)
v_3	impacting particle velocity ($m\,s^{-1}$)
v_{3bc}	critical velocity for fracture initiation for blunt projectile ($m\,s^{-1}$)
v_{3sc}	critical velocity for fracture initiation for sharp projectile ($m\,s^{-1}$)
v_r	local relative velocity ($m\,s^{-1}$)
V_a	activation volume ($m^3\,mol^{-1}$)
V_L	volume of workpiece (cm^3)
V_r	average relative velocity ($m\,s^{-1}$)
V_{rs}	relative velocity of septum ($m\,s^{-1}$)
w_o	initial width of Gaussian dome (FWHM) (μm)
w	width of Gaussian dome (μm)
w_c	width of crack (μm)
w_1	workpiece deflection (μm)
w_{max}	maximum workpiece deflection (μm)
w_p	width of the plastic nanoscratch (nm)
W_t	interaction energy between colloidal particles (J)
x_r	number fraction of rogue particles
x_L, y_L	point on the leading edge of the workpiece
Y	geometric constant for cracks
z	separation distance (nm)
z_c	valence of counter ion
z_o	local surface height above or below the mean height of the surface (nm)
z_g	thickness removed from workpiece during etching (μm)
z_p	Zernike polynomials
z_s	scaling factor for impact damage as function of projectile and workpiece hardness
dz_g/dt	thickness removal rate during etching ($\mu m\,h^{-1}$)
Z	interaction constant used in DLVO model ($J\,m^{-1}$)
Z_{sf}	constant relating fracture surface area to peak stress and workpiece volume ($cm^2\,psi^{-2}\,L^{-1}$)

α	stress parameter to determine between finite and infinite scratch length
α_1	thermal diffusivity of the workpiece ($m^2\ s^{-1}$)
α_{2-3}	combined thermal diffusivity of the lap and slurry ($m^2\ s^{-1}$)
α_{t1}	thermal expansion coefficient of the workpiece (K^{-1})
α_{Al}	thermal expansion coefficient of aluminum (K^{-1})
α_K	numerical factor in the range 0.03–0.04 used to determine k_{max}
α_r	radial crack initiation constant
α_e	edge angle on workpiece (°)
α_s	fraction of the kinetic energy that is used to plastically deform surface
α_g	groove height (μm)
β	geometric constant
γ	liquid media surface tension ($J\ m^{-2}$)
γ_{SG}	solid–gas interface surface energy ($J\ m^{-2}$)
γ_{SL}	solid–liquid interface surface energy ($J\ m^{-2}$)
γ_{LG}	liquid–gas interface surface energy ($J\ m^{-2}$)
γ_f	fracture surface energy ($J\ m^{-2}$)
Δ	removal increment amount (μm)
ΔPV	workpiece deflection due to PBB (nm)
Δz	AFM tip displacement (nm)
δ	RMS *surface roughness* (nm)
δ_i	partial charge of atom i
δ_{s-wp}	partial charge difference between slurry particle and workpiece
δ_s	partial charge of slurry particle
δ_{wp}	partial charge of workpiece material
δ_{PV}	PV *surface roughness* (μm)
δ_b	baseline PV roughness used in 2D etch model (μm)
δ_o	roughness constant for workpiece roughness – slope of PSD relationship (nm)
δ_e	strain generated upon cooling a glass bilayer with different compositions
ε	strain on lap
ε_c	elastic strain of lap at workpiece center
$\varepsilon(t')$	lap strain rate due to loading by workpiece (s^{-1})
ε_i	dielectric constant ($i = 3$ for colloidal particle, $i = 4$ for liquid medium; $i = r$ for interface media)
η_s	fluid viscosity (Pa s)
η_2	viscosity of viscoelastic lap material (Pa s)
θ	angle in polar coordinate system (°)
θ_a	angle of applied force on particle (°)

θ_c	contact angle of solvent droplet on a workpiece surface (°)
θ_x, θ_y	slopes of the workpiece in the x and y directions relative to the lap plane (°)
θ_L	solid angle of the lap covered by the workpiece (°)
ν_i	Poisson's ratio ($i = 1$ for workpiece, $i = 2$ for lap, $i = 3$ for polishing or abrasive particle)
ν	rate of desorption of an adsorbed particle (s^{-1})
κ	inverse Debye length (nm^{-1})
μ	interfacial friction coefficient
ρ	radial distance for Zernike polynomials (mm)
ρ_c	crack tip radius (nm)
ρ_i	mass density ($i = 1$ for workpiece, $i = 2$ for lap, $i = 3$ for polishing or abrasive particle) ($gm\,cm^{-3}$)
ρ_f	mass density of fluid ($gm\,cm^{-3}$)
ρ_L	radius of curvature of the lap surface (m)
ρ_o	x,y position coordinate on the workpiece relative to workpiece center (mm, mm)
σ, σ_i	local stress or pressure at workpiece–lap interface or local stress at pad asperity height i (Pa)
σ_f	critical failure stress for fracture (Pa)
σ_o	average applied stress on workpiece (Pa)
σ_{part}	standard deviation in PSD (nm)
σ_p	peak stress in workpiece (Pa)
σ_s	surface stress in workpiece (Pa)
σ_{sep}	applied pressure on septum (Pa)
σ_t	thermal stress on workpiece (Pa)
τ	shear stress (Pa)
τ_c	creep compliance time constant for viscoelastic lap (s)
τ_s	stress relaxation time constant for viscoelastic lap (s)
φ	azimuthal angle (°)
φ_{nd}	nanodiamond particle size (nm)
φ_{CI}	carbonyl iron particle size (nm)
χ_i	crack growth constant ($i = h$ for Hertzian fracture, $i = \ell$ for lateral fracture, $i = r$ for radial fracture) (N)
χ_v	term related to strain of a viscoelastic lap (unitless)
χ_i^o	electronegativity of neutral atom i
$\overline{\chi}$	mean electronegativity of the compound
Ψ_a	abrasive angle (°)
Ψ, Ψ_i, Ψ_j	Stern potential, Zeta potential, or surface potential (1 for workpiece, 2 for lap, and 3 for particle) (mV)
Ω	factor relating crack length to crack depth
ξ	geometric constant for Vickers indent

Abbreviations

2D	two dimensional
3D	three dimensional
AFM	atomic force microscopy
AR	aspect ratio
BOE	buffered oxide etch ($HF:NH_4F$)
CCOS	computer controlled optical surfacing
CCP	computer controlled polishing
CISR	convergent, initial-surface independent, rogue-particle free polisher
CMP	chemical mechanical planarization
CNC	computer numerical control
CP	continuous polisher
DC	diamond conditioning
DI	deionized water
DKDP	deuterated potassium dihydrogen phosphate
DLVO	Derjaguin, Landau, Verwey, and Overbeek
EDS	energy dispersive spectroscopy
EEM	elastic emission machining
EHMG	ensemble Hertzian multigap model
EUV	extreme ultraviolet
FEA	finite element analysis
FE-SEM	field emission scanning electron microscopy
FJP	fluid-jet polishing
FOM	figure of merit
FWHM	full width at half maximum
IBF	ion beam figuring
IDG	island distribution gap model
IEP	isoelectric point
IR	infrared
LDRD	Laboratory Directed Research and Development
LLNL	Lawrence Livermore National Laboratory
LRU	line replaceable units
MR	magnetorheological
MRF	magnetorheological finishing
ND	not detected or not determined
NIF	National Ignition Facility
NMR	nuclear magnetic resonance
OCT	optical coherent tomography
PACE	plasma-assisted chemical etching
PA-CVD	plasma-assisted chemical vapor deposition
PBB	pitch button blocking

PCVM	plasma chemical vaporization machining
PJCE	plasma jet chemical etching
PL	photoluminescence
ppb	part per billion
ppm	part per million
ppmw	part per million by weight
PSD	particle size distribution
PSDI	phase shifting diffractive interferometry
PV	peak to valley
PZC	point of zero charge
QCM	quartz crystal microbalance
RF	reflected wavefront
RIBE	reactive ion beam etching
RMS	root mean square
RT	room temperature
S1	Side 1
S2	Side 2
SEM	scanning electron microscopy
SI	standard units
SIMS	secondary ion mass spectroscopy
SPOS	single-particle optical sensing
SS	stainless steel
SSD	subsurface mechanical damage
TEM	transmission electron microscopy
TIRM	total internal reflection microscopy
UV	ultraviolet
WIWNU	within workpiece nonuniformity

Part I

Fundamental Interactions – Materials Science

1

Introduction

1.1 Optical-Fabrication Processes

Optical fabrication is the manufacture of optical components such as passive optics – e.g. lenses, transmission flats, mirrors, and prisms – and active optics – e.g. laser-gain media, frequency converters, polarizers, and adaptive optics. These optics are produced in a myriad of shapes and sizes, using a variety of materials.

A typical optical-fabrication process starts from a block of bulk material, or a workpiece, which is processed through a series of grinding and polishing steps, as shown in Figure 1.1 to become an optic. A tremendous number of grinding and polishing steps exist, of many different types [1–17]. Nevertheless, the overall objective of all optical-fabrication process steps is to remove material from the workpiece in a controlled fashion to meet the specifications of the next or final process step. Generally, any surface damage caused by a given step is removed in the subsequent step, as depicted in Figure 1.2.

Optical-fabrication processes often require many iterations of a given process step (as suggested by the circular arrows in Figure 1.1), which are accompanied by metrology and process modifications to achieve a desired *surface figure* and *surface quality*. Early process steps tend to remove material much faster, with less geometric control and more surface damage to the workpiece. The final steps are typically the opposite: slower *material removal rates*, but greater control of geometry and little (or, ideally, no) surface damage to the workpiece.

The size of the abrasive or polishing particle is a dominant factor controlling the *material removal rate* (see Figure 1.3). Because the *material removal rate* may vary by many orders of magnitude, an optimized number of process steps with appropriate between-step specifications is needed to determine the most

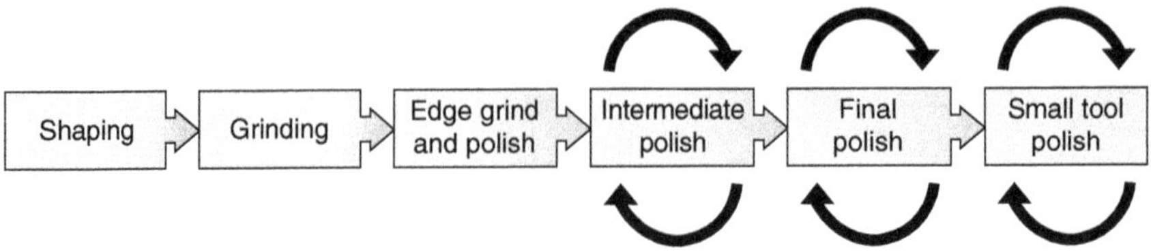

Figure 1.1 Typical steps in a conventional optical-fabrication process.

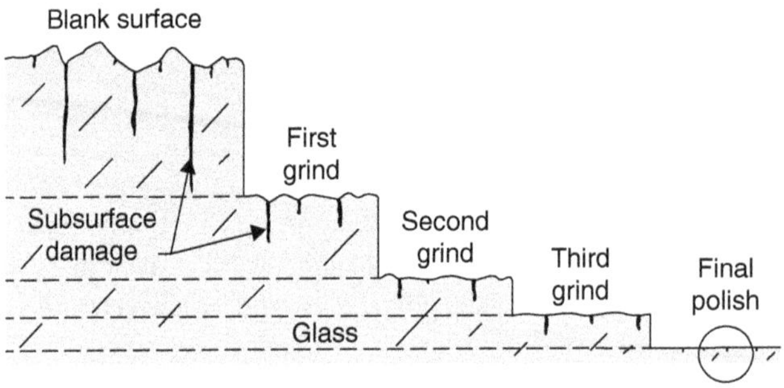

Figure 1.2 A schematic of material removal during various steps of the grinding/polishing process, illustrating surface-fracture removal.

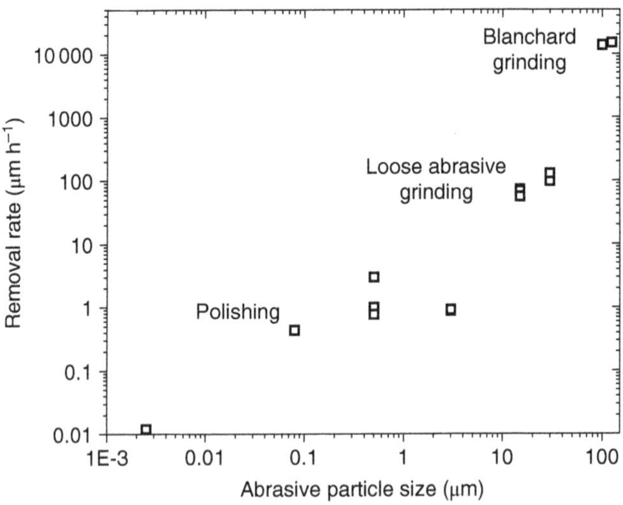

Figure 1.3 Log–log plot of material removal rate versus abrasive-particle size, determined for fused-silica glass.

economical fabrication process. Ultimately, the overall process time, and hence cost, of a given optical fabrication process is driven by a combination of

- the removal rates of the process steps (dominated by the latter, slower process steps)
- the number of process steps
- the number of iterations required in each step to yield an acceptable workpiece.

A common factor among the fabrication processes is the set of fundamental interactions at various spatial scale lengths between the

- workpiece, which will become the optic
- lap or tool, which leads to the time- and spatially dependent mechanical loading of the workpiece
- slurry or lubricant, which often contains particles to remove material from the workpiece.

The description and understanding of these phenomena and how processing affects workpiece structure, properties, and performance provide the materials-science basis of optical fabrication, as represented in Figure 1.4 https://en.wikipedia.org/wiki/Materials_science [11]. Defining these fundamental interactions or phenomena and the relationships among processing, structure, properties, and performance requires very controlled, systematic processing experiments and characterization of structures and properties, combined with quantitative modeling. Historically, this has been difficult to achieve, owing to a large number of simultaneous interactions and phenomena at various spatial scale lengths, as well as the large number of process variables.

In some ways similar to optical fabrication, chemical–mechanical planarization (CMP) is a fabrication method used for integrated circuits. CMP processing has received much study, because of its importance in the integrated-circuit industry. The key objective of CMP is to planarize the various layers of multilayer structures of a variety of materials (aluminum, silicon, silicon dioxide, copper, and tungsten, to name a few) with little or no defectivity. Though CMP

Figure 1.4 A schematic of the important relationships in performing materials science. Source: From Callister and Rethwisch 2010 [11] and https://enwikipediaorg/wiki/Materials_science.

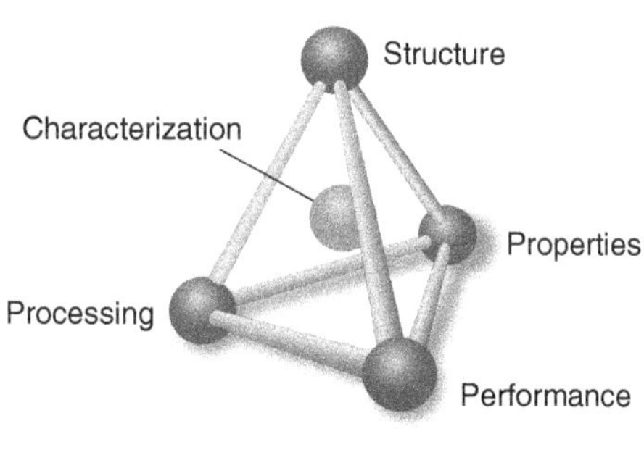

is governed by mechanics, tribology, and chemistry similar to that of optical fabrication, the technologies have distinct differences.

In CMP, the focus is planarizing and simultaneously polishing multimaterials, as depicted in Figure 1.5 [18–20]. By contrast, optical fabrication is focused on the *surface figure* and polishing of a single material at a time. The object of planarization is to reduce the entire multimaterial layer to the same thickness, while the object of achieving an optimal *surface figure* is to get a desired shape, whether flat, spherical, aspherical, or other. Hence with optical fabrication, the kinematics, or the relative motions of the tool with respect to the workpiece, is often more involved, and the slurry chemistry is quite different, as only one material (vs many) is utilized. CMP processes tend to be conducted at very high pressures (30–40 kPa), high relative velocities, and short polishing times – often minutes for final polishing and tens of minutes for stock removal. Optical fabrication generally uses lower pressures (0.7–7 kPa), lower velocities, and much longer polishing times (1 to >100 h). Even in making a flat optic, the *surface figure* differs from planarization. This is because an optical flat is specified based on optical path differences in reflection or transmission, while CMP planarization is specified on the thickness uniformity of the surface layer; this subtle, but important, difference implies that planarization allows for workpiece bending, which is generally undesirable in controlling *surface figure*. Finally, optical fabrication often requires slurry recirculation systems, owing to long polishing times, whereas CMP processes may make single-pass use of the polishing slurry.

Figure 1.5 Comparisons of integrated circuits fabricated with and without CMP. Source: From Li 2007 [18].

1.2 Major Characteristics of the Optical-Fabrication Process

The major characteristics of the optical-fabrication process and resulting optic are defined as follows:

(1) *Surface figure.* The long-range surface shape of the workpiece
(2) *Surface quality.* The characteristics of the surface, including subsurface mechanical damage (scratches and digs), clarity, particle cleanliness and impurities, and structural modification to the workpiece surface
(3) *Surface roughness.* The short-range surface topography of the workpiece
(4) *Material removal rate.* The rate at which material is removed from a workpiece surface at a given process step.

Figure 1.6 illustrates how these characteristics overlap. Each plays an important role in the performance and cost of the optic.

- *Surface figure* influences overall light-wavefront modification desired for that optic
- *Surface quality* affects the degree of scatter and use survivability (e.g. vulnerability to laser damage or operating strength)
- *Surface roughness* affects surface scatter and, for high-power laser applications, laser-beam contrast
- *Material removal rate* affects the processing time and cost of the optic

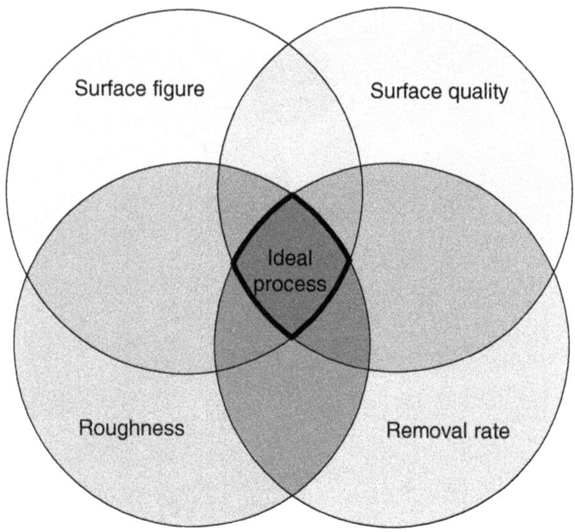

Figure 1.6 Venn diagram illustrating the four major characteristics of optical fabrication. Ideally, an optimal process is developed that meets all specifications simultaneously.

An ideal optical-fabrication process is one optimized for all four of these major characteristics. Yet a common challenge in making process changes is that improvements to one characteristic may occur at the expense of another. For example, achieving very low roughness usually involves processing conditions that yield low *material removal rates* and thus a costlier optic.

Surface figure is typically measured by interferometric techniques (e.g. Refs [21, 22]) that describe the long-scale shape of the final workpiece. *Surface figure* is often described by Zernike polynomials (such as power, astigmatism, coma, or irregularity, as indicated in Figure 1.7) and a power-spectral-density plot (as in Figure 1.8). Zernike polynomials are a convenient way of describing the final surface of a round optic as a series of components of various aspects of the surface in the form [23], as follows:

$$z_\mathrm{p} = a_n^m(R_n^m \cos m\varphi) + a_n^{-m}(R_n^{-m} \sin m\varphi) + a_n^0(R_n^0), \tag{1.1}$$

where $a_n{}^m$ is the coefficient associated with a particular term, $R_n{}^m$ are the radial polynomials, n is the radial order, m is the azimuthal frequency, and φ is the azimuthal angle. Note m and n are nonnegative integers with $n \geq m$. The sine–cosine terms in Equation (1.1) represent nonrotationally symmetric surfaces, and the last term represents rotationally symmetric surfaces. These polynomials have many useful characteristics, not least that they can be related directly to the classical aberrations and defined for circular, annular, and elliptical apertures. The *surface figure* specifications may involve transmitted or reflected wavefront, or both. In the case of transmitted wavefront, the homogeneity of the bulk material is important for many specifications [16, 24, 25].

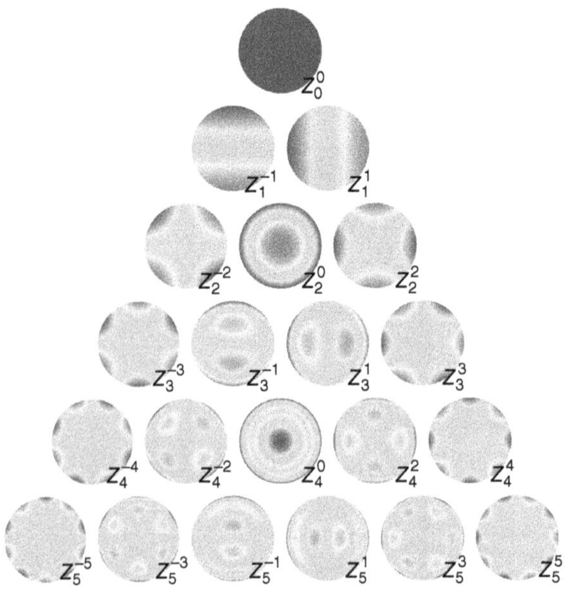

Figure 1.7 The first 21 Zernike polynomials describing the workpiece's surface figure, ordered vertically by radial degree and horizontally by azimuthal degree. Source: https://en.wikipedia.org/wiki/Zernike_polynomials.

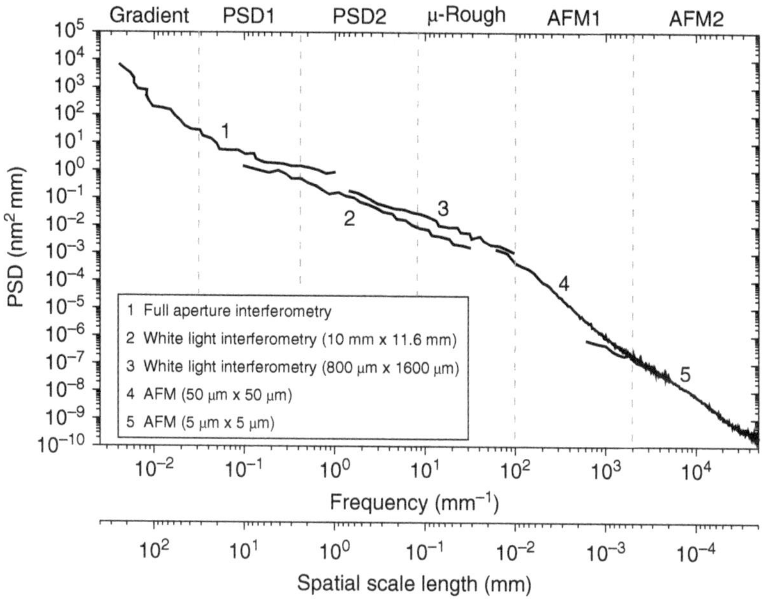

Figure 1.8 Example of a composite power spectral density of an optical surface over many spatial scale lengths, using different measurement techniques at different scale lengths. Typical techniques to measure spatial scale lengths are labeled. The left side of the plot represents long-range spatial scale lengths (referred to as surface figure). The right represents short-range spatial scale lengths (referred to as surface roughness).

Surface quality is a measure of the surface perfection achieved on a workpiece after finishing and cleaning. Examples of *surface quality* metrics include mechanical interactions such as surface microfractures or subsurface damage (SSD), plastic deformation, and densification; foreign particles or residue; and chemical interactions, such as surface molecular moieties and molecular impurities in the near-surface layer. SSD-type *surface quality*, commonly referred to scratch/dig, is usually specified using various standards [26–28]. For laser optics, an added requirement is to specify scratch/dig after a short chemical-etching process to reveal all such defects.

Surface roughness is a measure of the topographical relief of a surface. It is often used not only to describe ground surfaces (at μm-scale length) but also for optically smooth surfaces (at Angstrom- to nanometer-scale length). Roughness, δ, is typically described by the root mean square (RMS) of the surface topography:

$$\delta = \sqrt{\frac{1}{N}\sum_{i=1}^{N} {z_{oi}}^2} \tag{1.2}$$

where N is the number of discrete, equally spaced measured points along a surface, and z_0 is the local surface height above or below the mean height of the surface. Note that the computed RMS roughness will depend on the total length of the surface profile (the maximum spatial length), the surface area being averaged over (i.e. lateral resolution), and the distance between data points (the minimum spatial-scale length) [29, 30]. Hence, RMS is not a unique value and depends strongly on measurement technique.

Another important description of *surface roughness* is the power spectral-density function (also called power spectra), which is the spatial frequency spectrum of *surface roughness,* measured in inverse length units, calculated from the Fourier transform of the surface height data (z_0). The power spectra over the smaller spatial scale lengths in Figure 1.8 describe roughness [29]. From long- to short spatial-scale lengths, the various spatial bands are referred to as the RMS gradient, PSD-1, PSD-2, μ-roughness, and atomic force microscopy (AFM) roughness. Power spectra are a convenient way in which to describe optical surfaces, since they can be related to the scattering per unit solid angle from a surface. They are also powerful in identifying periodic structures on a surface, as manifested by a spike in the power spectrum. An example set of specifications, including roughness, for a high-power laser optic is shown in Table 1.1.

Table 1.1 An example set of high-level specifications on a typical laser optic, defined in terms of surface figure (here, specifically transmitted wavefront through Zernike terms), roughness bands (fine-scale power spectra), surface quality (subsurface damage or scratch/dig), and bulk properties.

Type	Property	Value
Surface	Peak-to-valley	211 nm ($\lambda/3$)
	Gradient	<7 nm cm^{-1} ($\lambda/90$/cm)
	PSD-1	1.8 nm rms
	PSD-2	1.1 nm rms
	Roughness	0.4 nm rms
	Scratch/Dig[a)]	20/10
Bulk	Index homogeneity	<5 ppm
	Absorbing inclusions (>5 μm)	0
	Transparent inclusions lenslets	0

Values typical for 3ω National Ignition Facility optics (40 cm aperture).

a) Post etch with number of scratches (width > 8 μm) <12.

1.3 Material Removal Mechanisms

At the macroscopic level, the *material removal rate* for a given process step is conventionally described by the Preston equation: [31]

$$\frac{\mathrm{d}h}{\mathrm{d}t} = k_{\mathrm{p}}\sigma_{\mathrm{o}}V_{\mathrm{r}}, \tag{1.3}$$

where $\mathrm{d}h/\mathrm{d}t$ is the average thickness *removal rate*, k_{p} is the Preston coefficient, σ_{o} is applied pressure, and V_{r} is the average relative velocity of the polishing particle, relative to the substrate. Simply stated the rate of removal increases linearly with pressure and velocity. It is clear, however, that despite its tremendous practical utility, the equation is unsatisfying from a materials-science perspective because it does not specifically identify the many microscopic and molecular phenomena that occur during processing.

Before dissecting the Preston equation in Chapter 2, it is important to describe the basic microscopic and molecular mechanisms by which a material can be removed from the surface of a workpiece. There is a robust literature on the various proposed removal mechanisms (e.g. Refs [9, 10, 32, 33]). A unifying consideration for most of these removal mechanisms is the load per particle or load per asperity. The four basic mechanisms for removing material from a surface are brittle removal, chemical/physical dissolution, plastic deformation, and chemical reaction, as follows:

- In *brittle removal* (i.e. fracturing), the load per particle is sufficiently large that fractures of various types are induced on the surface, leading to fractured-particle removal. The grinding processes used for glass and ceramics are governed by brittle removal.
- *Chemical or physical dissolution* is the process by which material is removed by etching, via an acid or a base, or by gas-phase ion bombardment, either by reacting or physically removing atoms from the surface. The *material removal rate* is determined by chemical reaction rate, largely independent of load. Etching behavior (isotropic vs anisotropic, congruent, or incongruent, reaction rate vs mass transport limited) has large implications in the evolution of the surface.
- In *plastic deformation*, the load per particle is in a regime where it is lower than the fracture limit, resulting in a plastic flow of material and removal from the surface. Clearly, metals and soft materials may demonstrate such behavior, but glasses and other brittle materials have also been shown to exhibit such removal, with plasticity at the nm level.
- In particle *chemical reaction*, the load per particle is low – below the plastic-deformation limit. The particle's chemical characteristics allow a

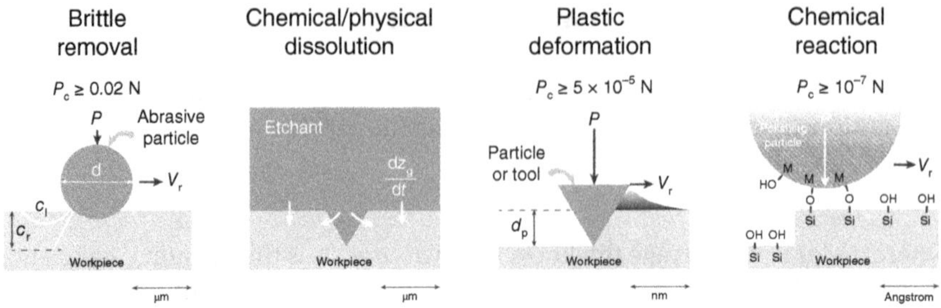

Figure 1.9 A schematic describing the various types of material removal mechanisms.

molecular-level reaction with the workpiece surface that causes material removal. The most common example in glass polishing is using cerium-oxide particles, where removal occurs via condensation and hydrolysis reactions. This phenomenon is known as the chemical tooth effect [10]. Figure 1.9 illustrates the material removal mechanisms given; further details are provided in subsequent chapters.

References

1 Karow, H.H. (1992). *Fabrication Methods for Precision Optics* (ed. J.W. Goodman), 1–751. New York: Wiley.

2 Williamson, R. (2011). *Field Guide to Optical Fabrication*, xii, 121. Bellingham, WA: SPIE.

3 Bass, M. and Mahajan, V.N. (2010). *Optical Society of America. Handbook of Optics*, 3e, vol. 1, 4–5. New York: McGraw-Hill.

4 Schwiegerling, J. (2014). *Optical Specification, Fabrication, and Testing*, xi, 203. Bellingham, WA: SPIE Press.

5 Twyman, F. (1952). *Prism and Lens Making; A Textbook for Optical Glassworkers*, 2e, viii, 629. London: Hilger & Watts.

6 Fynn, G.W., Powell, W.J.A., and Fynn, G.W. (1988). *Cutting and Polishing Optical and Electronic Materials*, 2e, xxiii, 229. Bristol, PA: A. Hilger.

7 De Vany, A.S. (1981). *Master Optical Techniques*, viii, 600. New York: Wiley.

8 Malacara, D. (2007). *Optical Shop Testing*, 3e, xx, 862. Hoboken, NJ: Wiley-Interscience.

9 Brown, N. (1981). *A Short Course on Optical Fabrication Technology.* Lawrence Livermore National Laboratory.

10 Cook, L. (1990). Chemical processes in glass polishing. *J. Non-Cryst. Solids* 120 (1–3): 152–171.

11 Callister, W.D. and Rethwisch, D.G. (2010). *Materials Science and Engineering: An Introduction*, 8e, xxiii, 885, 82. Hoboken, NJ: Wiley.
12 Parks, R.E. (1981). Traditions of optical fabrication. *Proc. Soc. Photo-Opt. Instrum. Eng.* 315: 56–64.
13 Oliver, M.R. (2004). *Chemical-Mechanical Planarization of Semiconductor Materials*, x, 425. Berlin, New York: Springer-Verlag.
14 Liang, H. and Craven, D.R. (2005). *Tribology in Chemical–Mechanical Planarization*, 185. Boca Raton, FL: Taylor & Francis.
15 Babu, S.V. (2000). Chemical–Mechanical Polishing – Fundamentals and Challenges: Symposium Held, San Francisco, CA, USA (April 5–7 1999). Warrendale, PA: Materials Research Society, ix, 281.
16 Izumitani, T. (1986). *Optical Glass*, x, 197. New York: American Institute of Physics.
17 Doi, T., Uhlmann, E., and Marinescu, I.D. (2015). *Handbook of Ceramics Grinding and Polishing*. William Andrew.
18 Li, Y. (2007). *Microelectronic Applications of Chemical Mechanical Planarization*. Wiley.
19 Landis, H., Burke, P., Cote, W. et al. (1992). Integration of chemical–mechanical polishing into CMOS integrated circuit manufacturing. *Thin Solid Films* 220 (1–2): 1–7.
20 Zantye, P.B., Kumar, A., and Sikder, A. (2004). Chemical mechanical planarization for microelectronics applications. *Mater. Sci. Eng., R* 45 (3): 89–220.
21 Robinson, D.W. and Reid, G.T. (1993). *Interferogram Analysis, Digital Fringe Pattern Measurement Techniques*. CRC Press.
22 Goodwin, E.P. and Wyant, J.C. ed. (2006). *Field Guide to Interferometric Optical Testing*. SPIE.
23 Evans, C.J., Parks, R.E., Sullivan, P.J., and Taylor, J.S. (1995). Visualization of surface figure by the use of Zernike polynomials. *Appl. Opt.* 34 (34): 7815–7819.
24 Bach, H. and Neuroth, N. (2012). *The Properties of Optical Glass*. Springer Science & Business Media.
25 Doremus, R.H. (1973). *Glass Science*. Wiley.
26 Kimmel, R.K., Parks, R.E., and OSA Standards Committee (1995). *ISO 10110 Optics and Optical Instruments: Preparation of Drawings for Optical Elements and Systems: a User's Guide*, 86. Washington, DC: Optical Society of America.
27 Salrin, J. and Gutlwin, G. (1945). Surface Quality Standards for Scratch and Dig. Picatinny Arsenal, NJ.
28 Aikens, D.M. ed. (2010). *The Truth About Scratch and Dig. Optical Fabrication and Testing*. Optical Society of America.
29 Bennett, J.M. and Mattsson, L. (1999). *Introduction to Surface Roughness and Scattering*, 2e, viii, 130. Washington, DC: Optical Society of America.

30 Duparre, A., Ferre-Borrull, J., Gliech, S. et al. (2002). Surface characterization techniques for determining the root-mean-square roughness and power spectral densities of optical components. *Appl. Opt.* 41 (1): 154–171.

31 Preston, F.W. (1922). The structure of abraded glass surfaces. *Trans. Opt. Soc.* 23 (3): 141.

32 Lawn, B.R. (1993). *Fracture of Brittle Solids*, 2e, xix, 378. Cambridge, New York: Cambridge University Press.

33 Evans, C.J., Paul, E., Dornfeld, D. et al. (2003). *Material Removal Mechanisms in Lapping and Polishing*. Elsevier.

2

Surface Figure

2.1 The Preston Equation

As described in Section 1.2, one of the major characteristics of the final optic is the *surface figure*, or long-range surface shape of the workpiece. Achieving the desired *surface figure* is a primary objective in fabricating an optic, because *surface figure* influences the wavefront modification of the incoming light, both in transmission and reflection. At the most basic level, the final *surface figure* of the workpiece is simply determined by its initial surface shape and the amount of material removed from each point on the workpiece surface, discounting residual stress changes. Hence, to quantitatively (i.e. deterministically) determine *surface figure* evolution in a given finishing process, one must understand all the phenomena that contribute to the *material removal rate* at each point and as a function of time. A useful approach to describe and organize these phenomena is to expand the traditional *material removal rate* Preston equation (Equation (1.3)). Preston's equation may be described in a more general form as follows: [1]

$$\frac{\mathrm{d}h_{\mathrm{i}}}{\mathrm{d}t}(x,y,t) = k_{\mathrm{p}}(x,y,t)\ \mu(v_{\mathrm{r}}(x,y,t))\,v_{\mathrm{r}}(x,y,t)\ \sigma(x,y,z,t), \tag{2.1}$$

$$\frac{\mathrm{d}h}{\mathrm{d}t}(x,y,t) = \frac{1}{t}\int_0^t k_{\mathrm{p}}(x,y,t')\mu(v_{\mathrm{r}}(x,y,t'))\,v_{\mathrm{r}}(x,y,t')\ \sigma(x,y,z,t')\ \mathrm{d}t', \tag{2.2}$$

where $\mathrm{d}h_{\mathrm{i}}/\mathrm{d}t$ and $\mathrm{d}h/\mathrm{d}t$ are the instantaneous and time-averaged removal rates, respectively, at some given time t and position x, y on the workpiece, μ is the friction coefficient, which is a function of the relative velocity (v_{r}) at the workpiece–lap interface, and σ is the pressure distribution resulting from the applied pressure (σ_{o}) and the nature of the workpiece–lap contact. Much of the literature starts with the Preston equation and modifies it for a given grinding or polishing system (e.g. [2–6]). In a review by Pal et al., [6] a number of modified versions of the Preston equation are described. Here, we start with the original, basic form of the Preston equation, as in (Equations (2.1) and (2.2)).

Figure 2.1 schematically describes the known phenomena that contribute to the general form of the Preston equation. The four major parameters in Equation (2.1) are Preston coefficient, interface friction, kinematics, and interface pressure distribution. The applied pressure distribution can vary significantly from the interface pressure distribution. The interface pressure distribution is not only a function of the applied pressure distribution but also influenced by the elastic-lap response, hydrodynamic forces, moment forces, viscoelastic-lap response, viscoplastic-lap response, and workpiece–lap shape mismatch. The workpiece–lap mismatch is influenced by a number of key phenomena, including the instantaneous *surface figure* of the workpiece, lap shape owing to pad wear or deformation, workpiece bending under applied load, workpiece bending owing to residual grinding stress, interface temperature distribution, global lap properties, slurry spatial distribution at the interface, and local nonlinear material deposits. Developing qualitative and quantitative understandings of these phenomena paves the way for improved fabrication processes that are more deterministic. Figure 2.1 provides a road map and reference for these phenomena, which are described in more detail in the following chapter.

2.2 The Preston Coefficient

For a given polishing process with a specific slurry, lap, and workpiece material, the Preston coefficient (k_p) quantifies the effective ability of slurry particles to remove material from the surface (Equation (2.1)). The Preston coefficient has units of thickness/(velocity pressure) and a value of $m^2\ N^{-1}$ (or $1\ Pa^{-1}$) in SI units. For polishing fused-silica glass using cerium-oxide slurry, k_p has typical values of $9–20 \times 10^{-13}\ m^2\ N^{-1}$ on a polyurethane pad [1] and $3 \times 10^{-13}\ m^2\ N^{-1}$ on pitch [8].

The value of the Preston coefficient is unique to a given set of polishing-process parameters, except for relative velocity and applied pressure. The extensive list of parameters includes slurry type, particle-size distribution (PSD), pH, slurry additives, lap-material properties, lap topography, lap thickness, shape, and groove pattern, and workpiece material. The utility of the Preston coefficient is that it can be used to compare the relative removal rates of polishing processes and workpiece materials. For example, for the same given polishing process (the same ceria slurry on a pitch lap at pH 7), the Preston coefficient is $\sim 3.3 \times 10^{-13}\ m^2\ N^{-1}$ for fused-silica glass, $10.4 \times 10^{-13}\ m^2\ N^{-1}$ for BK7 glass, and $1.6 \times 10^{-13}\ m^2\ N^{-1}$ for SF6 glass [8]. In other words, under the same polishing conditions, BK7 glass will polish ~3× faster than fused silica and SF6 will polish ~2× slower. Several references have tabulated values for the Preston coefficient for various polishing and grinding systems [8–10].

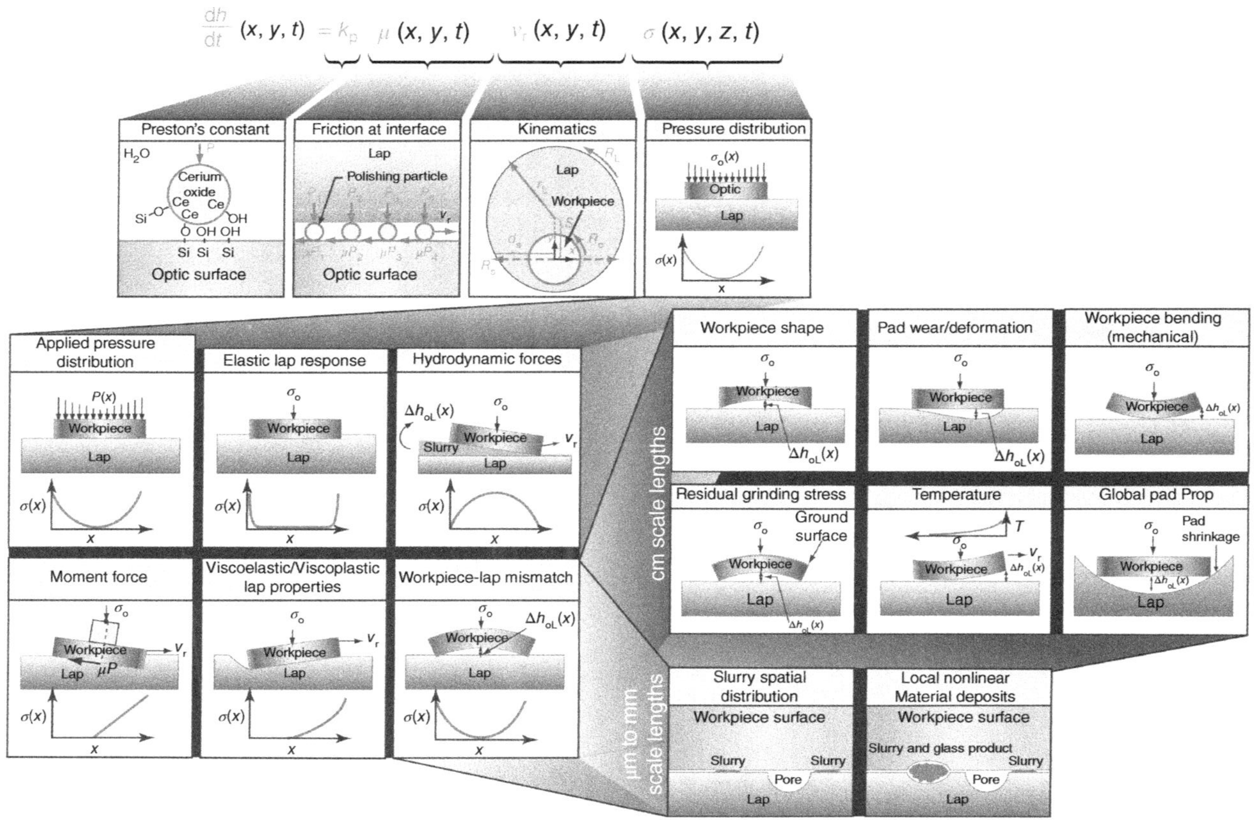

Figure 2.1 Dissection of the macroscopic Preston equation, illustrating the various phenomena that influence the spatial and temporal material removal rate during grinding and polishing. Source: Suratwala et al. 2014 [7]. Reproduced with permission from John Wiley & Sons.

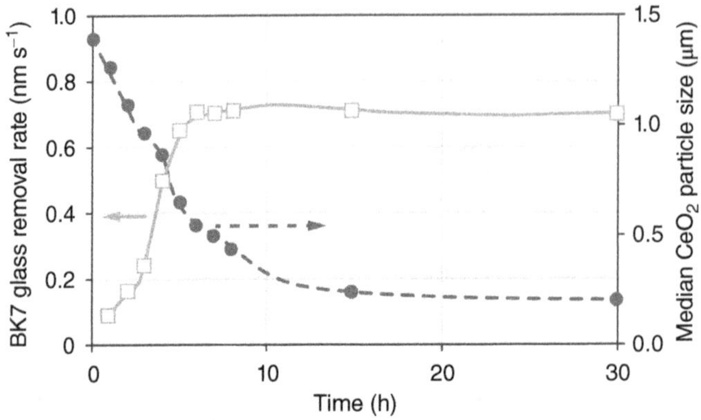

Figure 2.2 Changes in global material removal rate as a function of polishing time, using ceria slurry to polish BK7 glass. Source: Cumbo et al. 1995 [8]. Reproduced with permission from The Optical Society.

Despite the practical utility of the Preston coefficient at the macroscopic level, from a materials-science perspective, this constant alone is dissatisfying, because it does not specifically describe the influence of the fundamental material properties and process parameters that determine the Preston coefficient. In Chapters 4 and 5, microscopic-level interactions are explored to understand the contributions of individual slurry particles, their interactions with the workpiece, and how they build up to a macroscopic Preston coefficient, interface friction, *material removal rate*, and ultimate workpiece roughness.

The Preston coefficient is usually evaluated as a single-value constant for a given polishing scenario. Under certain conditions, however, the Preston coefficient may change as a function of polishing time and workpiece position. Cumbo et al. [8] showed a drop in the global *material removal rate* with increase in polishing time during ceria polishing of BK7 glass (Figure 2.2), attributing it to the change in slurry PSD with polishing time.

Another unique experiment showed that slurry reactivity can actually decrease with distance from the leading edge of a workpiece at the interface; that is, the Preston coefficient decreases with distance traveled through the workpiece–lap interface. The leading edge is defined as that portion of the workpiece edge where the points on the lap first make contact. Figure 2.3a illustrates a large difference in *surface figure* evolution during polishing with and without grooves on a polishing pad for otherwise identical polishing conditions, namely, fused-silica workpieces 25 cm in diameter, using cerium-oxide slurry on a polyurethane foam pad and employing the convergent polishing method described in Section 8.4. The grooves allow slurry to be replenished at the interface. In polishing using the pad without grooves, the *surface figure*

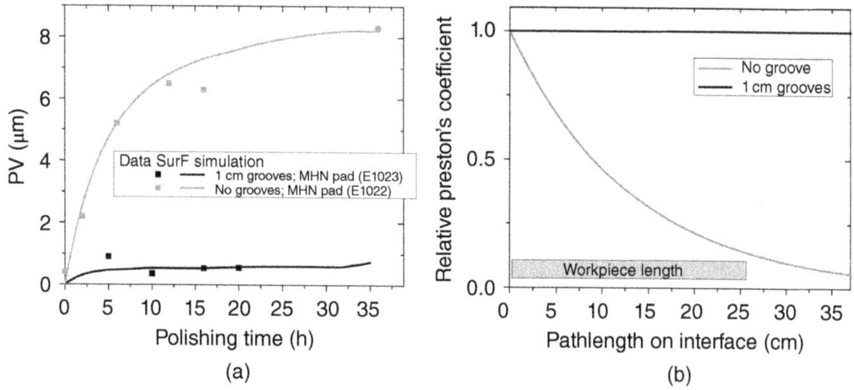

Figure 2.3 (a) Surface figure PV of an originally flat fused-silica workpiece after polishing with cerium-oxide slurry on a polyurethane foam pad (MHN, with and without grooves; (b) calculated effective slurry reactivity with distance across the workpiece.

changed dramatically, resulting in the workpiece becoming very convex, 8 μm peak to valley (PV). When the workpiece was polished using a pad with 1-cm-spaced grooves, however, *surface figure* migration was minimal. Based on these results, Figure 2.3b shows slurry reactivity as a function of distance that quantitatively explains *surface figure*. In the case of a pad with 1-cm grooves, slurry reactivity (or effective Preston coefficient) is largely unchanged; for a pad without grooves, slurry reactivity, and Preston coefficient decrease exponentially with a decay constant of ~8 cm. This experiment illustrates the polishing conditions that may arise when the Preston coefficient can change with distance, as well as the practical importance of groove designs that reduce this effect by replenishing fresh slurry at the interface.

2.3 Friction at Interface

The contribution of interfacial friction (μ) to material removal, as in Equation (2.1) can be thought of microscopically as proportional to the number of polishing particles making contact with the workpiece. The greater the number of particles making contact, the greater the friction and removal rate. At the macroscopic scale, the friction coefficient is defined as follows:

$$\mu = \frac{F}{P}, \tag{2.3}$$

where P is the applied load (N) and F is the tangential friction force (N). The magnitude of the F between the workpiece and lap is determined by the mode of contact between the two bodies, the applied load, the characteristics of

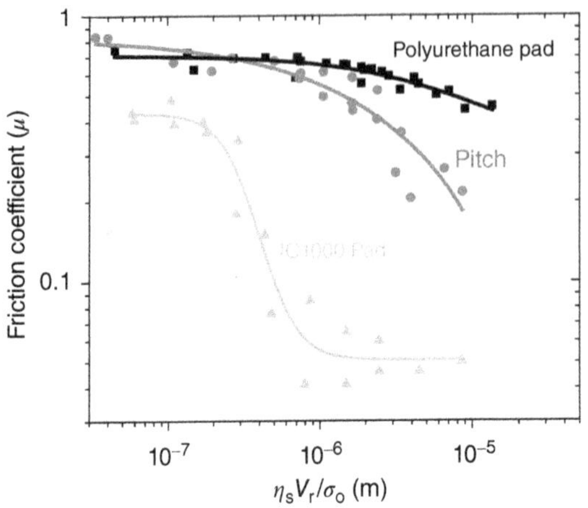

Figure 2.4 Measured friction coefficient as a function of applied pressure and relative velocity plotted in terms of the Stribeck curve. The line represents a sigmoidal fit to the data for each lap material. Source: Suratwala et al. 2010 [1]. Reproduced with permission from John Wiley & Sons.

the slurry (e.g. viscosity), and the workpiece–lap relative velocity [11, 12]. Experimentally, the friction force is typically measured as a function of applied load (P) and lap rotation rate (R_L) for a given polishing system [1]. It is common to describe the dynamic friction using the Stribeck curve (μ as a function of a figure-of-merit (FOM) $\eta_s V_r/\sigma_o$). η_s is the slurry fluid viscosity, V_r is the relative velocity between particle and workpiece, and σ_o is the applied pressure. This FOM is also known as the Hersey number [13]. An example of a measured Stribeck curve for polishing fused-silica glass on various lap materials is given in Figure 2.4 [1, 12]. There are three distinct regimes or modes of interface interactions, as shown schematically in Figure 2.5 [6, 11]. At low values of

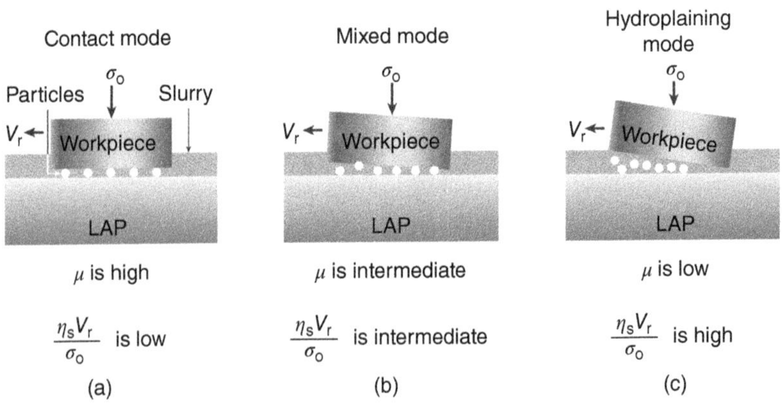

Figure 2.5 Types of interfacial contact between workpiece and polishing lap: (a) contact mode, (b) hydrodynamic mode, and (c) mixed mode. Source: Adapted from Pal et al. 2016 [6] and Hutchings 1992 [11].

$\eta_s V_r/\sigma_o$, the workpiece and lap make mechanical contact, referred to as contact mode, and the friction coefficient is high, typically 0.7–0.8; at high values of $\eta_s V_r/\sigma_o$, the fluid pressure of the slurry carries the workpiece off the lap, referred to as hydrodynamic mode, and the friction coefficient is low, <0.02. An example of hydrodynamic mode is when a car in the rain hydroplanes and slides after braking at high speeds. At intermediate values of $\eta_s V_r/\sigma_o$, the interface is in mixed mode, where the friction coefficient is between the contact and hydrodynamic modes. Here the friction coefficient can change significantly with small changes in relative velocity and applied pressure. It is important to note that the thickness of the film, or interface gap, increases from contact mode, to mixed mode, to hydrodynamic mode [13, 14].

Most conventional optical polishing is performed in contact mode, where the friction coefficient is large and does not significantly change with small changes in pressure and velocity. In chemical mechanical planarization (CMP), however, many processes operate in a mixed mode. Generally, hydrodynamic mode is avoided as essentially no removal occurs on the workpiece, but several novel optical-finishing processes, such as float polishing, intentionally operate in hydrodynamic mode, exploiting little friction and very low removal rates to achieve very low roughness [15–20]. Some of these methods are described in Chapter 8.

In Figure 2.4, the polyurethane pad and pitch laps follow the same basic behavior, with the friction coefficient on the Stribeck curve. The transition into hydrodynamic mode occurs at different values of $\eta_s V_r/\sigma_o$, depending on the properties of the lap material. The friction coefficient can be described quantitatively by a sigmoidal curve, which mimics the shape of the Stribeck curve. In the case of the polyurethane pad in Figure 2.4, the friction coefficient is

$$\mu = 0.7 - \frac{0.6}{1 + \left(7.7 \times 10^4\,\mathrm{m}^{-1} \frac{\eta_s V_r}{\sigma_o}\right)^{0.9}}. \tag{2.4}$$

Using the Stribeck curve and $\eta_s V_r/\sigma_o$ FOM has important utility in establishing the load and velocity conditions for polishing to maintain material removal, preferably high material removal. For example, take a system of polishing glass using a ceria slurry on a polyurethane pad. For a Baume 5 ceria slurry with viscosity of 25 cP and applied pressure of 0.3 psi, the relative velocity must stay below 0.785 m s^{-1} to stay within contact mode (with the criteria $\eta_s V_r/\sigma_o < 10^{-5}$ m from Figure 2.4) and to ensure that the *material removal rate* does not drop. This velocity is equivalent to a lap and workpiece rotation rate of $R_o = R_L = 100$ rpm, with a workpiece-center to lap-center offset of $s = 75$ mm. Note that the Stribeck effect is a bit counterintuitive to the expected behavior implied by the Preston equation, where increasing relative velocity results in an increased *material removal rate*. Accounting for the Stribeck friction effect, increasing relative velocity may actually decrease the removal rate.

2.4 Kinematics and Relative Velocity

Another major contributor to spatially dependent material removal affecting *surface figure* is the velocity of the lap or tool relative to the workpiece surface, defined as v_r (Equation (2.1)). Microscopically, this can be thought of as the relative velocity of the polishing particles to the workpiece surface in performing the work of removing material from the surface. The greater the velocity of the polishing particles relative to the workpiece, the greater the number of particles that will interact with the surface per unit time, thus yielding a greater *material removal rate.* Assuming the workpiece-particle relative velocity is roughly equivalent to the workpiece–lap relative velocity (i.e. the polishing particle is essentially stationary relative to the lap), the kinematic parameters of the polishing system can be used to calculate the relative velocity of the polishing particles. The kinematics of the polishing system is defined as the relative motions of the workpiece and lap over a defined area as a function of time.

The polishing of optical components can be divided into two major classes of geometry: full-aperture polishing, in which the workpiece is smaller than the lap, and subaperture polishing or small-tool polishing, where the lap contact zone is smaller than the surface of the workpiece. Kinematic analysis for various conventional full-aperture polishing scenarios has been previously summarized (e.g. see Brown [21] and Taylor et al. [22]). The kinematics during subaperture polishing for a multitude of polishing geometries has recently received much attention (see examples at [23, 24]).

A common full-aperture polishing system in terms of kinematic degrees of freedom and involving workpiece and lap rotation and workpiece stroke is shown in Figure 2.6. It is convenient to describe the local relative velocity in vector form as follows:

$$\vec{v}_r(x,y,t) = (\vec{R}_o \times \vec{\rho}_o(x,y,t)) - (\vec{R}_L \times (\vec{\rho}_o(x,y,t) - \vec{S}(t))) + \frac{d\vec{S}(t)}{dt}, \tag{2.5}$$

where ρ_o is a position on the workpiece given by coordinates x and y with the origin at the workpiece center, $\vec{R}_o$ and $\vec{R}_L$ are the rotation rates of the workpiece and lap in vector form, directed along the z-axis, and $\vec{S}$ is the vector describing the separation between the geometric centers of the workpiece and lap (see Figure 2.6).

Note that a similar set of equations can also be formulated for other polishing systems with a different set of kinematics and degrees of freedom. In this case, the first term on the right side of Equation (2.5) describes the rotational velocity of the workpiece for some given position at the workpiece-center frame of reference. The second term describes the rotational velocity of the lap at the workpiece-center frame of reference. The final term describes relative velocity

Figure 2.6 Top-view schematic of the polisher and workpiece, illustrating kinematic degrees of freedom and coordinate system of common full-aperture polishing systems. Source: Suratwala et al. 2010 [1]. Reproduced with permission from John Wiley & Sons.

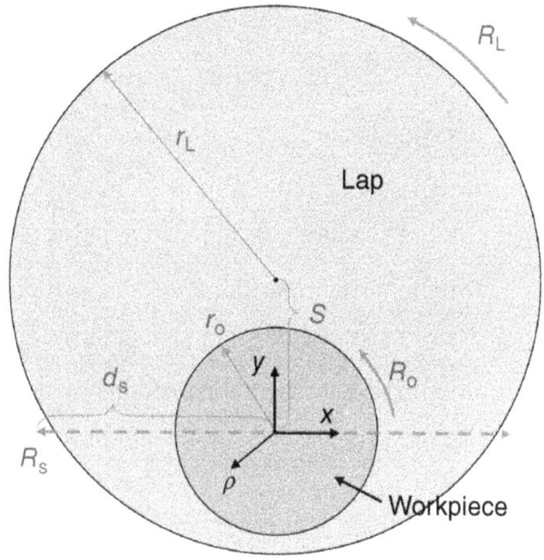

owing to the linear motion of the stroke. In the case of spindle polishing, each of the terms above can be described in vector form as follows:

$$\vec{R}_o = \begin{pmatrix} 0 \\ 0 \\ R_o \end{pmatrix}, \quad \vec{R}_L = \begin{pmatrix} 0 \\ 0 \\ R_L \end{pmatrix}, \quad \vec{S} = \begin{pmatrix} d_s \sin(R_s t) \\ s \\ 0 \end{pmatrix},$$

$$\vec{\rho}_o = \begin{pmatrix} \sqrt{x^2+y^2}\sin(\arctan(x/y) + 2\pi R_o t) \\ \sqrt{x^2+y^2}\cos(\arctan(x/y) + 2\pi R_o t) \\ 0 \end{pmatrix}, \tag{2.6}$$

where R_o is the optic rotation rate, R_L is the lap rotation rate, d_s is stroke amplitude, R_s is the stroke rate, and s is the separation distance between the centers of the workpiece and lap. To describe typical continuous-polisher (CP) type kinematics, one would set d_s equal to 0.

As relative velocity can lead to removal only when the lap and workpiece are in contact, an additional condition for a nonzero local relative velocity applies for the case of a circular lap:

$$|\vec{\rho}_o(x, y, t) - \vec{S}(t)| \leq r_L. \tag{2.7}$$

The time-averaged relative velocity (V_r) is then given by

$$V_r(x, y) = \frac{1}{t}\int_0^t \vec{v}_r(x, y, t')\, dt'. \tag{2.8}$$

Using Equations (2.4)–(2.8), the time-averaged velocity has been calculated for a variety of kinematics, where $r_o = 0.05$ m, $r_L = 0.10$ m, and $R_L = 28$ rpm, as shown in Figure 2.7 [1]. When V_r is higher on the edge relative to the center, the workpiece will become more convex, and when lower, more concave. Figure 2.7a suggests that as the workpiece rotation rate is mismatched with

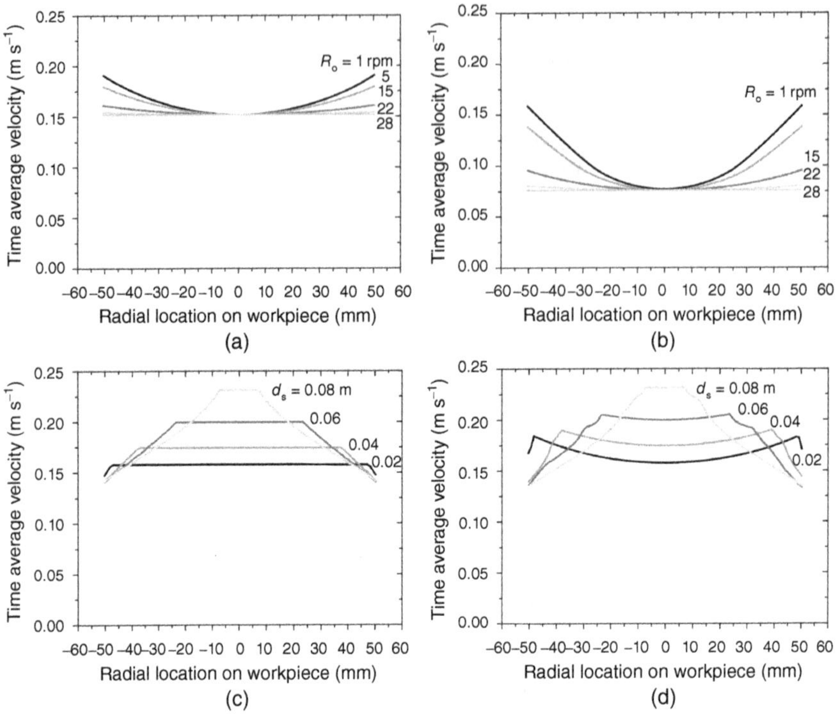

Figure 2.7 Calculated time-averaged relative velocity using kinematic equations described using Equations (2.4)–(2.7), where $r_o = 0.05$ m, $r_L = 0.1$ m, $R_L = 28$ rpm. (a) Effect of relative workpiece rotation rate with no stroke ($s = 0.025$ m, $R_s = 0$ rpm, $d_s = 0$ m, and R_o is varied). (b) Effect of relative workpiece rotation rate with larger separation distance ($s = 0.05$ m, $R_s = 0$ rpm, $d_s = 0$, and R_o is varied). (c) Effect of stroke with workpiece/lap rotation matched ($s = 0.05$, $R_s = 15$ rpm, $R_o = 28$ rpm, and d_s is variable). (d) Effect of stroke with workpiece–lap rotation mismatch ($s = 0.05$, $R_s = 5$ rpm, $R_o = 15$ rpm, and d_s is variable). Source: Suratwala et al. 2010 [1]. Reproduced with permission from John Wiley & Sons.

the lap rotation rate, the workpiece will become more convex. Comparing Figure 2.7a,b, shows that increasing the separation distance s increases the overall time-averaged velocity, and hence, removal rate. Figure 2.7c,d, illustrates that increasing the stroke distance generally leads to lower velocities at the edge, because the edges of the workpiece spend more time off the lap and the workpiece becomes more concave. These trends are consistent with those generally observed and practiced by opticians during conventional spindle polishing.

Another important kinematic case is CP mode where $R_o = R_L$ and there is essentially zero stroke. A unique condition is achieved wherein the relative velocity is spatially constant at all times and, hence, time-averaged relative velocity is constant across the workpiece. This case and concept is well described by Brown [21, 25]. Ignoring all other phenomena, CP mode provides a convenient set of kinematics with which to achieve uniform removal across a workpiece. In practice, however, the influence of other *material removal rate* nonuniformities from other phenomena, such as those in Figure 2.1, prevent true uniform removal. Nevertheless, kinematics is a major contributor to spatial material removal distribution during polishing.

The second major class of polishing geometries is subaperture grinding and polishing, which has become a very important capability in optical fabrication. Subaperture grinding and polishing includes techniques such as computer numerical control (CNC) grinding [26], computer-controlled optical surfacing (CCOS) [27], magnetorheological finishing (MRF) [28–30], stress lap polishing [31], and bonnet and fluid jet polishing [32, 33], to name a few. All these techniques have a tool invoking a load and relative velocity against the workpiece and computer control of the kinematics of the tool locally across the workpiece. The relative motion of the tool against the workpiece results in a removal function – that is, the amount removed locally against the surface of the workpiece as a function of time. These techniques use different small tools, whether stiff, compliant, adaptive, or other, and local kinematic degrees of freedom, for example, rotational or linear. A discussion of the kinematics involved with grinding and polishing subaperture tools is found in several valuable references [23, 24, 34]. These techniques have led to greater determinism and smaller spatial-scale lengths of the *surface figure* that can be controlled on the workpiece. Understanding and knowing the removal function as a function tool parameters and its stability are critical in determining how a workpiece will respond. For example, a removal function for MRF polishing has the characteristic shape shown in Figure 2.8. A number of the subaperture polishing techniques are reviewed in Chapter 8.

2.5 Pressure Distribution

Kinematics is clearly not the only factor that influences *surface figure*. This point is illustrated in Figure 2.9, in which a flat fused-silica workpiece is polished using ceria on a polyurethane pad. The time-averaged relative velocity is constant ($R_o = R_L$). If kinematics alone influenced the *surface figure*, workpiece should have stayed flat with polishing time. Instead, it changed significantly.

Many phenomena influence the pressure distribution outlined in Figure 2.1. These may occur simultaneously, which historically has led to problems in deciphering them quantitatively. These phenomena are described in more detail below. Where possible, controlled polishing experiments that isolate or illustrate a change in *surface figure* via a given phenomenon are described.

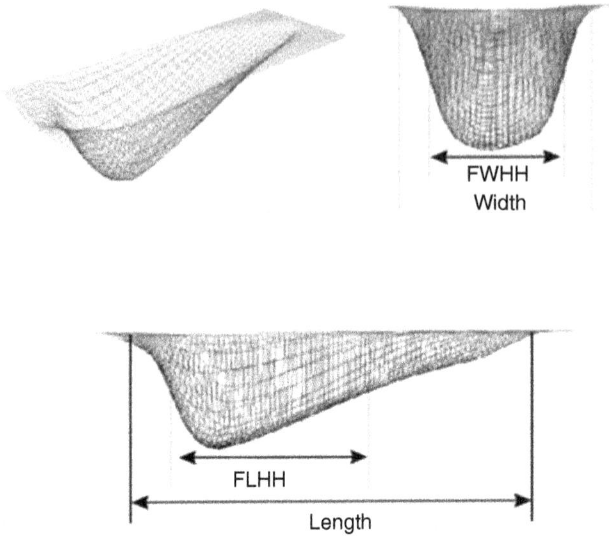

Figure 2.8 An example removal function using a subaperture MRF polishing technique. Source: Reproduced with permission of Menapace et al. 2003 [35].

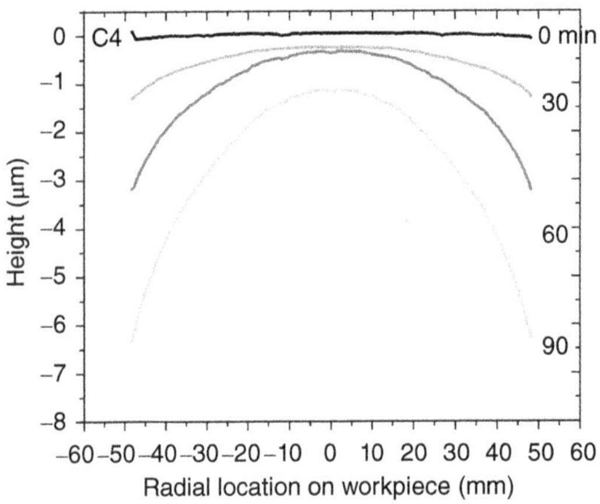

Figure 2.9 Surface figure of fused-silica glass surface as a function of polishing time using kinematics with constant time-averaged relative velocity (CP mode; $R_L = R_o$). Source: Suratwala et al. 2010 [1]. Reproduced with permission from John Wiley & Sons.

2.5.1 Applied Pressure Distribution

An obvious way to influence the pressure distribution on a workpiece is to alter the applied pressure distribution. This is a common, conventional optical-fabrication technique. For example, a wedge can be created on a flat plate by placing an applied load on one side of the workpiece face only, as oppose to uniformly loading the workpiece during polishing [34]. More involved approaches have been proposed by using spatially discrete and different mechanical loading at various locations on the workpiece face to alter the pressure distribution and consequently the rate of *surface figure* change [36–38].

In an analogous technique called stress lap polishing, differential loading is accomplished by spatially modifying the local pressure exerted by the lap or tool [31]. The challenge in using differential weighting is twofold: first, a quantitative correlation between applied pressure distribution (σ_o) and actual pressure-distribution experiences (σ) must be established for this method to be deterministic. This is not easily expressed analytically, as it typically requires finite element analysis (FEA). Second, the length scale of control of the actual pressure distribution is limited, based on the mechanical properties and geometry of the workpiece.

2.5.2 Elastic Lap Response

Even under uniform applied pressure, the elastic response of the lap can lead to nonuniform pressure distribution, because of the change in response of the lap at the edge of the workpiece–lap interface. This effect has been described (e.g. [39]) as the rigid-punch or edge effect. Rigid-punch, one-dimensional pressure distribution takes the form of

$$\sigma(r) = \frac{P}{\pi(r_o{}^2 - r^2)^{1/2}}, \tag{2.9}$$

where P is the uniform applied load and r_o is the radius of the workpiece. The resulting pressure distribution displays a significant rise in pressure at the edge of the workpiece. This effect is the likely reason for the edge roll-off in *surface figures* commonly observed after conventional polishing. A number of theoretical and experimental analyses have evaluated edge roll-off behavior. Kimura et al. [40], for example, measured the edge roll-off after Si-wafer polishing using different pads of various hardness (Figure 2.10). Fu and Chandra [41–43] evaluated the rigid-punch effect specifically for CMP wafer-pad polishing and developed an analytical expression for the interface stress profile accounting for workpiece shape. Figure 2.11 illustrates the large rise in pressure at the edge of the workpiece. The one-dimensional pressure distribution was shown to take the form

$$\sigma(r) = \frac{2}{\pi}\frac{E_2}{1-\nu_2{}^2}\frac{4a_2r^2 + (a_0 - 2a_2r_o{}^2)}{\sqrt{r_o{}^2 - r^2}}, \tag{2.10}$$

where E_2 is the pad modulus, ν_2 is the pad Poisson ratio, r_o is the workpiece radius, a_2 is workpiece curvature, and a_0 is the vertical displacement of the workpiece [43]. The larger the vertical displacement, the greater the discontinuity (a_0) between the workpiece and pad height, which results in a greater edge-stress rise. During CMP, a common technique to reduce the edge effect is to use a retaining ring (e.g. see [44]), where a large preload on the pad

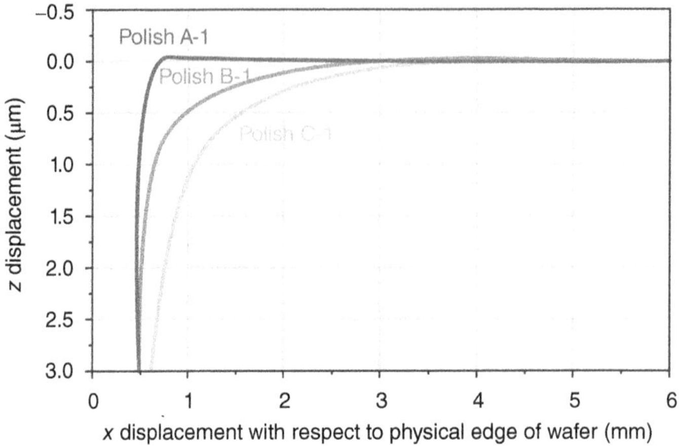

Figure 2.10 Measured edge roll-off after polishing Si wafer with pads of three hardnesses. Source: Kimura et al. 1999 [40]. Reproduced with permission from IOP Publishing.

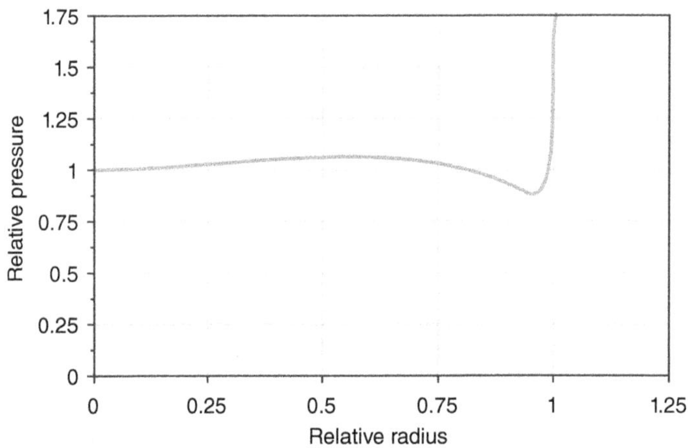

Figure 2.11 Wafer-pad interface pressure when wafer is under uniform backside loading. Source: Fu and Chandra 2005 [41]. Reproduced with permission from Elsevier.

reduces the height discontinuity, and therefore stress, at the workpiece edge. A similar technique has been shown to reduce edge roll-off during optical glass polishing [1].

2.5.3 Hydrodynamic Forces

As discussed in Section 2.3, under the right conditions of applied pressure (σ_o), relative velocity (V_r), and slurry viscosity (η_s), workpiece–lap frictional interactions may occur in the hydrodynamic regime (see Figure 2.4). Here, the slurry fluid trapped between the workpiece and the polishing pad forms a hydrodynamic film, where the friction coefficient is low, and there is larger interface fluid thickness (or gap) than in the contact regime. Fluid-film thickness will vary across the workpiece, resulting in a tilt or wedge of workpiece relative to the lap plane, with the leading edge of the workpiece having a larger thickness than the trailing edge (see Figure 2.5). This change in fluid thickness creates nonuniform pressure distribution across the workpiece. The resulting hydrodynamic pressure distribution can be determined by solving Reynold's equation derived for the incompressible fluid from the Navier–Stoke's equation [11, 12, 45–48]. In one dimension, it is described as follows:

$$\frac{\mathrm{d}(\sigma(x))}{\mathrm{d}x} = -6\eta_s V_r \frac{\Delta h_{oL}(x) - \Delta h_{oL}{}^*}{\Delta h_{oL}(x)^3}, \tag{2.11}$$

where σ is the resulting pressure distribution, η_s is the fluid viscosity, $\Delta h_{oL}(x)$ is the gap between the workpiece and lap (or workpiece–lap mismatch), and $\Delta h_{oL}{}^*$ is the gap at the point of maximum pressure on the workpiece.

While most optical fabrication processes do not operate in hydrodynamic mode, many CMP processes operate in mixed mode – hence hydrodynamic pressures make an important contribution to the overall pressure distribution on the workpiece. It is important that an optical fabricator recognize when process conditions result in these hydrodynamic effects.

Many CMP studies have calculated hydrodynamic pressure distribution and film thicknesses [46–51]. For example, Cho et al. [48] solved the three-dimensional Reynold's equation, determining that film thickness ranges from 15 to 110 μm, with variances in thickness across the workpiece owing to a workpiece tilt of 2–4× for typical CMP polishing conditions. In contrast, typical interface film thicknesses in contact mode are around 0.5–1 μm [52]. This study also found that film thickness increases with viscosity (η_s) and velocity (V_r), and decreases with applied pressure (σ_o), consistent with the Hersey number FOM, which measures the degree at which an interface is in hydrodynamic mode (see Section 2.3). Figure 2.12 shows calculated pressure distribution in terms of mean and shear-stress components as a function of applied pressure, workpiece rotation rate, and slurry viscosity. Mean pressure is highest at the workpiece center and lowest at wafer's edge, whereas shear stress has the opposite trend. Applied pressure increases the mean pressure (especially at the workpiece center), but rotation rate and slurry viscosity increases do not change mean pressure. As applied pressure decreases, mean pressure not only decreases but also becomes more uniform in the inner area of the workpiece.

While in hydrodynamic mode, the influence of this nonuniform pressure distribution on the spatial distribution of the *material removal rate* during polishing is complex to evaluate, compared to evaluation while in contact mode.

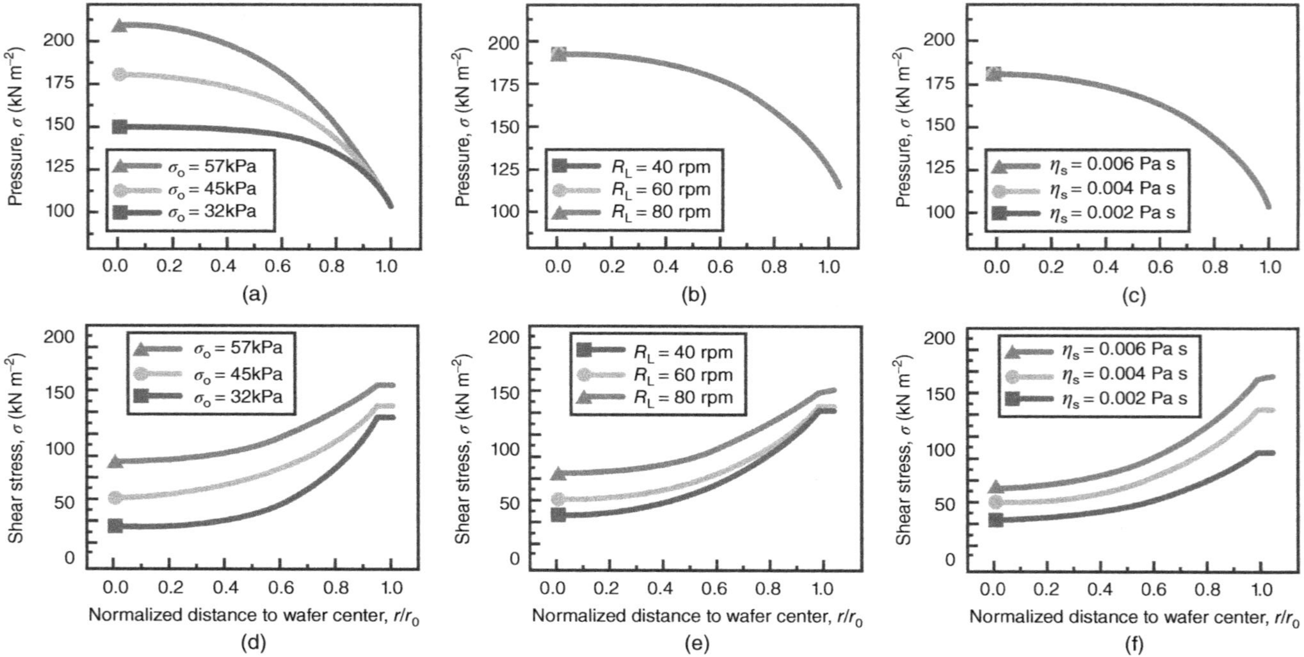

Figure 2.12 Calculated mean (a–c) and shear (d–f) stress due to hydrodynamic forces as a function of applied pressure (a,d), workpiece rotation rate (b,e), and slurry viscosity (c,f). Source: Cho et al. 2001 [48]. Reproduced with permission from Elsevier.

In contact mode, applied pressure is translated to active particles at the interface, as discussed in Section 2.3. In hydrodynamic mode, pressure is carried largely by the fluid – hence, the normal load per particle is small and shear load per particle dominates material removal (e.g. in MRF, as described in Section 8.1). Other effects have been proposed where fluid pressure distribution also changes fluid flow and replenishment at the interface, altering the *material removal rate* [53, 54].

2.5.4 Moment Forces

In addition to workpiece tilt or slope caused by fluid pressure in hydrodynamic mode, a moment force, driven by friction at the workpiece–lap interface while in contact mode, can also lead to a slope in the workpiece – and hence, a nonuniform spatial pressure distribution. Consider the spindle polishing setup shown in Figure 2.13, where the workpiece is held and mechanically loaded by a driving pin [38]. In this configuration, the workpiece has sufficient degrees of freedom to rotate about the driving-pin-spindle axis, which may result in a tilt of the workpiece relative to the lap plane. The driving force for this tilt is the frictional force at the interface and moment created by the spindle pivot point. Using a force and moment balance while at mechanical equilibrium, the total load (P) and moment (M) are given by

$$P = \int_{\text{optic}} \sigma(x, y)\, \mathrm{d}x\, \mathrm{d}y, \tag{2.12}$$

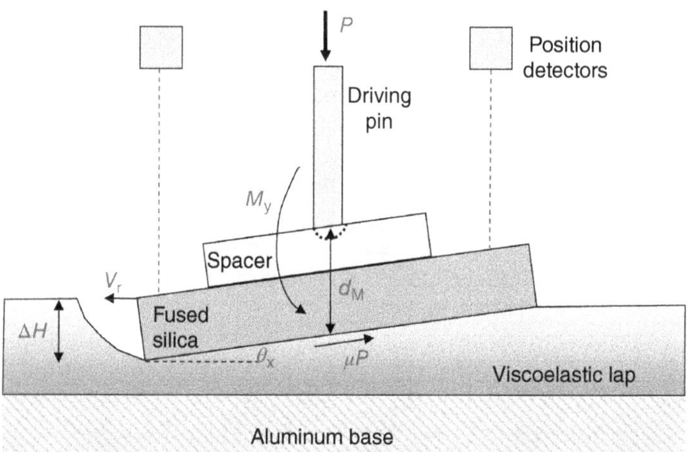

Figure 2.13 Schematic of a spindle polishing setup illustrating parameters affecting moment force and workpiece tilt. Source: Suratwala et al. 2010 [1]. Reproduced with permission from John Wiley & Sons.

$$M_x = \int_{\text{optic}} \sigma(x, y) y \, \mathrm{d}x \, \mathrm{d}y - F_y d_\mathrm{M} = 0, \tag{2.13}$$

$$M_y = F_x d_\mathrm{M} - \int_{\text{optic}} \sigma(x, y) x \, \mathrm{d}x \, \mathrm{d}y = 0, \tag{2.14}$$

where F_x and F_y are the friction force, M_x and M_y are moment in the x and y directions, and d_M is the moment-arm distance.

Figure 2.14 shows the measured optic tilt (or slope) during spindle polishing in the configuration shown in Figure 2.13. Shown are various applied pressures, moment-arm distances, and lap rotation rates, using cerium-oxide slurries on a polyurethane lap [1]. These results indicate a nonzero workpiece slope during polishing, where the leading edge of the workpiece is lower than the trailing edge. The slope increases with increase in moment-arm distance and applied pressure. This is qualitatively consistent with the above friction-induced moment formalism, both in terms of slope direction and scaling with moment-arm distance and applied pressure. Note the slope-direction from the friction-induced moment force is the opposite of the slope that occurs in hydrodynamic forces, where the leading edge of the workpiece is higher than the trailing edge.

The determination of moment and slope becomes more complicated with the addition of stroke in the kinematics, where the moment, and hence slope, becomes time-dependent – in other words, slope changes with the position of

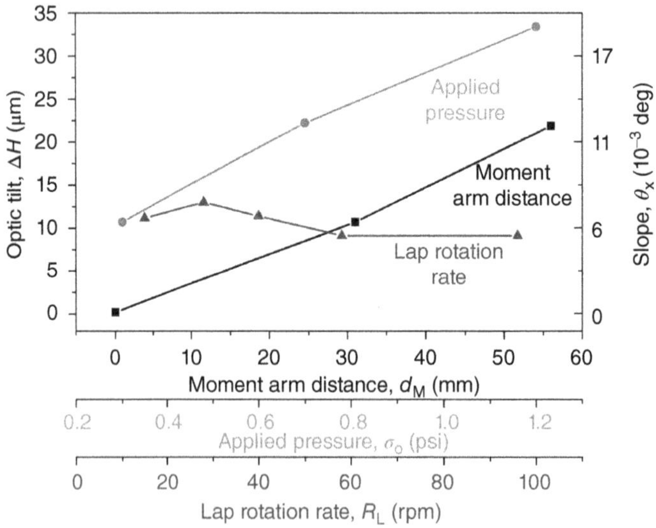

Figure 2.14 Measured workpiece tilt (ΔH) or slope as a function of moment-arm distance (d_M), applied pressure (σ_o), and lap rotation rate (R_L). Source: Suratwala et al. 2010 [1]. Reproduced with permission from John Wiley & Sons.

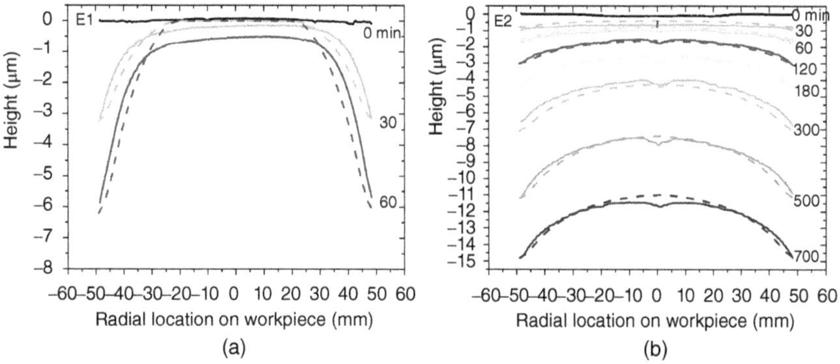

Figure 2.15 Comparison of surface figure evolution while polishing using (a) a spindle load (leading to workpiece slope) versus (b) a wheel loading (with zero workpiece slope). Source: Suratwala et al. 2010 [1]. Reproduced with permission from John Wiley & Sons.

the workpiece along the stroke trajectory. Any offset of the workpiece from the lap surface changes the pressure distribution over a smaller area of the workpiece, and any offset of the workpiece from the lap surface may lead to an additional slope, reflecting the center-of-gravity balance.

Figure 2.15 illustrates how polishing moment forces may change the resulting *surface figure* of a workpiece. Polishing using the spindle configuration, which has a large workpiece slope (as seen in Figure 2.13), was compared with polishing using a wheel configuration, which has essentially no workpiece slope [1]. The kinematics of the experiment was such that $R_o = R_L$. Thus, the time-averaged velocity was constant and material removal nonuniformities owing to kinematics were eliminated (see Section 2.4). In a wheel configuration, the workpiece is loaded using weight, there is no spindle, and wheels are located at the edge of the workpiece to aid workpiece rotation. With spindle loading (as in Figure 2.15a), an initially flat workpiece becomes very convex due to a large increase in material removal at the workpiece edge in a relatively short polishing time of 60 min. An approximately six-times differential in the center versus edge *material removal rate* was measured. Using a wheel configuration (Figure 2.15b), the center versus edge *material removal rate* differential was significantly reduced, closer to ~1.3×, because of the large reduction in the moment force.

The pressure distribution owing to moment force and workpiece slope results in a high-pressure rise at the leading edge of the workpiece and potentially zero pressure some distance from the leading edge, because the workpiece no longer contacts the lap. The larger the slope, the greater the nonuniformity in *material removal rate*. Since the workpiece is usually rotating, the overall surface figure observed would have the same pressure distribution that is rotationally averaged about the center of the workpiece. The convex *surface figure* observed in Figure 2.15a is consistent with the expected pressure distribution. A more quantitative treatment of the resulting pressure distribution owing to slope from the friction-induced moment combined with the viscoelastic-lap contributions is described in the next section.

2.5.5 Viscoelastic and Viscoplastic Lap Properties

2.5.5.1 Viscoelastic Lap

Even after the removal of moment forces via a wheel-driven workpiece, some level of nonuniform material removal is still observed causing the workpiece to go convex (see Figure 2.15b). This leads to the isolation of another source of nonuniform pressure distribution caused by viscoelastic-lap relaxation [1, 42]. The effective elastic modulus of a viscoelastic lap, which determines the stress or pressure response for a given strain, changes with time of loading. The pressure at the leading edge of the workpiece – the position on the lap that makes first contact with the workpiece – is higher than at the trailing edge – when the same position on the lap loses contact with the workpiece. This is because as the lap position travels through the workpiece interface, the lap material has a time-dependent relaxation. The influence of the viscoelastic properties of the polyurethane lap also correlates to the length of the scratches caused by rogue particles that are viscoelastically penetrating into the lap. This phenomenon is discussed in Section 3.1.3 [55].

For a viscoelastic lap loaded by an elastic workpiece, the pressure distribution on the workpiece ($\sigma(x, y)$) can be described by the heredity equation for a constant applied pressure as [1, 39]

$$\sigma(x,y) = \int_0^{t_L(x,y)} E_{rel}(t_L(x,y) - t')\dot{\varepsilon}(t')\,dt', \tag{2.15}$$

where $t_L(x, y)$ is the time of lap exposure at some point x, y on the workpiece for the corresponding point on the lap, E_{rel} is the stress-relaxation function for the viscoelastic-lap material, and $\dot{\varepsilon}(t')$ is the lap strain rate. These parameters are analyzed below.

The interface pressure on the workpiece depends on the time or duration of lap exposure to the viscoelastic lap. Hence the interface pressure distribution depends on the kinematics. The time of lap exposure can be determined using a line path of some point on the lap (x_L, y_L) at the leading edge of the workpiece as it travels to some given point on the workpiece (x, y), as illustrated in Figure 2.16a. The dashed line represents the line path. In the case of kinematics without stroke, as described in Figure 2.16, the time of lap exposure is given by

$$t_L(x,y) = \frac{1}{R_L}\arccos\left(\frac{x \cdot x_L(x,y) + (y+s)(y_L(x,y)+s)}{x^2+(y+s)^2}\right), \tag{2.16}$$

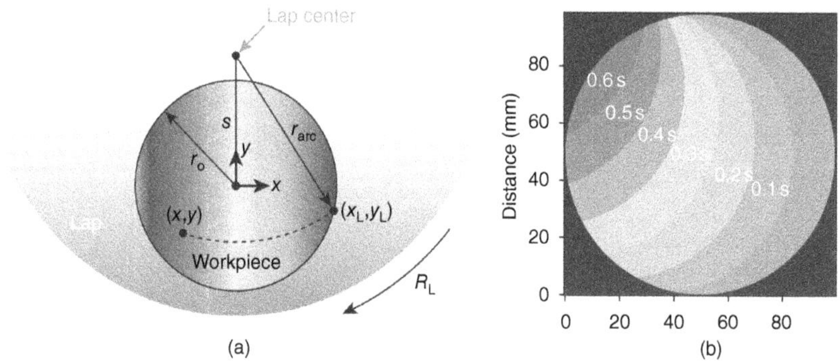

Figure 2.16 (a) Schematic illustrating a given point (x, y) on the workpiece at time of lap exposure $t_L(x, y)$, and the trace of the corresponding point on the lap while it was loaded, starting at the leading edge of the workpiece at point (x_L, y_L). (b) Contour plot of the calculated time of lap exposure $(t_L(x, y))$ for all points on the workpiece surface for the conditions ($r_o = 0.05$ m, $R_L = 20$ rpm, $s = 0.075$ m). The leading edge is at the right. Source: Suratwala et al. 2010 [1]. Reproduced with permission from John Wiley & Sons.

where

$$y_L(x, y) = \frac{x^2 + (y + s)^2 - {r_o}^2 - s^2}{2s}, \tag{2.17}$$

$$x_L(x, y) = \sqrt{r_o^2 - y_L(x, y)^2}, \tag{2.18}$$

and where s is the separation distance from the lap center to the workpiece center, R_L is lap rotation rate, and r_o is the radius of the workpiece. For every point selected on the workpiece (x, y), there is a corresponding point at the leading edge of the workpiece (x_L, y_L). Figure 2.16b shows the calculated time of lap exposure $t_L(x, y)$ for a specific set of kinematic conditions. The minimum time of lap exposure is at the leading edge of the workpiece, and the maximum is at the trailing edge, on the side of the workpiece closest to the lap center. The asymmetry of the time of lap exposure indicates that the velocity of a given point on the lap is lower closest to the lap center, which leads to longer lap exposure. For the case shown in Figure 2.16b, the maximum time of lap exposure is 0.6 s. A similar exercise can be performed for the case with stroke added, though the algebra is more complicated. The time of lap exposure would change along the stroke cycle, whereas without stroke, the time of lap exposure would stay constant.

The behavior of the viscoelastic lap material can be modeled using the delayed-elasticity viscosity model [1, 39]. Figure 2.17 illustrates this model, which comprises two moduli (springs) and one viscosity (the dashpot). The creep-compliance function $J(t)$ and the stress-relaxation function $E_{rel}(t)$

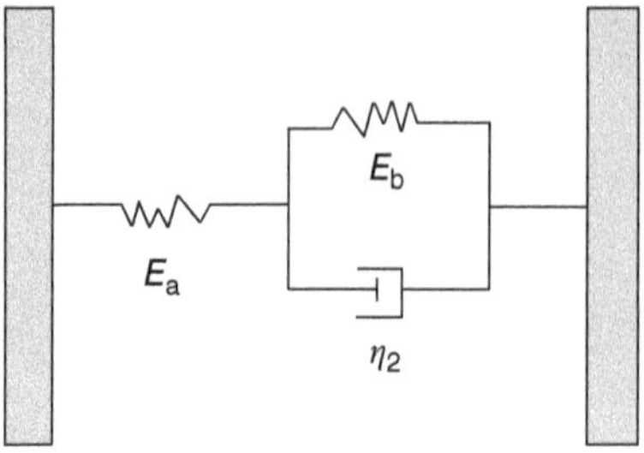

Figure 2.17 Schematic illustration of the delayed elasticity model to describe the viscoelastic behavior of a polyurethane polishing pad. Source: Suratwala et al. 2010 [1]. Reproduced with permission from John Wiley & Sons.

for the lap are described as [39]

$$J(t) = \frac{1}{E_a} + \frac{1}{E_b}\left(1 - e^{\frac{-t}{\tau_c}}\right), \tag{2.19}$$

$$E_{\text{rel}}(t) = \frac{E_a}{E_a + E_b}\left(E_b + E_a e^{\frac{-t}{\tau_s}}\right), \tag{2.20}$$

where τ_c is the creep-compliance time constant, τ_s is the stress-relaxation time constant, and E_a and E_b are modulus components for viscoelastic lap. The following self-similar relationships also apply: [39]

$$E_a + E_b = E_2,\ E_b = \frac{\eta_2}{\tau_c},\ \text{and}\ \tau_s = \frac{\eta_2}{E_2}, \tag{2.21}$$

where E_2 and η_2 are the bulk modulus and viscosity of the lap.

From constant loading experiments on the same polyurethane pad, the creep-compliance time constant was determined to be $\tau_c = 4.0$ s [1]. From dynamic mechanical analysis performed by Lu et al. [56] on the polyurethane pad, $E_2 = 100$ MPa and $\eta_2 = 9.7 \times 10^7$ P. Hence, using Equation (2.21), $E_a = 97.75$ MPa, $E_b = 2.25$ MPa, and $\tau_s = 0.1$ s. Note that the stress-relaxation time constant (τ_s) is less than the maximum time of lap exposure (see Figure 2.16b), suggesting that a significant amount of stress relaxation can occur under this set of kinematics with this pad. With all of these parameters quantitatively known, the stress-relaxation function, Equation (2.20), is now quantitatively defined.

The final component needed to determine the pressure distribution due to viscoelastic relaxation using Equation (2.15) is the strain rate ($\dot{\varepsilon}(t')$). The strain on the lap is constrained by the shape of the workpiece and its orientation with respect to the lap, or the slope. Note that workpiece slope contributions can also come from moment forces, as discussed in Section 2.5.4. For the cases in which the workpiece *surface figure* is flat, strain as a function of workpiece position can be defined as follows:

$$\varepsilon(x, y) = \frac{\tan(\theta_x)x}{t_2} + \frac{\tan(\theta_y)y}{t_2} + \varepsilon_c, \tag{2.22}$$

where θ_x and θ_y are the slopes of the workpiece in the x and y directions relative to the lap plane, ε_c is the elastic strain at the center of the workpiece, and t_2 is the thickness of the viscoelastic pad. It is convenient to describe the strain as a function of time ($\varepsilon(t)$) instead of position, which can be done using

$$x = r_{\mathrm{arc}} \cos\left(R_{\mathrm{L}} t + \left(\arccos \frac{x_{\mathrm{L}}}{r_{\mathrm{arc}}}\right)\right), \tag{2.23}$$

$$y = r_{\mathrm{arc}} \sin\left(R_{\mathrm{L}} t + \left(\arccos \frac{x_{\mathrm{L}}}{r_{\mathrm{arc}}}\right)\right) - s, \tag{2.24}$$

$$r_{\mathrm{arc}} = \sqrt{x^2 + (y+s)^2}, \tag{2.25}$$

where r_{arc} is the arc radius for a given point (x_{L}, y_{L}) at the leading edge of the workpiece (see Figure 2.16a) relative to the lap center. Substituting into Equation (2.22) and then differentiating gives the strain rate as follows:

$$\begin{aligned}\dot{\varepsilon}(t) = &-\frac{\tan(\theta_x)}{t_2} r_{\mathrm{arc}} R_{\mathrm{L}} \sin\left(R_{\mathrm{L}} t + \left(\arccos \frac{x_{\mathrm{L}}}{r_{\mathrm{arc}}}\right)\right)\\ &- \frac{\tan(\theta_y)}{t_2} r_{\mathrm{arc}} R_{\mathrm{L}} \cos\left(R_{\mathrm{L}} t + \left(\arccos \frac{x_{\mathrm{L}}}{r_{\mathrm{arc}}}\right)\right).\end{aligned} \tag{2.26}$$

Using the equations above, we can now determine the pressure distribution on a nonrotated workpiece. Figure 2.18a shows the calculated pressure distribution using the conditions where the workpiece does not rotate, but lap does. For comparison, the measured surface profile for a fused-silica workpiece polished under these conditions is shown in Figure 2.18b [1]. The leading edge of the workpiece in each image is marked by a star. The observed removal is qualitatively consistent with the calculated pressure distribution, where the

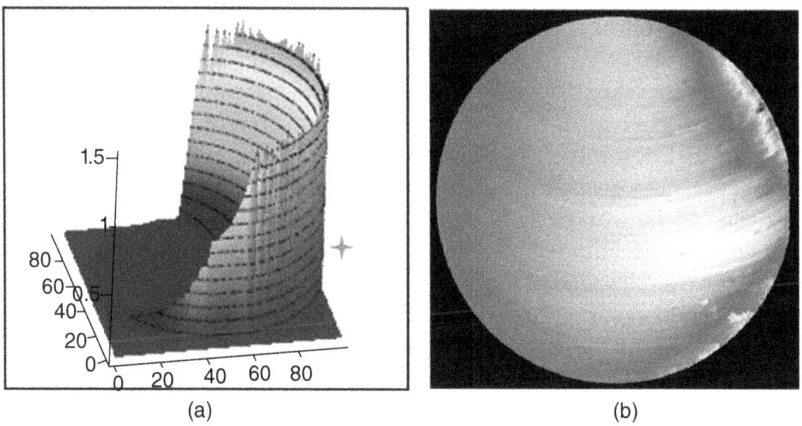

Figure 2.18 (a) Calculated stress profile on a nonrotated workpiece using Equation (2.15); (b) measured surface figure of a nonrotated workpiece during polishing. The leading edge of the workpiece in each image is marked by a star. Source: Suratwala et al. 2010 [1]. Reproduced with permission from John Wiley & Sons.

leading-edge experiences a much higher removal or pressure. During polishing, the workpiece is almost always rotated; hence, the overall pressure distribution on the workpiece would be a time-average of nonrotated pressure distribution rotated about the center of the workpiece, described as

$$\sigma(r) = \frac{1}{2\pi} \int_0^{2\pi} \sigma(r, \theta)\, \mathrm{d}\theta, \tag{2.27}$$

where $\sigma(r, \theta)$ is the pressure distribution determined by Equation (2.15) in cylindrical coordinates. As the slope of the workpiece is increased relative to the lap plane, the time-averaged rotated pressure distribution becomes more nonuniform, and thus material removal becomes more nonuniform. This is consistent with the observed increase in material removal nonuniformity owing to moment forces discussed in the previous section.

2.5.5.2 Viscoplastic Lap

The behavior of conventional pitch polishing is a bit different from pad polishing. For a pad lap with viscoelastic response, removal of workpiece load results in the lap returning to its original shape. By contrast, for a pitch lap with viscoplastic response, the workpiece or conditioner load results in irreversible change to the shape of the lap. Pitch has historically been the lap medium of choice for polishing optics because its viscoplastic response allows precise shaping of the lap, such that the workpiece can migrate to the shape of the lap – enabling very high *surface figure* control on the workpiece – and the ability of slurry particles to partially embed viscoplastically into the pitch lap – leading to low-roughness surfaces, as discussed in Chapter 4.

Pitch polishing, including characteristics such as viscosity, hardness, softening point, and the effect of temperature and additives for various compositions, is well documented [9, 21, 38, 57–60]. Here, we focus on viscoplastic response and impact on pressure distribution. Most commercial pitches have high viscosities (10^9–10^{10} P). Hence, the viscous response time of pitch is generally longer than that observed with polyurethane-pad polishing – many minutes versus fractions of a second. Hence, the pressure-distribution response at the leading edge of the workpiece is expected to be much less with pitch polishing, but continued loading of pitch by the workpiece or conditioner continues to modify the overall shape of the lap nevertheless, due to the plastic response. This leads to another source of pressure-distribution response to the workpiece: workpiece–lap mismatch.

2.5.6 Workpiece–Lap Mismatch

Another strong influence on pressure distribution is the surface mismatch at contact between the workpiece and lap, or workpiece–lap mismatch. If both surfaces are perfectly matched, the mismatch is zero and the pressure distribution is uniform (ignoring other effects contributing to pressure distribution). If the surfaces are not matched, pressure distribution will not be uniform, leading to lower pressures where the mismatch is larger and higher pressures where the mismatch is small or zero. The compliance, or modulus, of the workpiece and lap determine the magnitude of the pressure deviation.

To illustrate the mismatch effect quantitatively, two series of polishing experiments are compared using a polyurethane lap on either a flat base (±1 μm PV) or a convex-lap base (±20 μm PV) [1]. The average measured removal rates as a function of radial position on the workpiece are shown in Figure 2.19. In Figure 2.19a, polishing kinematics using no stroke is shown; Figure 2.19b shows the results of a large stroke (such that the workpiece

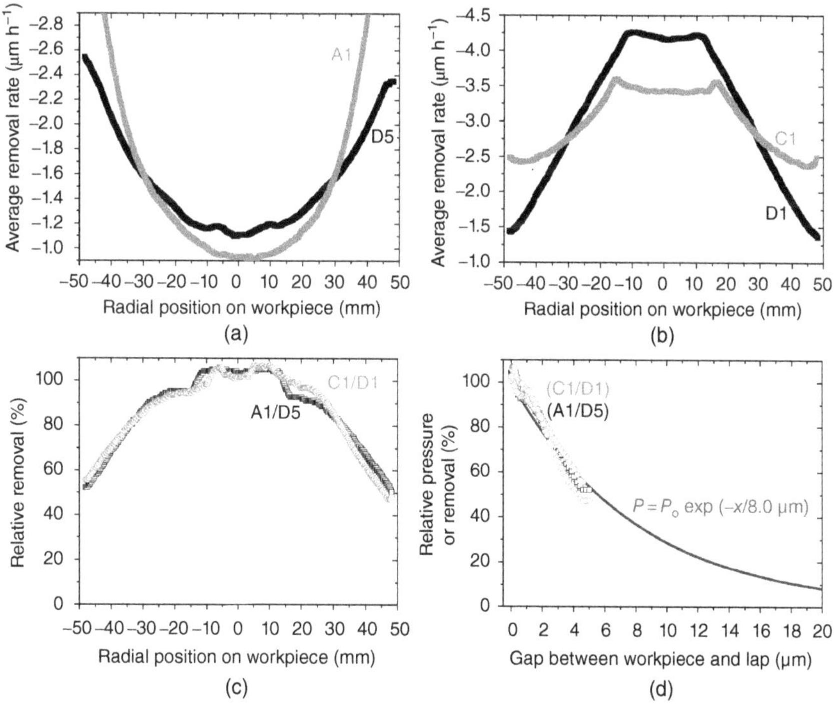

Figure 2.19 (a) Measured average removal rate as a function of radial position on a workpiece using a flat (A1) and convex lap (D5) and kinematics with no stroke. (b) Measured average removal rate using a flat (C1) and convex lap (D1) and kinematics with large stroke. (c) Relative removal (normalized at the center of the workpiece) between flat and convex lap using the measured average removal rate shown in panels (a) and (b). (d) Relative pressure or removal as a function of mismatch or gap between the contacting surface profiles between workpiece and lap. The solid line shows a single exponential fit. Source: Suratwala et al. 2010 [1]. Reproduced with permission from John Wiley & Sons.

comes partly off the lap during the stroke). Each plot compares two samples that were polished under identical conditions, except for the curvature of the lap metal base. The removal rate using the convex lap is greater at the center of the workpiece and lower at the edges, relative to the removal rate observed using the flat lap. This deviation in removal stems from the change in pressure distribution because of workpiece–lap mismatch. Taking the ratio of the removal rate at each radial position for each of the two series of samples in Figure 2.19a–b, the relative removal as a function of radial position on the workpiece is shown in Figure 2.19c. Note that the effect of lap curvature on removal for both polishing experiments was about the same, though they were conducted under very different kinematics. For both series of polishing experiments, the removal rate decreased by ~50% at the edge relative to the removal rate at the center.

The relative removal shown in Figure 2.19c can be described in terms of the workpiece–lap mismatch by describing the lap curvature as a sphere [1]. The height ($h_L(x)$) of the lap normalized to zero at the ends is given by

$$h_L(x) = \sqrt{{\rho_L}^2 - x^2} - (\rho_L - h_{PV}), \tag{2.28}$$

where ρ_L is the radius of curvature of the lap surface and h_{PV} is the PV height of the lap. For the convex lap used here, ρ_L is 250 m and $h_{PV} = 20$ μm. Allowing the workpiece to tilt to achieve best contact with the lap, the workpiece–lap mismatch (Δh_{oL}) is then

$$\Delta h_{oL}(x) = h_L(x) - h_o(x), \tag{2.29}$$

where the $h_o(x)$ is the height of the workpiece described by

$$h_o(x) = \tan(\theta_x)x + h_c, \tag{2.30}$$

where h_c is height at the center of the workpiece. For the convex lap used here, $\tan(\theta_x) = 20$ μm/100 mm $= 2 \times 10^{-4}$ and $h_c = 20$ μm. Applying Equations (2.28)–(2.30), the data in Figure 2.19c was converted to the relative pressure as a function of the workpiece–lap mismatch in Figure 2.19d. The data shows that the removal rate (and hence pressure) on the workpiece drops linearly; at a 5-μm mismatch, pressure, and removal drop ~50%. The belief is, however, that for larger mismatch, the pressure dependence on the mismatch will become nonlinear. Hence, relative pressure as a function of mismatch $\Delta h_{oL}(x)$ is quantitatively described as

$$\frac{\sigma}{\sigma_o}(\Delta h_{oL}) = e^{-\Delta h_{oL}/\overline{h}}, \tag{2.31}$$

where $\overline{h}$ is a constant describing the rate at which pressure drops with increase in workpiece–lap mismatch. A reasonable fit to the data is obtained using $\overline{h} = 8$ μm, which is shown as a solid line in Figure 2.19d.

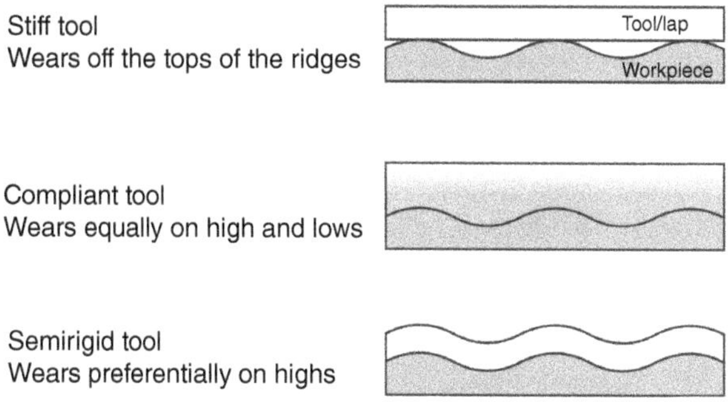

Figure 2.20 The effect of tool stiffness on the workpiece–lap mismatch and ultimate wear on the workpiece. Source: Burge et al. 2001 [61]. Reproduced with permission from SPIE.

This polishing example was for a relatively stiff polyurethane pad. However, the mechanical properties of the workpiece, and more so the lap, determine the value of $\bar{h}$. A low-modulus lap (or more compliant lap) would lead to a larger value of $\bar{h}$ and less convergence (or smoothing) of the workpiece to the shape of the lap. Conversely, a higher modulus (less compliant lap) would lead to a lower value of $\bar{h}$ and greater convergence, or smoothing, of small-scale errors of the workpiece to the shape of the lap. This concept is qualitatively illustrated by Burge et al. [61] as shown in Figure 2.20. The example in Figure 2.18 is a useful technique to determine $\bar{h}$ for a given polishing system.

In the example above, the source of workpiece–lap mismatch was the curvature of the lap base. There are many potential sources for mismatch, notably workpiece shape, lap shape, mechanical bending of the workpiece, residual stress, and temperature-gradient-induced bending of the workpiece, slurry spatial distribution, and local nonlinear material deposit on the lap. These are described in detail below.

2.5.6.1 Workpiece Shape

A primary objective in optical polishing is to modify the workpiece shape, that is, the *surface figure*. Since the workpiece shape is changing over time and the shape itself also contributes to pressure distribution, there will be a time-dependent, pressure-distribution response to workpiece shape. The effects of workpiece shape on pressure distribution behave like those discussed earlier – they are governed by Equation (2.31). In Section 8.3, a novel polishing technique called Convergent polishing is described, where all the sources of material removal nonuniformity (including those that affect pressure distribution) have been eliminated or reduced, except mismatch caused by workpiece shape.

2.5.6.2 Pad Wear/Deformation

The influence of pad wear and changes in pad properties during polishing are well documented [62–65]. In these studies, the workpiece and/or a conditioner – typically a fixed diamond abrasive to treat the pad or sacrificial workpiece used in pad optical polishing – is loaded against the pad surface. As a result, spatial and temporal changes in pad properties such as thickness profile, roughness, and slurry-charging level may occur. Since the size of the workpiece or conditioner is typically not the same as the lap, and since the kinematic controls are largely focused on achieving uniform removal on the workpiece, the pad usually wears sacrificially and nonuniformly, resulting in workpiece–lap mismatch, nonuniform pressure distribution, and nonuniform material removal [1]. Chang et al. [63] modeled the typical kinematics of the conditioner and showed that it will wear the pad nonuniformly, resulting in a concave pad surface. Park et al. [65] showed that the wafer (or workpiece) also contributes to the nonuniform roughness and thickness of a polyurethane foam pad.

Based on the above mentioned facts, pad wear may evolve temporally and spatially in a complex way. It is useful to describe it quantitatively. The time-averaged pad wear may be described in a manner similar to that of material removal from the workpiece, with a Preston-type equation in the form of

$$\frac{\mathrm{d}h_{\mathrm{lap}}}{\mathrm{d}t}(x,y,t) = \frac{f_{\mathrm{o}}(r)}{t}\int_0^t k_{\mathrm{lap}}(t')\,\mu(x,y,t')\,v_{\mathrm{r}}(x,y,t')\,\sigma(x,y,t')\,\mathrm{d}t', \tag{2.32}$$

where $f_{\mathrm{o}}(r)$ is the fraction of the lap circumference loaded by the workpiece as a function of radial distance (r) from lap center and k_{lap} is the lap Preston coefficient. μ, v_{r}, and σ are synonymous with those in Equation (2.2) [66].

Even though the lap Preston coefficient is observed to be generally spatially constant, it may vary with time. Figure 2.21 shows the results of the measured pad wear rate of a fibrous polyurethane polishing pad (specifically, a Suba 550), which was determined by measuring the thickness of the pad after various polishing iterations [66]. The wear rate was initially high, but decreased with polishing time and then stabilized. From the data, the lap Preston coefficient (k_{lap}) can be empirically described as a single exponential decay as a function of time (t) in the form

$$k_{\mathrm{lap}}(t) = k_{\mathrm{olap}}e^{-t/t_{\mathrm{lap}}}, \tag{2.33}$$

where $k_{\mathrm{olap}} = 1.65\times10^{-4}$ μm m N^{-1} and $t_{\mathrm{lap}} = 23$ h. Note the initial value of the lap Preston coefficient is ~100× greater than the Preston coefficient for the fused-silica workpiece ($k_{\mathrm{p}} = 1.9\times10^{-6}$ μm m N^{-1} using $\mu = 0.8$ on Suba 550 and $k_{\mathrm{p}} = 1.5\times10^{-6}$ μm m N^{-1} using $\mu = 0.5$ on an IC1000 pad) [1]. However, the aged lap shows wear that is equivalent to or less than that of the workpiece. The nature of this change in wear rate is believed to stem from the fibrous structure

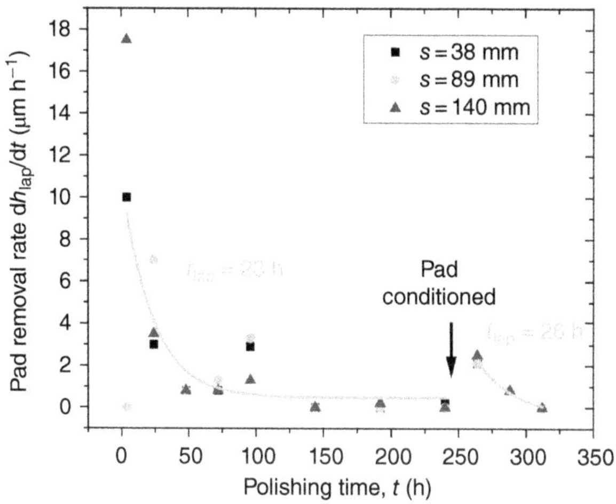

Figure 2.21 Measured pad removal rate after polishing fused-silica coupons. The lines are single exponential fits to the measured data. Source: Suratwala et al. 2012 [66]. Reproduced with permission from John Wiley & Sons.

of the Suba pad, in which loose fibers are initially removed and subsequent removal slows down – possibly from fiber compaction or the exposure of more strongly held fibers. In contrast, the solid-foam IC1000 pad appears to show little change in pad wear rate with time, as inferred by a similar measured workpiece *surface figure* for polishing conducted on a new or aged IC1000 pad (data not shown) [66].

To evaluate the overall spatial lap wear rate for the case of a round workpiece (Equation (2.32)), $f_o(r)$ can be described as follows:

$$f_o(r) = \frac{\theta_L(r)}{2\pi} = \frac{\arcsin\left(\frac{x_L(r)}{r}\right)}{\pi}, \tag{2.34}$$

where θ_L is the solid angle of the lap covered by the workpiece, and $x_L(r)$ is the x component of a point on the leading edge of the workpiece (see Figure 2.22). Using the geometric relations of

$$x_L{}^2 + y_L{}^2 = r^2, \tag{2.35}$$

$$x_L{}^2 + (y_L + s)^2 = r_o{}^2, \tag{2.36}$$

results in

$$x_L(r) = \sqrt{-\left(\frac{r_o{}^2 - s^2 - r^2}{2s}\right)^2 + r^2}. \tag{2.37}$$

With a uniform applied pressure and matched rotation ($R_o = R_L$), k_{lap}, μ, σ, and v_r are spatially constant, and hence, the nonuniformity of the lap wear due

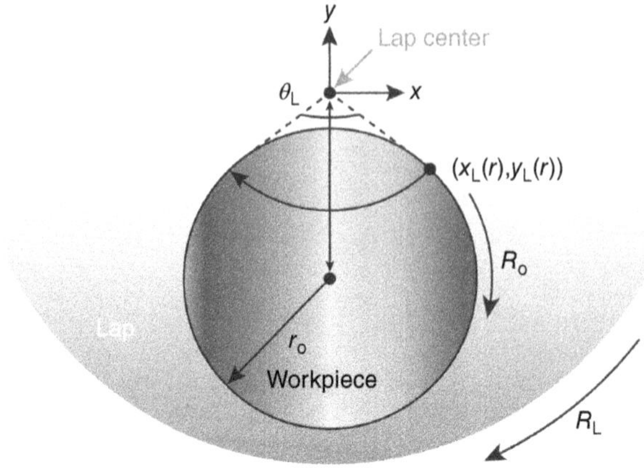

Figure 2.22 Schematic of the polishing setup illustrating the coordinate system and geometry associated with a point (x_L, y_L) at the leading edge of the workpiece and its corresponding path on the lap. Source: Suratwala et al. 2010 [1]. Reproduced with permission from John Wiley & Sons.

to the workpiece given by Equation (2.32) is dominated by $f_o(r)$. In this case, the pad wear rate simplifies to

$$\frac{dh_{lap}(r)}{dt} = 2\arcsin\left(\frac{x(r)}{r}\right) k_{lap}\,\mu R_o s \sigma_o, \tag{2.38}$$

where the time-averaged relative velocity is $V_r = 2\pi R_o s$. Note that while k_{lap} may be nonlinear with time, it does not vary spatially [66, 67].

Using Equation (2.38) for typical polishing conditions, the calculated pad wear-rate radial profile is shown in Figure 2.23a,b. The calculated profile is qualitatively consistent with the pad profile measured experimentally by Park et al. [65] after polishing many Si wafers under similar kinematic conditions. Because pad wear is greatest near the workpiece center, the *surface figure* of the workpiece is expected to become convex, owing to the greater workpiece–lap mismatch and pressure drop in the center versus the edges of the workpiece.

The rate of evolution of the *surface figure* because of pad wear is illustrated in Figure 2.24a,b, which shows the workpiece going convex at different rates under a new pad versus used. On a new Suba 550 pad, the rate of turning convex is higher because of the large Preston coefficient of the lap; see Equation (2.38). Whether these pads are old or new, the within workpiece nonuniformity (WIWNU) has high values of 92% and 27%, respectively [66].

2.5.6.3 Workpiece Bending

Mechanical bending of or residual stress in a high-aspect-ratio (AR) (diameter to thickness) workpiece or lap can also contribute to a change in workpiece–lap mismatch. Depending on the type of removal needed, workpiece bending can

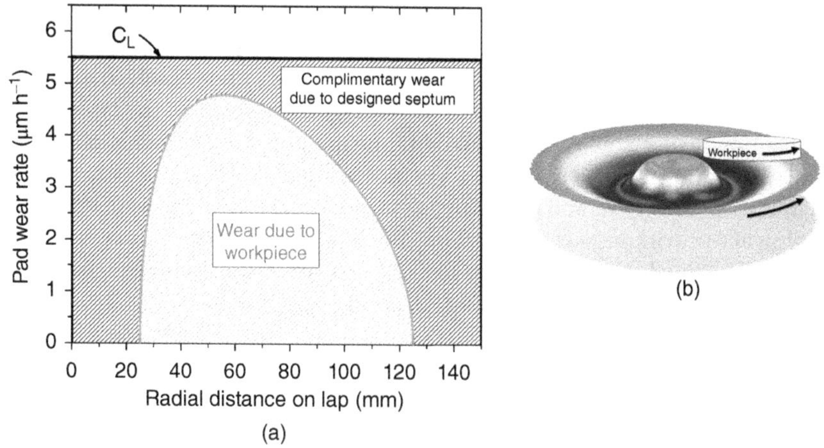

Figure 2.23 Calculated pad wear rate using Equation (2.38) for a round workpiece with polishing conditions of $\mu = 0.7$, $\sigma_o = 2068$ Pa (0.3 psi), $R_o = 20$ rpm, $s = 75$ mm, $r_o = 50$ mm, $r_{lap} = 150$ mm: (a) wear-rate as a function of radial distance on lap and (b) 3D contour illustrating the shape of lap wear due to workpiece, with scale exaggerated for illustration. Source: Suratwala et al. 2010 [1]. Reproduced with permission from John Wiley & Sons. Suratwala et al. 2012 [66]. Reproduced with permission from John Wiley & Sons.

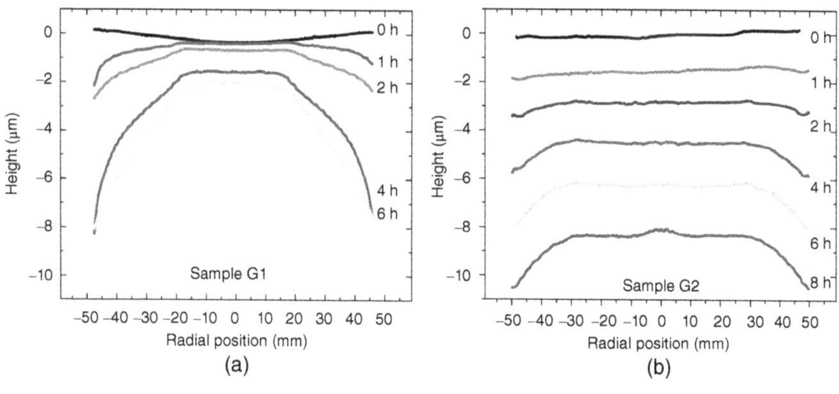

Figure 2.24 Radial surface profile of fused-silica workpieces after each polishing iteration, using (a) a new (G1) and (b) aged (G2) Suba 550 polishing pad. Source: Suratwala et al. 2010 [1]. Reproduced with permission from John Wiley & Sons. Suratwala et al. 2012 [66]. Reproduced with permission from John Wiley & Sons.

be a desired or undesired characteristic of the polishing process. For example, when polishing Si wafers during CMP, uniform removal over the whole workpiece is desired, and hence bending may be advantageous. Ng et al. [68] measured the degree of silicon-wafer bending during CMP and compared it to FEA calculations. Uniform removal is typically aided by using a compliant subpad that reduces the workpiece–lap mismatch and results in improved WIWNU. However, in the case of optical finishing where the workpiece is initially not at the desired *surface figure*, workpiece bending is usually undesirable, because it would interfere with the workpiece's converging to the shape of the lap.

The amount that a workpiece will bend during polishing depends on many factors, including workpiece geometry – for example, shape, AR, elastic modulus, initial *surface figure*, and mounting conditions. One method to estimate the bending of the workpiece is the Kirchoff–Love extension to the Euler–Bernouli bending theory, which describes the deflection of an elastic, homogeneous thin plate, or beam with no shear bending, given by

$$\sigma_o = \frac{E_1 t_1^3}{12(1 - {v_1}^2)} \nabla^4 w_1 + \frac{E_2}{t_2} w_1, \tag{2.39}$$

where w_1 is the workpiece deflection, E_1 is the workpiece modulus, t_1 is workpiece thickness, v_1 is workpiece Poisson's ratio, E_2 is the lap modulus, and t_2 is lap thickness [69]. The second term on the right accounts for the elastic response of the lap. For thinner workpieces, the second term dominates, suggesting that the plate will largely bend to the lap. For thicker workpieces, the first term dominates, in the solution of which the maximum deflection takes the form of

$$w_{\max} = 0.813\sigma_o \frac{{r_o}^4}{E_1 {t_1}^3}, \tag{2.40}$$

where r_o is the workpiece radius [70]. Note that $w_{\max}$ is proportional to $\sim\mathrm{AR}^3$ (AR = aspect ratio or $2r_o/t_1$), which suggests that degree of deflection is strongly dependent on the AR of the workpiece.

The amount of workpiece bending can be solved more precisely to account for shear affects and surface figure of the workpiece using FEA. In these calculations, a fused-silica workpiece (r_o = 75 mm) of varying thicknesses and with an initially concave surface figure (PV = 15.8 μm) was uniformly loaded (σ_o = 2068 Pa or 0.3 psi) onto a polyurethane pad. The results, shown in Figure 2.25, indicate that a thick workpiece (t_2 = 10 mm or AR < 15) will not bend significantly, whereas a thin workpiece (t_2 < 3.3 mm or AR > 45) will. The solid line is the freestanding initial surface profile; the dashed lines are the surface profile that results under applied load. There is a large increase in bending going from 10- to 3.3-mm thick workpieces, then little change after reducing the thickness to 1 mm.

The method of mounting or blocking the workpiece may control the amount of workpiece bending during polishing. Where workpiece bending is desired, a compliant interface with a low elastic modulus between the workpiece and loading weight can be used, such as foam. Achieving a mounting condition for a high-AR workpiece that prevents bending is more challenging. One can mount

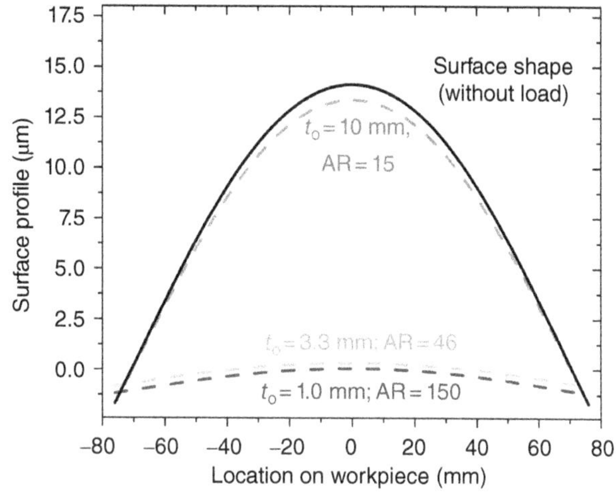

Figure 2.25 Calculated surface profile using FEA of fused-silica workpieces (r_o = 75 mm, initial PV = 15.8 μm) with different thicknesses (t_1) or aspect ratios (ARs) after being loaded at 0.3 psi. Source: Suratwala et al. 2012 [66]. Reproduced with permission from John Wiley & Sons.

to a stiff mounting block, but the degree of workpiece bending will depend on the surface figures of both the opposing surface of the workpiece (i.e. the side not being polished) and the surface of the mounting block. To reduce this dependency, an optical-fabrication technique known as pitch-button blocking (PBB) can be used (see Section 7.1.1 for details) [71]. This blocking technique uses small islands of pitch between the workpiece and mount, which are cooled from the pitch softening temperature. At room temperature, the workpiece–pitch-button–mount system is stiff and the workpiece largely maintains its initial surface figure. Figure 2.26a,b, shows the results of polishing a high-AR workpiece with a concave surface using PBB. After polishing, the workpiece converges to the shape of the lap because of a lack of workpiece bending and a dominance of the workpiece–lap mismatch (due to the workpiece's surface figure leading to convergence with the lap's flat shape). In contrast, Figure 2.26c,d, shows the results of polishing a concave surface on a high-AR workpiece, which remains largely concave after polishing due to the compliant mounting. The bending of the workpiece allows minimizing the workpiece–lap mismatch and the spatial variation in pressure distribution. This results in uniform removal on the workpiece and little or no change in the surface figure.

2.5.6.4 Residual Grinding Stress

In any optical-fabrication process, there is a transition from grinding (removal by fracture) to polishing (removal by chemical or nanoplastic means). During polishing, the removal of the ground surface layer of the workpiece, which is known to be in high compression [34, 72], changes the equilibrium stress distribution in the workpiece. An unconstrained workpiece ground on one face

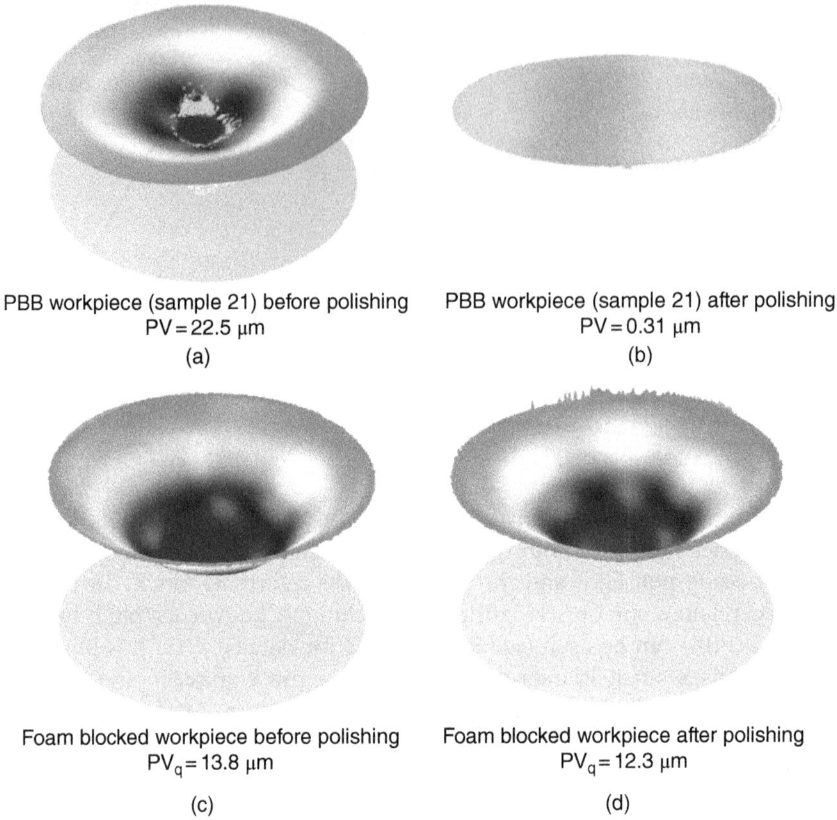

Figure 2.26 Surface figure of a high-AR workpiece (100-mm × 2.2-mm thick) using (stiff) PBB (a) before and (b) after polishing. The full scale is −18 to 5 μm. Below, surface figure of a high-AR workpiece (100 mm × 2.2 mm thick) using foam (compliant) blocking (c) before and (d) after polishing. The full scale is −8.0 to 6.0 μm. Source: Feit et al. 2012 [71]. Reproduced with permission from Optical Society of America.

will bend, changing the *surface figure*, a result referred to as the Twyman effect [38, 73]. Lambropoulos et al. [74] quantified the Twyman stress for a variety of grinding processes, where the magnitude of the stress is scaled with the size of the abrasive used in grinding (see Figure 2.27). A similar study was performed by Chen and De Wolf [77] on Si wafers.

A simple experiment can be used to quantify the effect of grinding residual stress on the change in surface figure. Fused-silica samples were ground on one side of a polished workpiece and the change in the surface figure of the opposing surface was measured [1]. Views a through d in Figure 2.28 show the impact of loose-abrasive grinding with 9 and 30 μm alumina, which caused the opposing surface to deform significantly (PV = 3.51 and 6.5 μm concave, respectively).

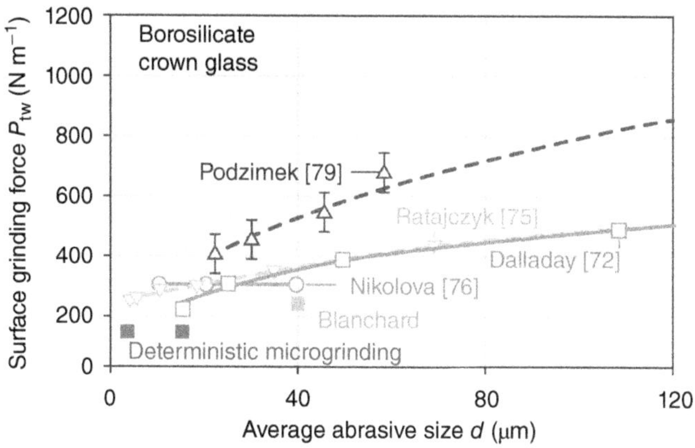

Figure 2.27 Grinding-surface Twyman stress on a workpiece as a function of abrasive for borosilicate crown glass. Source: Lambropoulos et al. 1996 [74]. Reproduced with permission from The Optical Society.

The magnitude of the stress can be inferred from the following form of the Twyman stress equation [78], given by

$$PV = \frac{3}{4}\frac{P_{tw}(1-\nu_1)}{E_1}\left(\frac{2r_o}{t_1}\right)^2, \tag{2.41}$$

where P_{tw} is the Twyman stress and ν_1 is the workpiece Poisson's ratio. Based on these deflections, $P_{tw} = 270\,\mathrm{N\,m^{-1}}$ using 9-μm alumina and $P_{tw} = 420\,\mathrm{N\,m^{-1}}$ using 30-μm alumina [66]. These values match well and scale similarly with the particle size of those reported by Lambropoulos et al. [74] and Podzimek and Heuvelman [79] (see Figure 2.27).

The residual stress on the ground workpiece will be additive to the applied pressure (σ_o) during polishing. Hence, the residual stress may influence workpiece bending – and therefore, the workpiece–lap mismatch and removal uniformity. However, since a ground surface cannot be measured interferometrically, it is difficult to quantify this effect. Instead, two methods to remove the residual stress have been explored: (i) polishing away the ground surface in a separate step or (ii) chemically etching the surface to remove residual stress [66]. The results for incrementally polishing a ground surface and its impact on the final workpiece shape of the opposing surface are shown in Figure 2.29. As polishing occurs, the residual stress layer is incrementally removed and the resulting Twyman stress-induced *surface figure* change is reduced. The depth of the residual stress layer is <10 μm. Hence, by prepolishing away at least 10 μm, one can remove the residual stress and avoid potential effects on removal uniformity. Alternatively, one may can etch the ground surface using, for example, NH_4F:HF for a fused silica workpiece. After removing

Figure 2.28 Surface figure of a high AR workpiece (100 mm × 2.2 mm) (a) before and (b) after grinding the opposing surface with 9 μm alumina abrasive. Surface figure of a high-AR workpiece (100 mm × 2.2 mm thick) (c) before and (d) after grinding the opposing surface with 30 μm alumina abrasive. (e) Surface figure of a high-AR workpiece (100 mm × 2.2 mm thick) first ground with 9 μm alumina abrasive and then etched in NH_4F:HF for 30 min. All surface figures are plotted at same full scale, −3.5 to 3.5 μm. Source: Suratwala et al. 2012 [66]. Reproduced with permission from John Wiley & Sons.

~10 μm of material by etching, the surface figure of the opposing surface returns nearly to its preground state (see Figure 2.28e), confirming the ability to remove residual stress and its potential effects on removal uniformity. Note that etching a ground surface has other advantages, such as exposing scratches for *surface quality* assessment and reducing the amount of material removal needed to polish through the subsurface mechanical damage from the grinding process [80, 81].

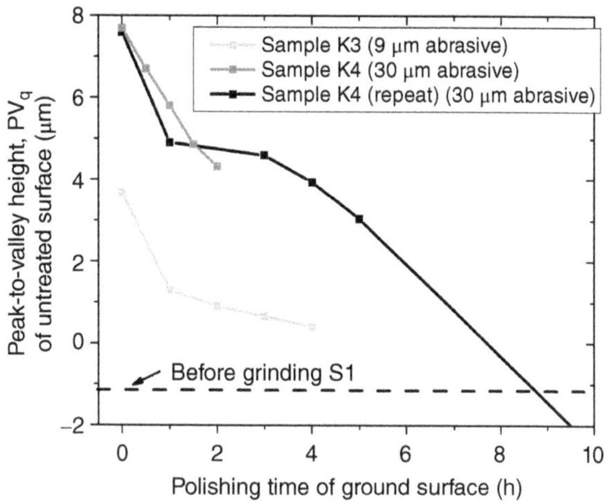

Figure 2.29 Surface figure of the opposing surface in terms of PV of high-AR fused-silica workpieces (100 mm × 2.2 mm thick) initially ground with 9- or 30-μm alumina abrasive and then polished for various times. Source: Suratwala et al. 2012 [66]. Reproduced with permission from John Wiley & Sons.

2.5.6.5 Temperature

The temperature at the interface between the workpiece and lap influences the *material removal rate* magnitude by changing the chemical reactivity of the slurry. The major source of thermal heating is the frictional work of the removal by the individual polishing particles at the workpiece–lap interface. A number of models describing this temperature rise have been proposed, using kinematics [82], frictional-energy generation [83], asperity micro-contact [84, 85], and frictional heating plus enhanced chemical reactivity [86, 87]. Most of these studies focused on the global temperature rise of the polishing system [83, 84, 87, 88]. For example, Kim et al. [87] and Horng et al. [84] showed a linear temperature rise with increase in rotation rate.

It has been shown in certain polishing systems that *material removal rate* has an Arrhenius temperature dependence as follows:

$$\frac{\mathrm{d}h}{\mathrm{d}t} = \left(\frac{\mathrm{d}h}{\mathrm{d}t}\right)_{\mathrm{o}} \exp\left(-\frac{\Delta E}{RT}\right), \tag{2.42}$$

where $(\mathrm{d}h/\mathrm{d}T)_{\mathrm{o}}$ is a constant removal rate, T is temperature, and ΔE is the activation energy. Figure 2.30 shows an Arrhenius temperature dependence of the polishing removal rate with an activation energy of 21–58 kJ mol^{-1} [34]. Sugimoto et al. [89] also showed an increased removal rate with temperature (a sixfold increase with a 10 °C temperature rise).

In addition to global temperature rise, it has been shown that there is also a radial lap temperature distribution with a temperature maximum located at the maximum workpiece radial contact with the lap (f_{o}) (see Equation (2.34)) [7, 82, 90]. Such spatial temperature differences may contribute to nonuniform spatial material removal on the workpiece. In a recent study, the temperature distribution on a lap and workpiece was evaluated in detail using an infrared

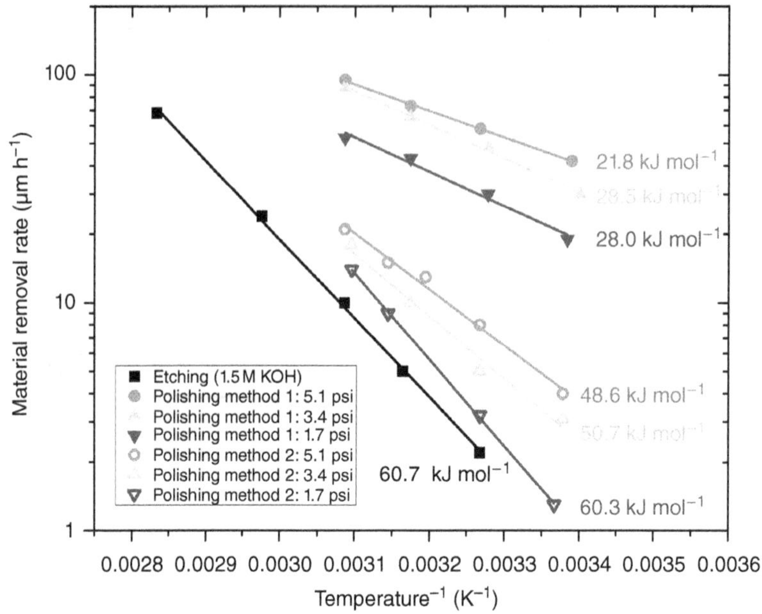

Figure 2.30 Effect of temperature on polishing removal rate and chemical etching. Source: Data from Doi et al. 2015 [34].

imaging camera [7]. For measuring the workpiece interface surface, the workpiece was removed from the polisher, the slurry was quickly wiped off, and the workpiece surface temperature, on the side facing the polisher, was imaged. Due to the relatively large heat capacity of the workpiece, the interface temperature was largely unchanged during this procedure. The measured thermal images of the lap and a nonrotated workpiece (at the workpiece–lap interface) as a function of lap rotation rate are shown in Figure 2.31. As expected, as rotation rate increases, lap and workpiece temperatures increase, owing to greater frictional heating per unit time. The temperature distribution had a largely radially symmetrical profile relative to the lap, both on the lap and workpiece. The temperature maximum corresponded to the radial location of the maximum workpiece path length on the lap.

Examples of the measured lap thermal profiles of various polishing configurations, using a nonrotated workpiece, a rotated workpiece, and a rotated workpiece plus compensating septum [7], are shown in Figure 2.32a. Rotating the workpiece reduced the temperature variation across the lap due to a radial (relative to the workpiece) time averaging. Also, the addition of a compensating septum further reduced temperature variation. This is more clearly illustrated in Figure 2.32b, which shows the maximum temperature difference across the lap as a function of lap rotation rate for the same three polishing

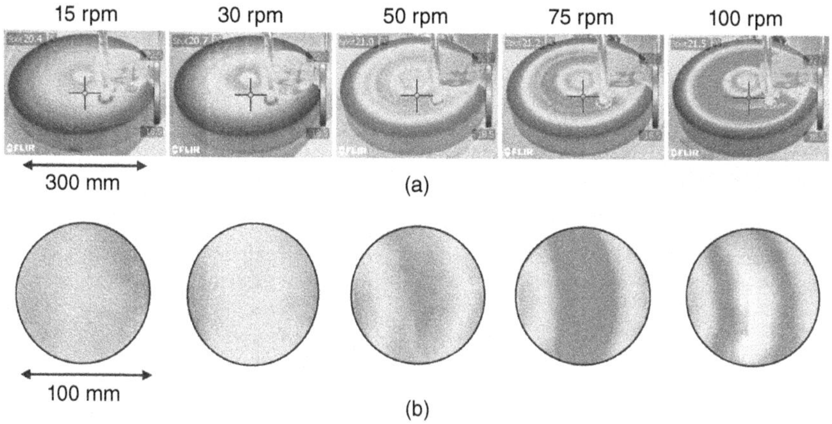

Figure 2.31 IR thermal images of the lap surface (a) and workpiece surface at the workpiece–lap interface (b) as a function of lap rotation rate. For the workpiece surface image, the lap center is at the left of each image. Each image has the same thermal-imaging color scale, from 18.5 °C (dark) to 22.5 °C (light). Source: Suratwala et al. 2014 [7]. Reproduced with permission from John Wiley & Sons.

configurations. Having a smaller temperature gradient in the polishing system is beneficial because it reduces material removal spatial nonuniformity, which increases the likelihood of greater material removal determinism during polishing.

A relatively simple thermal model, where the major source of thermal heating during polishing is the frictional work occurring at the workpiece–lap interface, can be used to predict temperature magnitude and distribution during polishing [7]. Consider a static, round workpiece on a rotating lap, as depicted in Figure 2.33. The thermal balance of the system can be described by the steady-state heat equation with rotational convective transport as follows:

$$\alpha_1 \nabla^2 T = 0_{\text{(workpiece)}}, \tag{2.43}$$

$$\alpha_{2-3} \nabla^2 T = R_\text{L} \left(y\frac{\partial T}{\partial x} - x\frac{\partial T}{\partial y} \right)_{\text{(Lap)}}, \tag{2.44}$$

where T is the spatial temperature distribution, α_1 is the thermal diffusivity of the workpiece glass, α_{2-3} is the combined thermal diffusivity of the lap and slurry, R_L is the lap rotation rate, and x and y are coordinate positions on the lap. Owing to the relatively low slurry-flow rate used in the experiment, heat transport due to slurry flow is considered small and is treated as a combined, effective thermal diffusivity of the lap and slurry.

The rate of heat generation per unit area (in W m^{-2}) at the workpiece–lap interface for a given lap radius (r) is defined as follows:

$$q(r) = f_\text{o}(r)\, \mu\, \sigma_\text{o}\, v_\text{r}(r), \tag{2.45}$$

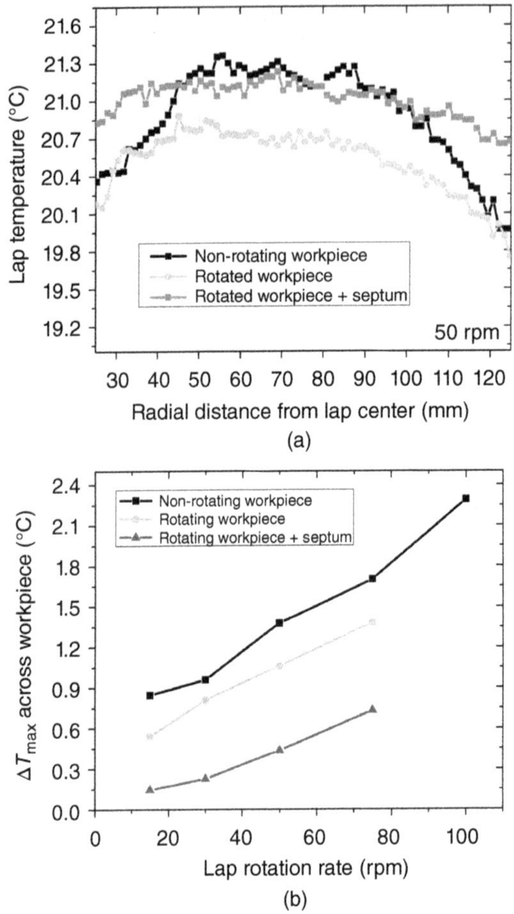

Figure 2.32 (a) Temperature profile of lap surface in a region where the workpiece contacts the lap for cases where the workpiece is not rotating, is rotating, and rotating with a supporting septum: (b) maximum temperature difference across the lap surface in a region where the workpiece contacts the lap, as a function of lap rotation rate for the same configurations in (a). Source: Suratwala et al. 2014 [7]. Reproduced with permission from John Wiley & Sons.

where μ is the friction coefficient, σ_o is the applied pressure on the workpiece, v_r is the relative velocity between the workpiece and lap, and $f_o(r)$ is the fraction of the lap circumference covered by workpiece, given by Equation (2.34) and discussed earlier [7, 66]. For a nonrotated workpiece, $v_r(r)$ is simply $R_L r$. Note in Equation (2.34), f_o increases and then decreases with increasing radial lap position (r); at the same time, relative velocity increases linearly with radial lap position (r). The combination of the two gives a skewed maximum in the radial frictional-heating distribution.

The heat-source interface boundary condition is given by

$$k_{wp}\frac{\partial T}{\partial z} = -q(r), \tag{2.46}$$

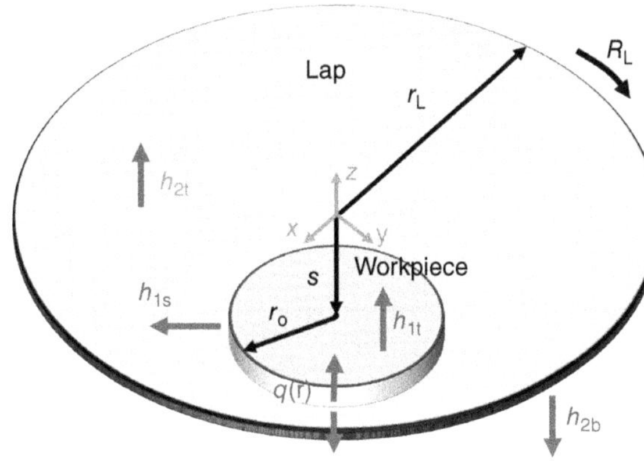

Figure 2.33 Schematic of polishing thermal model identifying geometric, kinematic, and thermal parameters. Source: Suratwala et al. 2014 [7]. Reproduced with permission from John Wiley & Sons.

and the effective heat transfer from the main surfaces of the system are used to formulate the remaining boundary conditions as follows:

$$-k_i \frac{\partial T}{\partial n^*} = h_{ij} \Delta T, \tag{2.47}$$

where k_i is the thermal conductivity of the lap ($i = 2$) or workpiece ($i = 1$), n^* is normal direction of the boundary condition surface, and h_{ij} is the heat transfer coefficient (W m^{-2} K^{-1}) for each surface of the system: top of the lap (h_{2t}), bottom of the lap (h_{2b}), side of the workpiece (h_{1s}), and top of the workpiece (h_{1t}).

Using the thermal conductivity, heat capacity, and density of the fused-silica workpiece ($k_1 = 1.38$ W m^{-1} K^{-1}; $C_{p1} = 740$ J kg^{-1} K^{-1}; $\rho_1 = 2200$ kg m^{-3}) and of the combined pad and slurry ($k_2 = 0.3$ W m^{-1} K^{-1}; $C_{p2} = 4186$ J kg^{-1} K^{-1}; $\rho_2 = 1000$ kg m^{-3}) for the experiment shown in Figure 2.31, the thermal diffusivity of the workpiece and lap/slurry were determined using $\alpha_i = k_i/(\rho_i C_{pi})$. The heat transfer coefficient from the bottom of the lap was determined using the thermal conductivity of the granite lap base and its thickness, giving $h_{2b} = 60$ W m^{-2} K^{-1}. Similarly, the heat transfer coefficient for the top of the workpiece, which is attached to a stainless-steel weight, is calculated as $h_{1t} = 648$ W m^{-2} K^{-1}. The heat transfer coefficient from the top of the lap was based on typical values, accounting for the room-temperature evaporation of the water in the slurry ($h_{2t} = 60$ W m^{-2} K^{-1}). The heat transfer from the side of the workpiece was based on typical air-solid values ($h_{1s} = 20$ W m^{-2} K^{-1}). Finally, the effective friction coefficient was adjusted to get the best fit to the data ($\mu = 0.012$). The small resulting value of this coefficient may be attributable to the fact that only a fraction of the work done by friction between surfaces manifests as heat [91].

The calculated thermal distributions on the workpiece are compared with the experimental data in Figure 2.34a,b. The thermal model captures the general

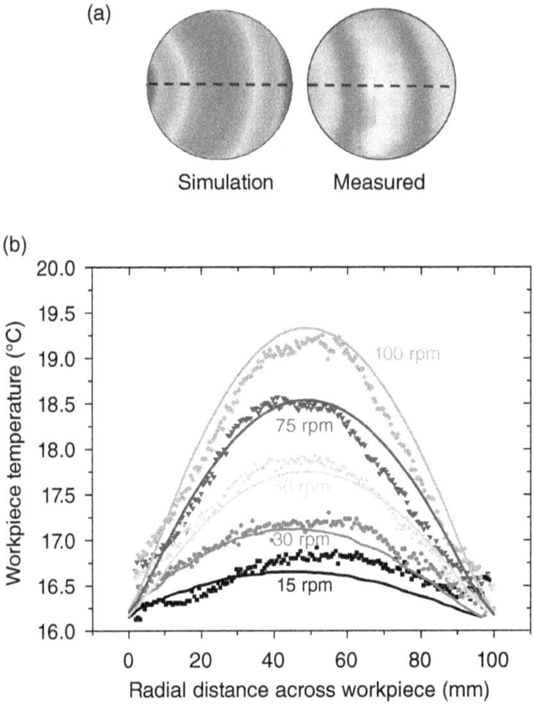

Figure 2.34 (a) Comparison of simulated and measured temperature distribution on a nonrotated, 100 mm diameter, fused-silica workpiece on rotated lap (100 rpm). The workpiece is oriented with its left side toward lap center. (b) Comparison of measured (points) and calculated (lines) temperature lineouts across the workpiece at various lap-rotation rates. Source: Suratwala et al. 2014 [7]. Reproduced with permission from John Wiley & Sons.

spatial shape of the observed thermal profile and compares quantitatively well to the radial data lineouts at various lap-rotation rates. Though the proposed thermal model greatly simplifies some of the known complex thermal behavior expected with the slurry fluid flow – such as at the workpiece interface, on the lap, and in the grooves of the pad – it captures the salient features of the observed thermal behavior.

The thermal model presented earlier, combined with some benchmark temperature measurements, provides a useful tool for understanding the thermal characteristics of a given polishing process and may suggest strategies to reduce temperature spatial nonuniformities and corresponding material removal spatial nonuniformities. The use of a compensating septum is a good example of a strategy to reduce temperature nonuniformities [7, 66, 67].

2.5.6.6 Global Pad Properties

In addition to pad wear (as discussed in Section 2.5.5.2), several other pad properties may influence the workpiece–lap mismatch. The overall modulus of the lap determines lap compliance and hence, the magnitude of the pressure differential established for a given workpiece–lap mismatch (see Figure 2.20). During polishing and repeated use of the lap, global lap properties can

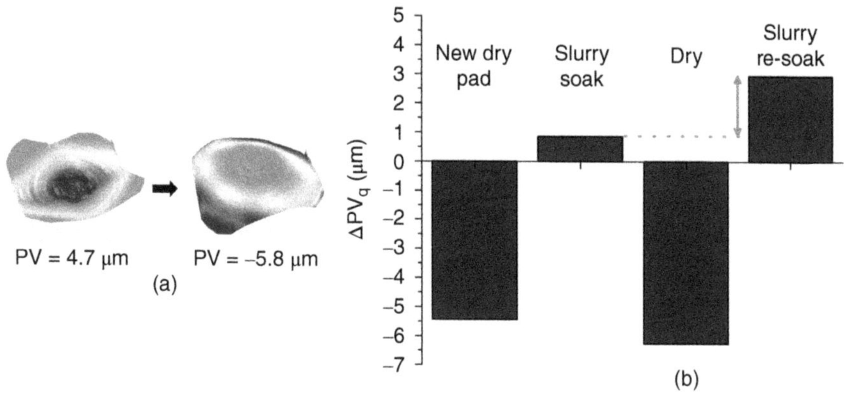

Figure 2.35 (a) Reflected wavefront measured on square fused-silica workpiece measuring 265 × 265 × 10 mm^3 after polishing on an MHN polyurethane pad before and after allowing the pad to dry completely between polishing iterations. (b) Reflected wavefront of the backside of a round fused-silica plate measuring 150 mm in diameter ×2.2 mm) after adhering an MHN polyurethane pad to the front surface and exposing to various environmental conditions.

change, owing to, for example, the effects of water penetration [92–94], pad shrinkage after drying between polishing runs, and polishing particles [95] or workpiece-removal products [96] embedded in the lap. Hence, the effective modulus of the lap may change with the use and treatment of the lap.

Figure 2.35a shows the *surface figure* of a 265 × 265 mm^2, fused-silica workpiece polished using the convergent-polishing method [66] (see Section 8.4) after polishing on a polyurethane foam pad that was not allowed to dry after adhesion to the lap, and then subsequent to polishing on the same pad after it was allowed to dry between polishing iterations. The results show a large change in the *surface figure* of the workpiece from PV = +4.7 to PV = −5.8 μm, a total change of 10.5 μm. A proposed mechanism is a permanent shrinkage of the pad after application to the lap base and drying after polishing, resulting in an irreversible change in the lap-height profile, workpiece–lap mismatch, and the observed change in workpiece shape. This behavior is consistent with the standard optical-finishing rule that the pad should not be allowed to dry between polishing iterations.

The pad-shrinkage hypothesis was tested using Twyman deflection measurements, with results in Figure 2.35b. The plot shows that the PV height of the reflected wavefront on the backside of the fused-silica plate adhered to the polyurethane pad after the pad was exposed to various chronological environments. Soaking the pad in slurry resulted in a change in the figure of the workpiece. After drying and resoaking, the workpiece deflection relative to the previous soaking differed by ~2 μm due to pad shrinkage. This suggests that if a pad is allowed to dry between polishing iterations and then resoaked, permanent change to the pad dimensions will result. This effect is analogous to that observed on the polisher and workpiece for the data shown in Figure 2.35a.

2.5.6.7 Slurry Spatial Distribution

How slurry particles are spatially distributed on the lap during polishing can influence the uniformity of material removal across the workpiece. This effect has been shown to occur both at mm-scale and μm-scale lengths [7, 96]. The latter is discussed in Chapter 4.

After polishing a nonrotated fused-silica workpiece on a rotated lap using cerium-oxide slurry on a polyurethane lap, a sub-mm-scale ripple pattern that lines up circumferentially, relative to the lap, is observable visually on the workpiece (see Figure 2.36), indicating a small spatial-length-scale material removal nonuniformity [7]. Changes in the stroke (the tangential or radial linear motion of the workpiece relative to the lap), diamond conditioning, and pad materials affected the characteristics of the ripple pattern (see Figure 2.37a,b) [7]. Without stroke on the baseline 50-mil-thick MHN polyurethane pad, the workpiece surface showed a ripple pattern with a high degree of nonuniformity and an average spacing of ~ 270 μm. Using PV height as a measure of material removal nonuniformity, the no-stroke case had a value of 2.78 μm. Adding simultaneous diamond conditioning result in a ripple pattern with a similar magnitude (PV = 2.32 μm). On the other hand, the addition of stroke kinematics led to significant improvement in uniformity; using a radial stroke result in PV = 0.38 μm. Also, as shown in Figure 2.37b, uniformity can be greatly improved by changing the pad material. Material removal nonuniformity improved in the following order: polyurethane-foam pad (50-mil MHN and 25-mil MHN) with PV = 0.81–2.78 μm, polyurethane-fiber pad (Suba 550) with PV = 0.52 μm, and soft rubber-foam pad (Chem Pol) with PV = 0.23 μm.

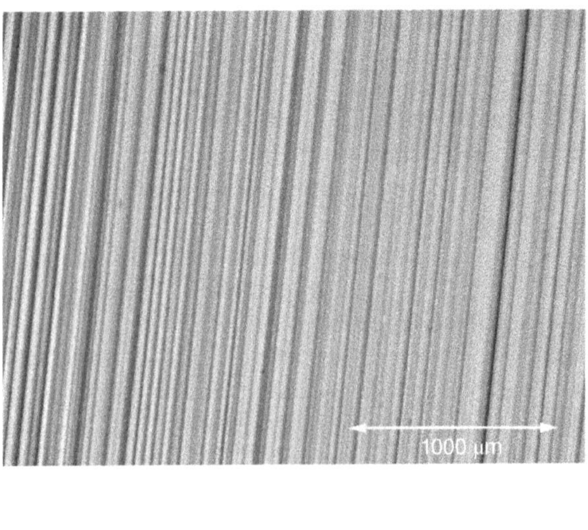

Figure 2.36 Optical micrograph of ripples observed on a fused-silica surface after polishing (nonrotated and without stroke). Source: Suratwala et al. 2014 [7]. Reproduced with permission from John Wiley & Sons.

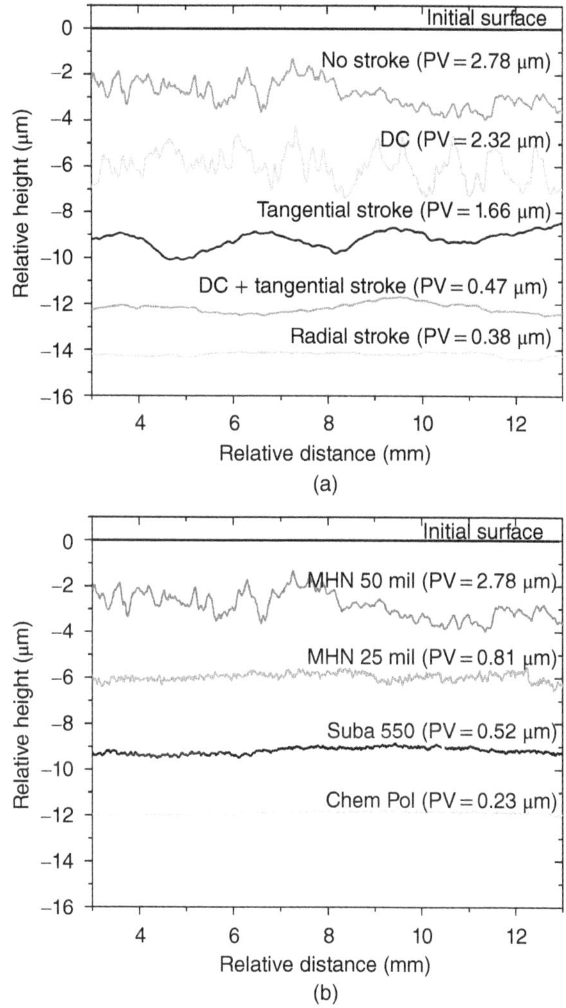

Figure 2.37 (a) Surface lineouts of fused-silica surfaces after polishing (nonrotated) under various conditions. (b) Surface lineouts of fused-silica surfaces after polishing (nonrotated) on different pads. DC = diamond conditioning. Source: Suratwala et al. 2014 [7]. Reproduced with permission from John Wiley & Sons.

The results suggest that this fine-scale material nonuniformity increased with pad hardness. Note that for improved convergence control with convergent polishing, however, a pad of higher hardness is typically desired [52, 66]. This is another example of how improvement in one polishing characteristic may come at the expense of another (see Figure 1.6).

It is proposed that the sub-mm-scale ripple pattern observed on the workpiece is caused by randomly located, microscopic islands of slurry on the pad where greater removal occurs [7]. The radial movement of the lap with respect to the workpiece results in a time-averaged, one-dimensional spatial radial material removal nonuniformity, thus leading to the ripple

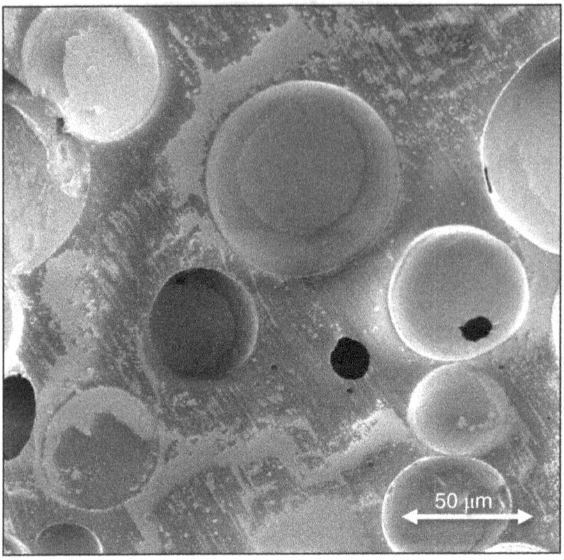

Figure 2.38 Scanning electron microscopy (SEM) image of 50-mil MHN pad surface after polishing, showing distribution islands of ceria particles on the plateaus. Circular regions are the open pores of the pad. Source: Suratwala et al. 2014 [7]. Reproduced with permission from John Wiley & Sons.

pattern's lining up with the lap radius on the workpiece (see Figure 2.36). A scanning electron-microscopic image of the top surface of a used 50-mil MHN polishing pad shows the presence of slurry islands (Figure 2.38) with spacing like that of ripple spacing. Under the proposed mechanism, one would expect a radial stroke (that is, movement perpendicular to the ripple direction) to optimally, spatially time-average the nonuniformity, leading to a reduction in the magnitude of the ripple pattern. This is consistent with experimental results showing largest reduction in the PV when using a radial stroke from 2.78 to 0.38 μm (Figure 2.37a).

2.5.6.8 Local Nonlinear Material Deposits

With longer polishing times – typically tens of hours – and using a rotated workpiece on a pad lap, another midscale-length material removal nonuniformity is observed, as illustrated in Figure 2.39a. Here the workpiece *surface figure* developed a relatively deep depression near the center [7]. No special pad treatment during polishing was performed between polishing iterations. Upon polishing the same workpiece, initially with poor *surface figure*, on the same pad, but this time after a large amount of diamond conditioning (DC), the midrange structure – and hence, this source of material removal nonuniformity – was shown to be removed (see Figure 2.39b).

The development of the center depression on the workpiece surface with continued polishing is due to preferential slurry- and glass-products material deposition at a specific radial location on the lap. This was confirmed in three

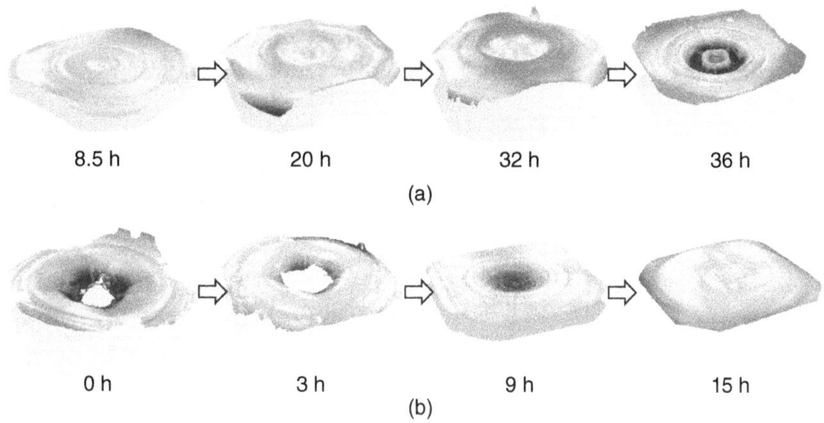

Figure 2.39 Evolution of the surface figure of a 265 mm square, fused-silica workpiece with polishing time (a) without any pad treatment (full scale −2 to 1 μm) and (b) after pad treatment (DC) (full scale −3 to 3.5 μm). Source: Suratwala et al. 2014 [7]. Reproduced with permission from John Wiley & Sons.

ways: by measuring the profile of the lap, showing an elevated height of material deposit at a fixed radial distance on the lap (see Figure 2.40a); by measuring a nonrotated workpiece on the same pad that showed a very strong preferential material removal increase (~7×) at the same radial location (see Figure 2.40b); and by removal of the temperature rise on the lap during polishing (caused by friction heating from increased material removal) by using DC on the same pad to remove the material deposit (see Figure 2.41).

Material deposit can be simulated by incorporating a source term, in addition to the removal term for pad wear [1, 7]. Material removal on the workpiece as a function of position and time ($\mathrm{d}h/\mathrm{d}t$) is determined by the Preston equation, given by Equation (2.2). Similarly, material removal on the lap ($\mathrm{d}h_{\mathrm{lap}}/\mathrm{d}t$) as a function of position (Equation (2.32)) can be expanded with a term to include material deposit using

$$\frac{\mathrm{d}h_{\mathrm{lap}}}{\mathrm{d}t}(x,y,t) = \frac{f_{\mathrm{o}}(r)}{t}\int_0^t [k_{\mathrm{lap}}(t')\,\mu(x,y,t')\,v_{\mathrm{r}}(x,y,t')\,\sigma(x,y,t')] - [f_{\mathrm{redep}}k_{\mathrm{p}}\,\mu(x,y,t')\,v_{\mathrm{r}}(x,y,t')\,\sigma(x,y,t')]\,\mathrm{d}t, \tag{2.48}$$

where f_{redep} is the fraction of the glass products removed from the workpiece that deposits on a particular position of the lap. The first term in the integral of Equation (2.48) describes the wear of the pad; the second describes the material deposition rate on the lap. Note the deposition rate is proportional to $f_{\mathrm{o}}(r)f_{\mathrm{redep}}k_{\mathrm{p}}$, which means the rate is proportional to workpiece radial path-length on the lap.

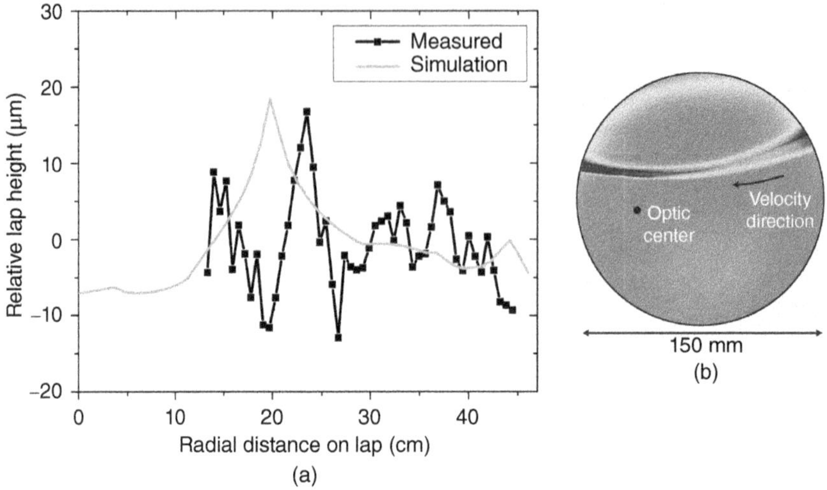

Figure 2.40 (a) Measured (data points) and calculated (red line) profiles of pad surface after polishing for ~50 h, showing a ring of material deposited on the pad surface at a lap radius of ~20–22 cm; (b) Measured surface figure (PV = 1.4 μm) of a portion of a nonrotated, 265 mm square, fused-silica workpiece after 30 min. of polishing on pad shown in (a). Note that measurement was made near the workpiece center only). Source: Suratwala et al. 2014 [7]. Reproduced with permission from John Wiley & Sons.

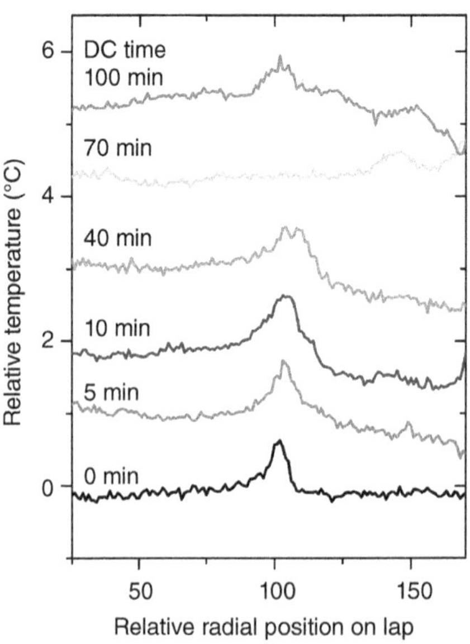

Figure 2.41 Measured radial temperature distribution on the lap, observed during polishing after DC for various amounts of time. Source: Suratwala et al. 2014 [7]. Reproduced with permission from John Wiley & Sons.

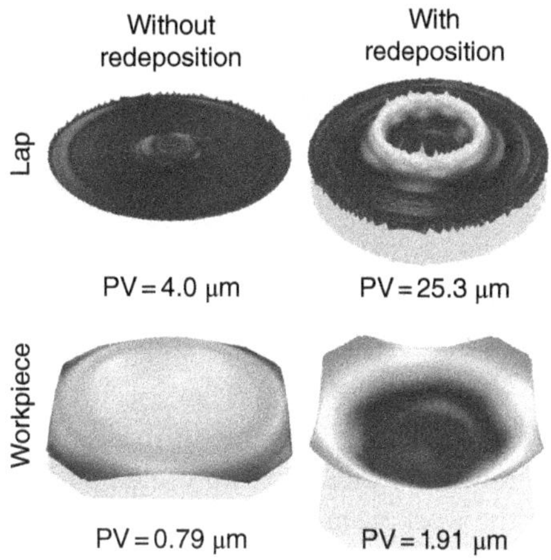

Figure 2.42 Simulated surface figure, using SurF code and Equation (2.48), of the lap (top) and workpiece (bottom) – at left, without ($f_{redep} = 0$) and at right, with ($f_{redep} = 1.5$). (Full scale for the lap is −8 to 30 μm; full scale for the workpiece is −1.2 to 1.2 μm.) Source: Suratwala et al. 2014 [7]. Reproduced with permission from John Wiley & Sons.

Figure 2.42 shows simulation results using the SurF code of the lap (top) and workpiece surface (bottom) with no material deposition ($f_{redep} = 0$) and with material deposition ($f_{redep} = 1.5$, representing best match to experimental data)[7]. SurF code is described in Section 2.6. With material deposition, a fixed location on the lap develops a large material deposit (~20 μm high) at a lap radius of 20 cm, which is fairly close in magnitude and location to experimental results (see Figure 2.39a). The workpiece surface establishes a center depression in a manner also similar to experimental data. Using small values of f_{redep} ($\ll$1) in the simulation resulted in very little deposition variation on the lap; however, using a larger-value f_{redep} (>1), a specific location on the lap deposited material nonlinearly. The nonlinear aspect of the deposition physically stems from the fact that if a particular radial location starts to deposit material on the lap, the workpiece–lap mismatch will result in enhanced pressure at that location – further increasing the *material removal rate* and hence the material deposition rate. Note that $f_{redep} > 1$ implies that more material than that removed from the workpiece is deposited on the surface of the pad. Energy-dispersive spectroscopy (EDS) measurements of the deposit on the lap showed a Si:Ce ratio of 2.3 : 1, indicating the material deposit is a combination of glass products (from the fused-silica workpiece) and polishing slurry (cerium oxide). Thus, the best fit value of $f_{redep} > 1$ is reasonable.

2.6 Deterministic *Surface Figure*

In this chapter, the *material removal rate* Preston equation has been dissected for the large number of phenomena that influence material removal, both spatially and temporally, and ultimately determine the final *surface figure* of the polished workpiece (see Figure 2.1). Historically, the ability to predict *surface figure* for full-aperture polishing involved polishing iterations, where polishing-process changes, such as kinematics and lap shaping with a conditioner, are made with *surface figure* measurements at each iteration. The number of iterations and quality of the final *surface figure* achieved depends greatly on the skill of the master optician.

With an improved understanding of the above-discussed phenomena, several general approaches can be made to increase determinism in controlling *surface figure.*

1. *Subaperture polishing.* In this approach, a subaperture tool with a defined removal function is combined with a CNC machine to scan the tool across the workpiece – with different dwell times at each position – to determine the amount removed at each point on the surface. An example is using MRF (see Section 8.1). This method has aroused much interest in the field of optical fabrication, especially in the fabrication of free-form optics (i.e. optics with complex shapes or surface deviating from conventional spheres, aspheres, and flats). However, as with full-aperture polishing, the challenge still remains of understanding all phenomena that affect the removal function. How deterministic and repeatable the removal function is as a function of position and time will largely drive the determinism of the overall process. Expanding the principles outlined in this chapter to the small-tool removal function and the resulting *surface figure* is a fruitful area for optical-fabrication research.
2. *Surface figure simulator.* Another approach to increase determinism is to combine all the phenomena outlined in this chapter and quantitatively incorporate them in a polishing simulator that allows *surface figure* prediction. One approach is a recently developed code called *Surface Figure* (or SurF), which incorporates a large number of phenomena that simultaneously account for spatial and temporal removal on the workpiece and lap and material deposit. Figure 2.43 illustrates the high-level algorithm for SurF; for a more detailed description see Ref. [1]. Figure 2.44, along with Figures 2.3, 2.15, 2.40, and 2.42, shows SurF simulation results compared with experimental data for a number of polishing scenarios. Because of the large number of phenomena that occur simultaneously, as well as input parameters that may not be known in advance, this type of code is more valuable today in identifying trends of *surface figure* evolution versus providing a given set of polishing conditions for achieving a final *surface figure.* However, with an improved grasp of all these phenomena, it is foreseeable that a code may be truly deterministic and useful in quantitative prediction of the *surface figure.*

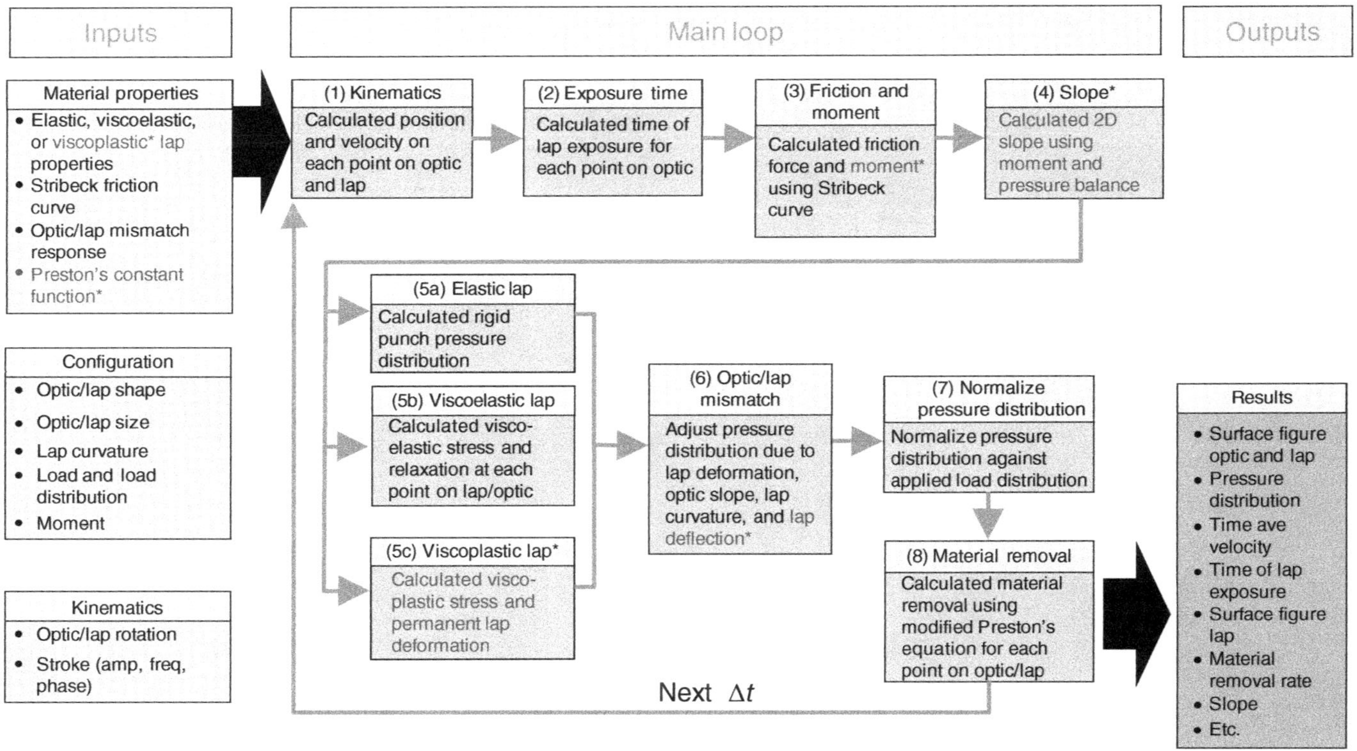

Figure 2.43 Flow chart illustrating how SurF code incorporates all the contributions to the Preston material removal rate equation. Items marked with an asterisk are modules under development. Source: Suratwala et al. 2010 [1]. Reproduced with permission from John Wiley & Sons.

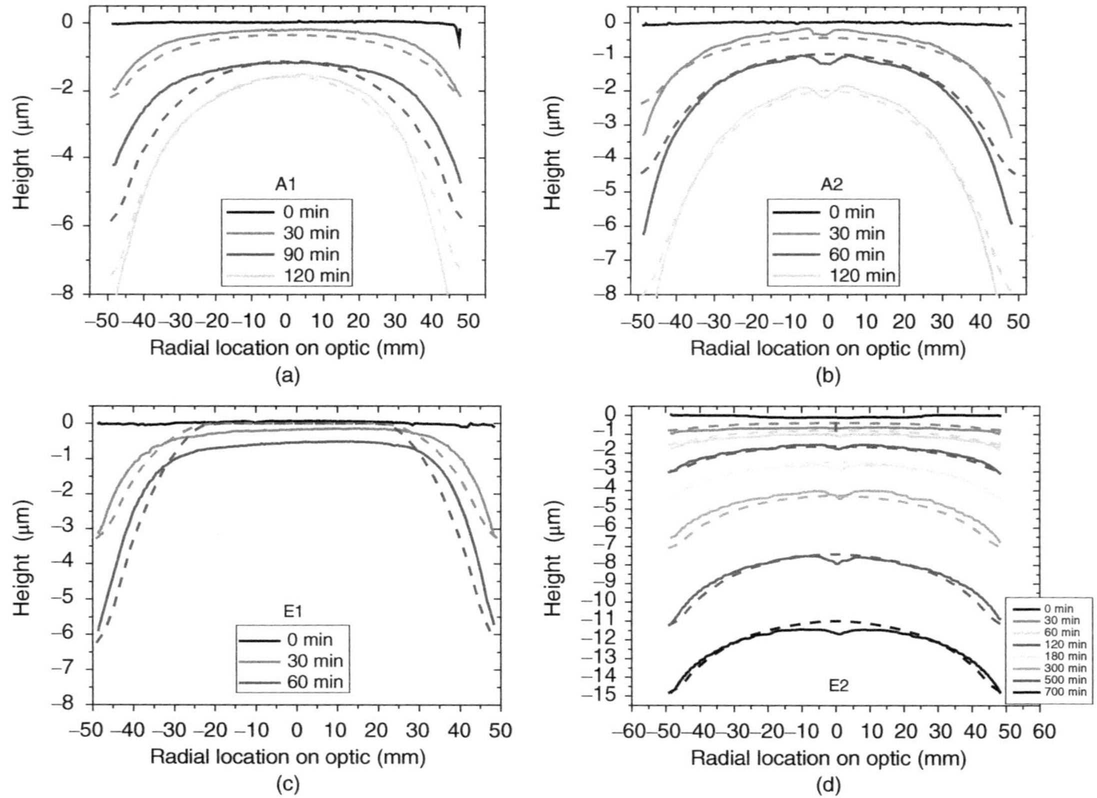

Figure 2.44 (a–h) Comparison of measured data (solid lines) versus SurF simulations (dashed lines) of the surface figure lineouts for fused-silica workpieces polished under various conditions. Source: Suratwala et al. 2010 [1]. Reproduced with permission from John Wiley & Sons.

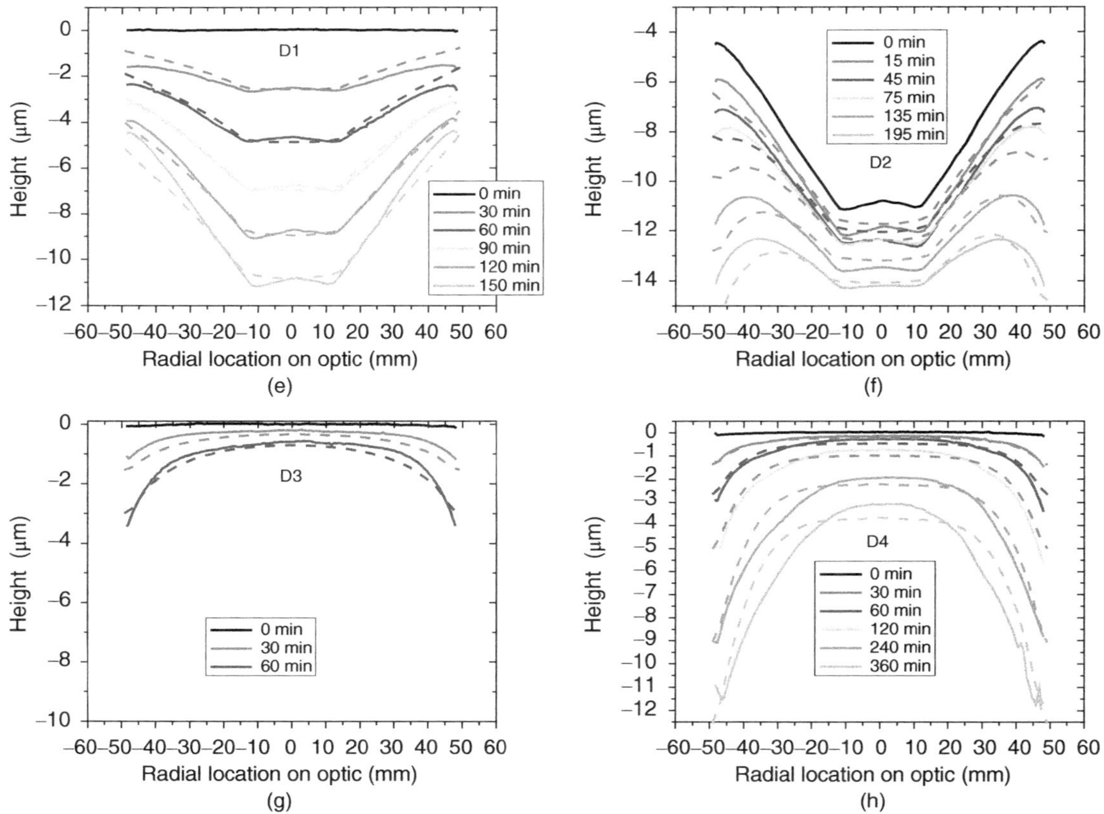

Figure 2.44 *(Continued)*

3. *Reduction of material removal nonuniformities.* Another approach in increasing determinism is systematically reducing or eliminating material removal nonuniformities, either by changing the design of the polishing system or by controlled processing parameters. In this way, the number of polishing-process variables that must be quantitatively understood or monitored to control the *surface figure* may be reduced – which would lead to a greater chance of achieving determinism in polishing, as there would be fewer variables influencing the process. An example is incorporating a septum into the polisher that counteracts the nonuniform pad wear that the workpiece incorporates onto the pad, with the result that pad wear no long contributes to material removal nonuniformity (see Section 7.1.3). Convergent polishing takes full advantage of this concept by removing all sources of material removal nonuniformity except for workpiece–lap mismatch owing to workpiece shape. A full description of Convergent polishing is found in Section 8.4.

References

1 Suratwala, T.I., Feit, M.D., and Steele, W.A. (2010). Toward deterministic material removal and surface figure during fused silica pad polishing. *J. Am. Ceram. Soc.* 93 (5): 1326–1340.

2 Luo, Q., Ramarajan, S., and Babu, S. (1998). Modification of the Preston equation for the chemical–mechanical polishing of copper. *Thin Solid Films* 335 (1): 160–167.

3 Cordero-Davila, A., Gonzalez-Garcia, J., Pedrayes-Lopez, M. et al. (2004). Edge effects with the Preston equation for a circular tool and workpiece. *Appl. Opt.* 43 (6): 1250–1254.

4 Lin, S.-C. and Wu, M.-L. (2002). A study of the effects of polishing parameters on material removal rate and non-uniformity. *Int. J. Mach. Tools Manuf.* 42 (1): 99–103.

5 Tseng, W.T. and Wang, Y.L. (1997). Re-examination of pressure and speed dependences of removal rate during chemical-mechanical polishing processes. *J. Electrochem. Soc.* 144 (2): L15–L17.

6 Pal, R.K., Garg, H., and Karar, V. (2016). Full aperture optical polishing process: overview and challenges. In: *CAD/CAM, Robotics and Factories of the Future*, 461–470. Springer.

7 Suratwala, T., Feit, M.D., Steele, W.A., and Wong, L.L. (2014). Influence of temperature and material deposit on material removal uniformity during optical pad polishing. *J. Am. Ceram. Soc.* 97 (6): 1720–1727.

8 Cumbo, M., Fairhurst, D., Jacobs, S., and Puchebner, B. (1995). Slurry particle size evolution during the polishing of optical glass. *Appl. Opt.* 34 (19): 3743–3755.

9 Izumitani, T. (1986). *Optical Glass*, 197. New York: American Institute of Physics.
10 Lambropoulos, J.C., Xu, S., and Fang, T. (1997). Loose abrasive lapping hardness of optical glasses and its interpretation. *Appl. Opt.* 36 (7): 1501–1516.
11 Hutchings, I.M. (1992). *Tribology: Friction and Wear of Engineering Materials*, viii, 273. Boca Raton, FL: CRC Press.
12 Lai, J.-Y. (2001). *Mechanics, Mechanisms, and Modeling of the Chemical Mechanical polishing Process*. Massachusetts Institute of Technology.
13 Hersey, M.D. (1966). *Theory and Research in Lubrication: Foundation for Future Developments*. Wiley.
14 Moon, Y. (1999). Mechanical aspects of the material removal mechanism in chemical mechanical polishing (CMP). PhD Thesis. University of California Berkeley, 193 pages.
15 Soares, S., Baselt, D., Black, J.P. et al. (1994). Float-polishing process and analysis of float-polished quartz. *Appl. Opt.* 33 (1): 89–95.
16 Namba, Y., Ohnishi, N., Yoshida, S. et al. (2004). Ultra-precision float polishing of calcium fluoride single crystals for deep ultra violet applications. *CIRP Ann. Manuf. Technol.* 53 (1): 459–462.
17 Namba, Y., Tsuwa, H., Wada, R., and Ikawa, N. (1987). Ultra-precision float polishing machine. *CIRP Ann. Manuf. Technol.* 36 (1): 211–214.
18 Namba, Y. and Tsuwa, H. (1979). A chemo-mechanical ultrafine finishing of polycrystalline materials. *Ann. CIRP* 28 (1): 425–429.
19 Namba, Y. and Tsuwa, H. (1978). Mechanism and some applications of ultra-fine finishing. *Ann. CIRP* 27 (1): 511–516.
20 Namba, Y. and Tsuwa, H. (1977). Ultra-fine finishing of sapphire single crystal. *Ann. CIRP* 26 (1): 325.
21 Brown, N. (1981). *A Short Course on Optical Fabrication Technology*. Lawrence Livermore National Laboratory.
22 Taylor, J., Davis, C., Demiris, A. et al. (1995). High speed lapping of flat optics. UCRL-ID-119056, pp. 4–13.
23 Marinescu, I.D., Hitchiner, M.P., Uhlmann, E. et al. (2006). *Handbook of Machining with Grinding Wheels*. CRC Press.
24 Marinescu, I.D., Uhlmann, E., and Doi, T. (2006). *Handbook of Lapping and Polishing*. CRC Press.
25 Brown, N. (1977). *Optical Polishing Pitch*. Livermore, CA: California University, Lawrence Livermore Laboratory.
26 Jones, R.A. (1979). Grinding and polishing with small tools under computer control. *Opt. Eng.* 18 (4): 184–390.
27 Jones, R.A. and Rupp, W.J. ed. (1990). Rapid optical fabrication with CCOS. In: *Advanced Optical Manufacturing and Testing*, vol. 1333, 34–44. International Society of Optics and Photonics.

28 Golini, D., Jacobs, S., Kordonski, W., and Dumas, P. (1997). Precision optics fabrication using magnetorheological finishing. Proceedings of SPIE CR67-16, pp. 1–23.

29 Pollicove H., Golini D. Deterministic manufacturing processes for precision optical surfaces. *InKey Eng. Mater.* 2003 (Vol. 238, pp. 53–58) Trans Tech Publications.

30 Pollicove, H. and Golini, D. (2002). Computer numerically controlled optics fabrication. *Int. Trends Appl. Opt.* 5: 125–144.

31 West, S., Martin, H., Nagel, R. et al. (1994). Practical design and performance of the stressed-lap polishing tool. *Appl. Opt.* 33 (34): 8094–8100.

32 Walker, D., Brooks, D., King, A. et al. (2003). The 'Precessions' tooling for polishing and figuring flat, spherical and aspheric surfaces. *Opt. Exp.* 11 (8): 958–964.

33 Beaucamp, A., Namba, Y., and Freeman, R. (2012). Dynamic multiphase modeling and optimization of fluid jet polishing process. *CIRP Ann. Manuf. Technol.* 61 (1): 315–318.

34 Doi, T., Uhlmann, E., and Marinescu, I.D. (2015). *Handbook of Ceramics Grinding and Polishing*. William Andrew Publishing.

35 Menapace, J.A., Dixit, S.N., Génin, F.Y., and Brocious, W.F., ed. (2004). Magnetorheological finishing for imprinting continuous-phase plate structures onto optical surfaces. XXXV Annual Symposium on Optical Materials for High Power Lasers: Boulder Damage Symposium. International Society for Optics and Photonics.

36 Heynacher, E., Beckstette, K., and Schmidt, M. (1989). Apparatus for lapping and polishing optical surfaces. Google Patents.

37 Nagahara, R.J. and Lee, D.M. (2001). Method and apparatus for using across wafer back pressure differentials to influence the performance of chemical mechanical polishing. Google Patents.

38 Twyman, F. (1952). *Prism and Lens Making; A Textbook for Optical Glassworkers*, 2e, viii, 629. London: Hilger & Watts.

39 Johnson, K.L. and Johnson, K.L. (1987). *Contact Mechanics*. Cambridge University Press.

40 Kimura, M., Saito, Y., Daio, H., and Yakushiji, K. (1999). A new method for the precise measurement of wafer roll off of silicon polished wafer. *Jpn. J. Appl. Phys.* 38 (1R): 38.

41 Fu, G. and Chandra, A. (2005). The relationship between wafer surface pressure and wafer backside loading in chemical mechanical polishing. *Thin Solid Films* 474 (1): 217–221.

42 Fu, G. and Chandra, A. (2002). A model for wafer scale variation of material removal rate in chemical mechanical polishing based on viscoelastic pad deformation. *J. Electron. Mater.* 31 (10): 1066–1073.

43 Fu, G. and Chandra, A. (2001). A model for wafer scale variation of removal rate in chemical mechanical polishing based on elastic pad deformation. *J. Electron. Mater.* 30 (4): 400–408.

44 Guthrie, W.L., Cheng, T., Ko, S.-H. et al. (1998). Method and apparatus for using a retaining ring to control the edge effect. Google Patents.
45 Terrell, E.J. and Higgs, C.F. (2006). Hydrodynamics of slurry flow in chemical mechanical polishing a review. *J. Electrochem. Soc.* 153 (6): K15–K22.
46 Runnels, S.R. and Eyman, L.M. (1994). Tribology analysis of chemical-mechanical polishing. *J. Electrochem. Soc.* 141 (6): 1698–1701.
47 Park, S.-S., Cho, C.-H., and Ahn, Y. (2000). Hydrodynamic analysis of chemical mechanical polishing process. *Tribol. Int.* 33 (10): 723–730.
48 Cho, C.-H., Park, S.-S., and Ahn, Y. (2001). Three-dimensional wafer scale hydrodynamic modeling for chemical mechanical polishing. *Thin Solid Films* 389 (1): 254–260.
49 Runnels, S.R. (1994). Feature-scale fluid-based erosion modeling for chemical-mechanical polishing. *J. Electrochem. Soc.* 141 (7): 1900–1904.
50 Chen, J. and Fang, Y.-C. (2002). Hydrodynamic characteristics of the thin fluid film in chemical–mechanical polishing. *IEEE Trans. Semicond. Manuf.* 15 (1): 39–44.
51 Bullen, D., Scarfo, A., Koch, A. et al. (2000). In situ technique for dynamic fluid film pressure measurement during chemical mechanical polishing. *J. Electrochem. Soc.* 147 (7): 2741–2743.
52 Suratwala, T., Feit, M., Steele, W. et al. (2014). Microscopic removal function and the relationship between slurry particle size distribution and workpiece roughness during pad polishing. *J. Am. Ceram. Soc.* 97 (1): 81–91.
53 Sundararajan, S., Thakurta, D.G., Schwendeman, D.W. et al. (1999). Two-dimensional wafer-scale chemical mechanical planarization models based on lubrication theory and mass transport. *J. Electrochem. Soc.* 146 (2): 761–766.
54 Yu, T.-K., Yu, C., and Orlowski, M. ed. (1994). Combined asperity contact and fluid flow model for chemical-mechanical polishing. Proceedings of International Workshop on Numerical Modeling of Processes and Devices for Integrated Circuits, 1994 NUPAD V, 1994. IEEE.
55 Suratwala, T., Steele, R., Feit, M.D. et al. (2008). Effect of rogue particles on the sub-surface damage of fused silica during grinding/polishing. *J. Non-Cryst. Solids* 354 (18): 2023–2037.
56 Lu, H., Obeng, Y., and Richardson, K. (2002). Applicability of dynamic mechanical analysis for CMP polyurethane pad studies. *Mater. Charact.* 49 (2): 177–186.
57 Karow, H.H. (1992). *Fabrication Methods for Precision Optics* (ed. J.W. Goodman), 1–751. New York: Wiley.
58 Rupp, W.J. (1971). Conventional optical polishing techniques. *J. Mod. Opt.* 18 (1): 1–16.
59 DeGroote, J.E., Jacobs, S.D., Gregg, L.L. et al. ed. (2001) Quantitative characterization of optical polishing pitch. International Symposium on Optical Science and Technology. International Society for Optics and Photonics.

60 Gillman, B.E. and Tinker, F. ed. (1999). Fun facts about pitch and the pitfalls of ignorance. SPIE's International Symposium on Optical Science, Engineering, and Instrumentation. International Society for Optics and Photonics.

61 Burge, J.H., Anderson, B., Benjamin, S. et al. ed. (2001). Development of optimal grinding and polishing tools for aspheric surfaces. International Symposium on Optical Science and Technology. International Society for Optics and Photonics.

62 Chen, C.Y., Yu, C.C., Shen, S.H., and Ho, M. (2000). Operational aspects of chemical mechanical polishing polish pad profile optimization. *J. Electrochem. Soc.* 147 (10): 3922–3930.

63 Chang, O., Kim, H., Park, K. et al. (2007). Mathematical modeling of CMP conditioning process. *Microelectron. Eng.* 84 (4): 577–583.

64 Zhou, Y.-Y. and Davis, E.C. (1999). Variation of polish pad shape during pad dressing. *Mater. Sci. Eng., B* 68 (2): 91–98.

65 Park, B., Lee, H., Park, K. et al. (2008). Pad roughness variation and its effect on material removal profile in ceria-based CMP slurry. *J. Mater. Process. Technol.* 203 (1): 287–292.

66 Suratwala, T., Steele, R., Feit, M. et al. (2012). Convergent pad polishing of amorphous silica. *Int. J. Appl. Glass Sci.* 3 (1): 14–28.

67 Suratwala, T., Steele, R., Feit, M. et al. (2014). Convergent polishing: a simple, rapid, full aperture polishing process of high quality optical flats & spheres. *J. Vis. Exp.* 94: doi: 10.3791/51965.

68 Ng, S.H., Yoon, I., Higgs, C.F., and Danyluk, S. (2004). Wafer-bending measurements in CMP. *J. Electrochem. Soc.* 151 (12): G819–G823.

69 Love, A.E.H. (2013). *A Treatise on the Mathematical Theory of Elasticity*. Cambridge University Press.

70 Roark, R.J. and Young, W.C. (1975). *Formulas for Stress and Strain*. McGraw-Hill.

71 Feit, M.D., DesJardin, R.P., Steele, W.A., and Suratwala, T.I. (2012). Optimized pitch button blocking for polishing high-aspect-ratio optics. *Appl. Opt.* 51 (35): 8350–8359.

72 Dalladay, A. (1922). Some measurements of the stresses produced at the surfaces of glass by grinding with loose abrasives. *Trans. Opt. Soc.* 23 (3): 170.

73 Twyman, F. ed. (1905). Polishing of glass surfaces. In: *Proceedings of the Optical Convention*, 78. London: Northgate & Williams.

74 Lambropoulos, J.C., Xu, S., Fang, T., and Golini, D. (1996). Twyman effect mechanics in grinding and microgrinding. *Appl. Opt.* 35 (28): 5704–5713.

75 Ratajczyk, F. (1966). Die Abhangigkeit des Twymaneffekts von den Scheifbedingungen des optischen Glases. *Feingeratetechnik* 15: 445–449.

76 Nikolova, E.G. (1985). Review: on the Twyman effect and some of its applications. *J. Mater. Sci.* 20: 1–8.

77 Chen, J. and De Wolf, I. (2003). Study of damage and stress induced by backgrinding in Si wafers. *Semicond. Sci. Technol.* 18 (4): 261.
78 Rupp, W. ed. (1987). Twyman effect for ULE. Optical Fabrication and Testing Workshop.
79 Podzimek, O. and Heuvelman, C. (1986). Residual stress and deformation energy under ground surfaces of brittle solids. *CIRP Ann. Manuf. Technol.* 35 (1): 397–400.
80 Wong, L., Suratwala, T., Feit, M.D. et al. (2009). The effect of HF/NH_4F etching on the morphology of surface fractures on fused silica. *J. Non-Cryst. Solids* 355 (13): 797–810.
81 Suratwala, T., Wong, L., Miller, P. et al. (2006). Sub-surface mechanical damage distributions during grinding of fused silica. *J. Non-Cryst. Solids* 352 (52–54): 5601–5617.
82 Hocheng, H., Huang, Y.L., and Chen, L.J. (1999). Kinematic analysis and measurement of temperature rise on a pad in chemical mechanical planarization. *J. Electrochem. Soc.* 146 (11): 4236–4239.
83 White, D., Melvin, J., and Boning, D. (2003). Characterization and modeling of dynamic thermal behavior in CMP. *J. Electrochem. Soc.* 150 (4): G271–G278.
84 Horng, J.H., Chen, Y.Y., Wu, H.W. et al. ed. (2011). Study of contact temperature in polishing surfaces. *Adv. Mater. Res.* 189–193: 1527–1531.
85 Horng, J.-H., Jeng, Y.-R., and Chen, C.-L. (2004). A model for temperature rise of polishing process considering effects of polishing pad and abrasive. *J. Tribol.* 126 (3): 422–429.
86 Oh, S. and Seok, J. (2008). Modeling of chemical – mechanical polishing considering thermal coupling effects. *Microelectron. Eng.* 85 (11): 2191–2201.
87 Kim, H., Kim, H., Jeong, H. et al. (2002). Friction and thermal phenomena in chemical mechanical polishing. *J. Mater. Process. Technol.* 130: 334–338.
88 Kim, N.-H., Ko, P.-J., Seo, Y.-J., and Lee, W.-S. (2006). Improvement of TEOS-chemical mechanical polishing performance by control of slurry temperature. *Microelectron. Eng.* 83 (2): 286–292.
89 Sugimoto, F., Arimoto, Y., and Ito, T. (1995). Simultaneous temperature measurement of wafers in chemical mechanical polishing of silicon dioxide layer. *Jpn. J. Appl. Phys.* 34 (12R): 6314.
90 Sampurno, Y.A., Borucki, L., Zhuang, Y. et al. (2005). Method for direct measurement of substrate temperature during copper CMP. *J. Electrochem. Soc.* 152 (7): G537–G541.
91 Chen, Q. and Li, D. (2005). A computational study of frictional heating and energy conversion during sliding processes. *Wear* 259 (7): 1382–1391.
92 Castillo-Mejia, D., Gold, S., Burrows, V., and Beaudoin, S. (2003). The effect of interactions between water and polishing pads on chemical mechanical polishing removal rates. *J. Electrochem. Soc.* 150 (2): G76–G82.

93 McGrath, J. and Davis, C. (2004). Polishing pad surface characterisation in chemical mechanical planarisation. *J. Mater. Process. Technol.* 153: 666–673.
94 Lu, H., Fookes, B., Obeng, Y. et al. (2002). Quantitative analysis of physical and chemical changes in CMP polyurethane pad surfaces. *Mater. Charact.* 49 (1): 35–44.
95 Charns, L., Sugiyama, M., and Philipossian, A. (2005). Mechanical properties of chemical mechanical polishing pads containing water-soluble particles. *Thin Solid Films* 485 (1): 188–193.
96 Suratwala, T., Steele, R., Feit, M. et al. (2017). Relationship between surface μ-roughness and interface slurry particle spatial distribution during glass polishing. *J. Am. Ceram. Soc.* 100: 2790–2802.

3

Surface Quality

The terms "*surface quality*" and "subsurface mechanical damage" (SSD) are often loosely used. Here we define *surface quality* as a measure of the level of perfection a workpiece surface exhibits after finishing and cleaning. A perfect surface is defined as a surface free of mechanical, structural, and chemical modification relative to bulk. Note *surface quality* does not include *surface roughness,* which is treated separately in Chapter 4. In practice, there is no such thing as a perfect surface, because a variety of microscopic and molecular surface modifications may occur on or just below the surface of the workpiece. As illustrated in Figure 3.1, surface modifications include the following:

- Subsurface mechanical interactions (SSD), which may lead to fracturing, plastic flow, or densification at the surface.
- Foreign particles or residue that may be deposited on the surface as particles land or precipitate during drying.
- Chemical and structural interactions that may result from changes in surface molecular moieties or by altering the near-subsurface (Beilby) layer.

These factors affecting *surface quality* may vary significantly in scale length, from tens of μm to Angstrom level.

3.1 Subsurface Mechanical Damage

Subsurface Mechanical Damage consists of surface microcracks created on the processed workpiece during the grinding or polishing of brittle-material surfaces, such as glass – indeed, the generation of microfractures is the very mechanism by which material is removed (see Figure 1.9). After polishing, any remaining or newly created surface cracks are identified macroscopically as scratches and digs. For optics, reducing detectable SSD is strongly desired, because SSD contributes to surface scatter, degrading optical performance [1]. Some SSD may be hidden below an index-matched Beilby layer or may have closed (i.e. healed); for this reason, not all SSD can be detected by visual

Materials Science and Technology of Optical Fabrication, First Edition. Tayyab I. Suratwala.

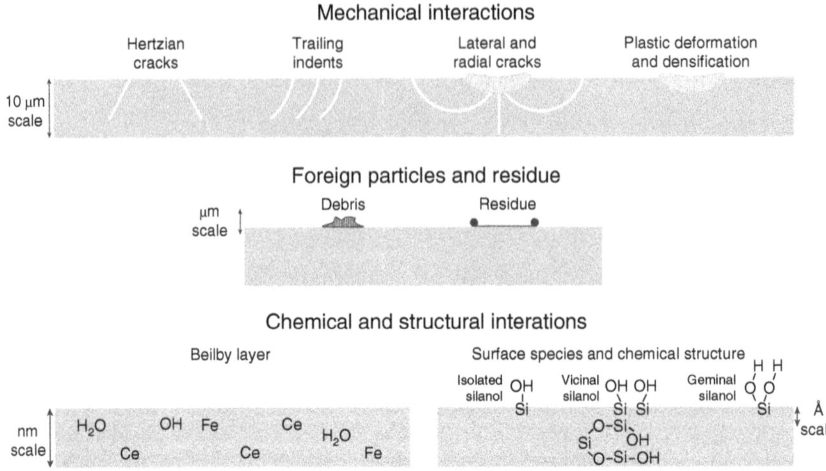

Figure 3.1 A variety of mechanical, chemical, and structural surface modifications may affect surface quality.

inspection or standard optical microscopy, until exposed by chemical etching [2]. In some high-end optical applications, further removal of SSD, even hidden SSD, is required for improving material strength – for example, in spacecraft, underwater windows and barriers, and military applications in which the surface flaws determine the ultimate strength. Removing SSD also reduces laser-induced damage in high-power, high-energy lasers [3, 4]. Of special concern in laser-optic applications, surface microcracks are known to contain intrinsic defect states on the fracture surface and serve as a reservoir for absorbing precursors that will heat up and explode upon irradiation with high-fluence laser light (see Chapter 9 for discussion) [5–8]. As a result, the fabrication of SSD-free optics and windows has been a quest for the optical-fabrication industry for many years [9–11].

3.1.1 Indentation Fracture Mechanics

3.1.1.1 Static Indentation

The grinding of glass can be described microscopically as the removal of glass particles created from an ensemble of intersecting brittle fractures, which are caused by an ensemble of normally loaded, hard indenters (abrasives) sliding or rolling across the surface of the workpiece. These brittle fractures lead both to material removal and the development of an SSD depth distribution in the processed workpiece. Comprehension of the basic fracture mechanical relationships, starting with a single particle or asperity loaded mechanically on a material, is vital in developing useful correlations for understanding the complex processes of grinding and polishing.

Figure 3.2 illustrates the cracks that may occur upon static indentation on brittle solids. The three basic types are Hertzian [12], radial [12–14], and lateral [12–14].

- Hertzian cracks are conical, created by a spherical indenter.
- Radial cracks are semicircular and perpendicular to the glass surface, created by a sharp indenter.
- Lateral cracks propagate more parallel to the glass surface and are typically created by a sharp indenter.

Fractures in brittle materials propagate when the stress intensity (K_I) at the crack tip is greater than the fracture toughness of the material (K_{Ic}) [15]. Fractures almost always propagate normal to the Mode I tension created in the material (see Chapter 6). The direction of crack propagations for the types presented in Figure 3.2 are determined by the stress distribution imparted by the indenter on the workpiece during loading and unloading [12].

Consistent with their geometry, the formation of lateral cracks largely results in material removal and contributes significantly to the observed *surface roughness* during grinding. Hertzian and radial cracks, on the other hand, largely

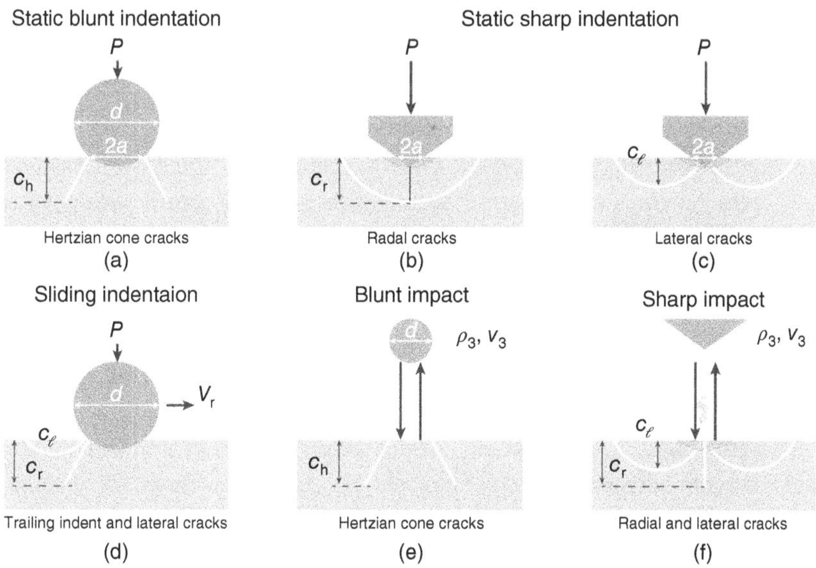

Figure 3.2 Schematic illustration of the fracture geometry of idealized fractures created by static (a–c) and dynamic (d–f) indentation: (a) Hertzian cone from a blunt, static indenter; (b) radial cracks from a sharp, static indenter; (c) lateral crack from a sharp, static indenter; (d) trailing indent and lateral cracks from a sliding, blunt indenter; (e) Hertzian conical crack from the impact of a blunt indenter; and (f) radial and lateral cracks from the impact of a sharp indenter.

contribute to SSD depth distribution. They may also contribute to material removal owing to intersection with other cracks. The relationships that govern the initiation load (P_c) to first observe the fracture on the surface have previously been described as follows [12–14]:

$$P_{ch} = Ar, \tag{3.1}$$

$$P_{cr} = \alpha_r \frac{{K_{Ic}}^4}{{H_1}^3}, \tag{3.2}$$

$$P_{c\ell} = \text{constant}, \tag{3.3}$$

where A is Auerbach's constant (N m^{-1}), α_r is the radial-crack initiation constant, r is the local radius of the indenting particle or sphere (m), K_{Ic} is the workpiece-material fracture toughness ($\text{MPa m}^{1/2}$), and H_1 is the workpiece-material hardness (GPa). The subscripts h, r, and ℓ are designations for Hertzian, radial, and lateral cracks, respectively. The initiation load for Hertzian cracks depends greatly on the radius curvature of the indenter, where it is more difficult to initiate a fracture using a larger radius of curvature (blunter) indenter. The initiation load for the radial-crack and lateral-crack scale with the properties of the substrate, namely the ratio ${K_{Ic}}^4/{H_1}^3$ of fracture toughness and hardness, is described as the index of brittleness or the ratio previously discussed by Lawn et al. [12, 16, 17].

The relationships that govern the extent of the fractures generated after initiation are given by [12–14].

$$c_h = \left(\frac{\chi_h P}{K_{Ic}}\right)^{2/3}, \tag{3.4}$$

$$c_r = \left(\frac{\chi_r P}{K_{Ic}}\right)^{2/3}, \tag{3.5}$$

$$c_\ell = \frac{\chi_\ell \left(\frac{E_1}{H_1}\right)^{2/3} P^{1/2}}{{H_1}^{1/2}}. \tag{3.6}$$

c_h, c_r, and c_ℓ are the crack-growth depths, χ_h, χ_r, and χ_ℓ are the crack-growth constants (unitless), E_1 is the elastic modulus for the brittle workpiece (GPa), and P is the applied normal load per particle (N). Again, the subscripts h, r, and ℓ designate Hertzian, radial, and lateral cracks. The crack depths for Hertzian and radial cracks scale as $P^{2/3}$, and the crack depth for lateral cracks scales as $P^{1/2}$. The growth rate of the crack is known to scale with the material properties of the substrate, namely ${E_1}^{1/2}/H_1$ for lateral cracks and $(E_1/H_1{K_{Ic}}^2)^{1/3}$ for radial cracks [18].

To determine the load of crack initiation and subsequent growth depth using the relationships above, the initiation constants (A, α_r, $P_{c\ell}$) and growth constants (χ_h, χ_r, χ_ℓ) must be determined. These are material properties not typically found in the literature; the constants may be determined by applying controlled static indents at various loads and then measuring the crack properties. Figure 3.3a,b shows the measured depth of the Hertzian and radial cracks as a function of $P^{2/3}$ for the cracks on various optical materials created by standard Hertzian and Vickers indentation. Similarly, Figure 3.3c shows the depth of lateral cracks as a function of $P^{1/2}$ for same materials. A reasonable linear fit is obtained for all three sets of data, suggesting that the observed load dependence is largely consistent with the indentation-growth expressions above. The x-intercept values from the linear fits are used to determine α_r and A, $P_{c\ell}$. The slopes of the linear fits are used to calculate χ_ℓ, χ_r, and χ_h using Equations (3.4)–(3.6). Table 3.1 shows the determined values of these constants. Fang and Lambropoulos also measured anisotropy in the crack growth of some anisotropic single crystals [19].

The applied normal load (P) and material properties of the workpiece are the key parameters that determine the extent of fracture. The growth constants for

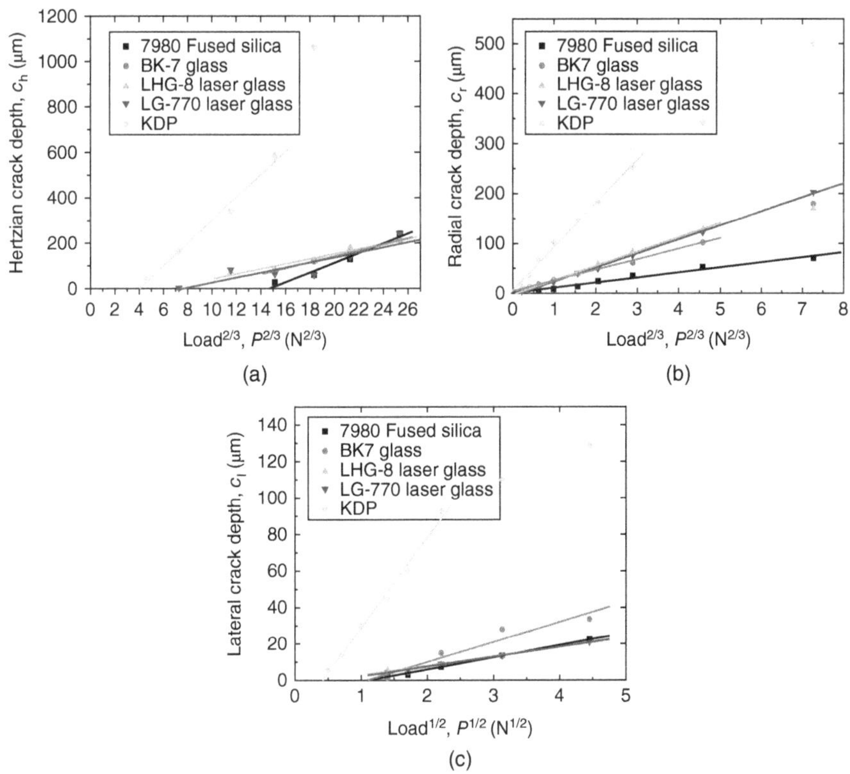

Figure 3.3 Measured (a) Hertzian, (b) radial, and (c) lateral cracks dimensions on selected optical glass and single crystals as a function of scaled load.

Table 3.1 Determined crack initiation and crack growth constants for a series of optical glasses and crystals. nm = not measured.

Name	Variable	Units	Fused silica (dry)	Fused silica (wet)	Borosilicated glass (BK7)	Phosphate glass (LHG8)	Phosphate glass (LG770)	KDP
Initiation constant for Hertzian cracks	A	10^4 N m^{-1}	9.8 ± 2.0	—	9.8 ± 2.0	5.9 ± 2.0	2.9 ± 1.0	1.0 ± 1.0
Growth constant for Hertzian cracks	χ_h	Unitless	0.050 ± 0.008	—	0.034 ± 0.005	0.019 ± 0.003	0.017 ± 0.003	0.028 ± 0.004
Initiation constant for radial cracks	α_r	$\times 10^5$	10.0 ± 3.3	5.0 ± 1.7	6.0 ± 2.0	2.2 ± 0.7	4.1 ± 1.4	1.1 ± 1.1
Growth constant for radial cracks	χ_r	Unitless	0.022 ± 0.003	0.03 ± 0.005	0.085 ± 0.013	0.083 ± 0.013	0.067 ± 0.011	0.081 ± 0.013
Initiation constant for lateral cracks	$P_{c\ell}$	N	1.5 ± 0.5	0.73 ± 0.24	1.5 ± 0.5	1.5 ± 0.5	1.5 ± 0.5	0.17 ± 0.07
Growth constant for lateral cracks	χ_ℓ	Unitless	0.19 ± 0.03	0.14 ± 0.02	0.35 ± 0.06	0.10 ± 0.02	0.12 ± 0.02	0.38 ± 0.06
Edge toughness	T_e	N mm^{-1}	109	nm	110	58	133	nm

a given material serve as the basis for quantifying grinding rates, SSD depth, and scratching characteristics, as discussed in subsequent sections.

3.1.1.2 Edge Chipping and Bevels

Another type of fracture in brittle materials is edge chipping. It is a static indentation fracture that occurs near the edge of the workpiece. Because of the edge boundary condition, the stress distribution due to static indentation is different than on a continuous planar surface, leading to crack propagation to the edge of the workpiece and dislodgement of a chip (often called a conchoidal fracture) from the workpiece (see Figure 3.4a). A key characteristic for an edge chip is that the normal load (P_{ce}) required for its formation on a given material scales linearly with the distance (d_e) that the load is applied from the edge of the workpiece (Figure 3.4b). Hence,

$$P_{ce} = T_e d_e, \tag{3.7}$$

where T_e is the edge toughness.

The edge toughness for a given material can be determined by performing a series of indentations at various loads and distances away from an edge of the workpiece. This is illustrated in Figure 3.5 that shows indents at 9.8 N applied with decreasing distances from the edge. Notice that the edge chip is formed once the edge distance decreases to 105 μm. This process is repeated for various loads and then the edge toughness can be determined as the slope in the P_{ce} versus d_e plot shown in Figure 3.4b.

Also, an approximate value of the edge toughness for a given workpiece material can be determined using the material's strain energy release rate

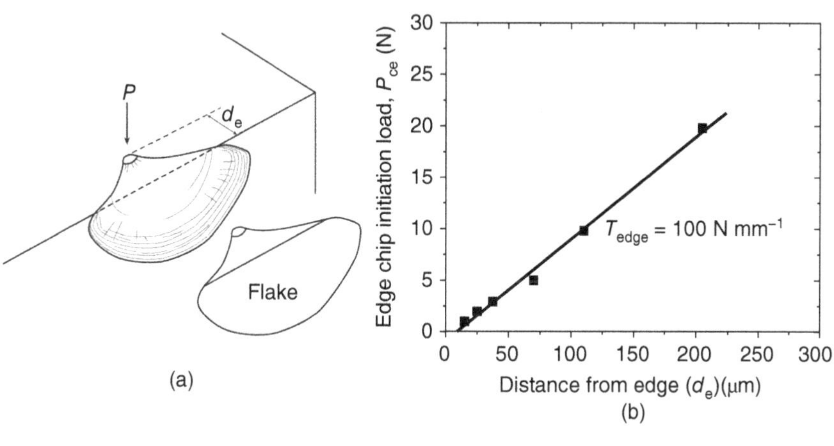

Figure 3.4 (a) Schematic illustration of the creation of an edge chip (after [20]); (b) Correlation between critical load to generate an edge chip and distance from edge of the workpiece for fused-silica glass.

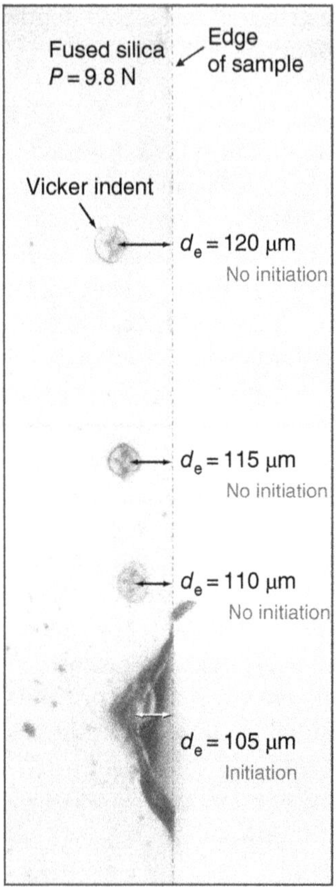

Figure 3.5 Optical micrograph of the surface of a fused-silica workpiece showing edge chip creation. Static indentations at 9.8 N are placed at various distances from the edge of the workpiece.

($G_{Ic} = K_{Ic}{}^2/E_1$) [21]. Figure 3.6 shows a reasonable correlation of the edge toughness with G_{Ic} for a large variety of different workpiece materials. Empirically, the edge toughness is given by

$$T_e \approx 120\left(\sqrt{\frac{K_{Ic}{}^2}{E_1}} - 1\right) \tag{3.8}$$

where T_e has units of N mm^{-1} and $K_{Ic}{}^2/E_1$ has units of J m^{-2}.

This expression has practical utility to estimate the edge toughness using basic material properties when direct experimental data of the edge toughness is not available. Some determined values of edge toughness of typical optical materials are shown in Table 3.1.

The geometry of the edge (angle and bevel design) can dramatically influence the resistance of the workpiece to edge chipping. Figure 3.7a,b shows the effect of edge angle on relative toughness of the edge, where the load required to edge chip increases significantly with edge angle [22, 23]. Having an edge

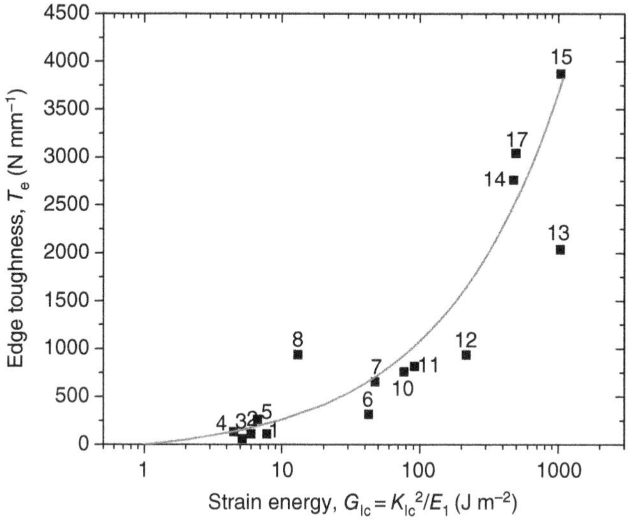

Figure 3.6 Correlation of the edge toughness of various materials. (1) Fused silica, (2) BK7, (3) LHG8 laser glass, (4) LG770 laser glass, (5) soda lime glass; (6) RBSN, (7) Vitox alumina, (8) SiC, (9) Sintox alumina, (10) Si_3N_4, (11–14) Various grades of WC, (15) various grades of stainless steel. Source: Data from Morrell and Gant 2001 [21].

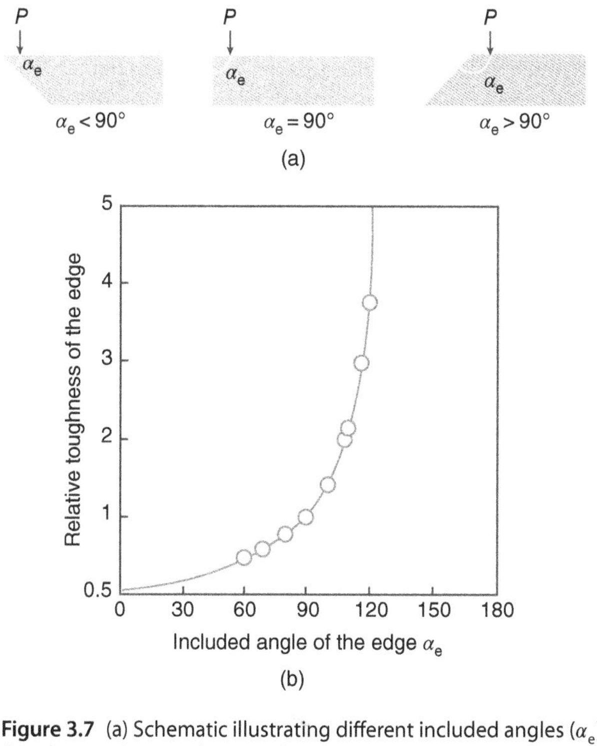

Figure 3.7 (a) Schematic illustrating different included angles (α_e) on the workpiece; (b) relative edge toughness as function of included edge angle on the workpiece. Source: From Hangl 1996 [22] and Almond 1990 [23].

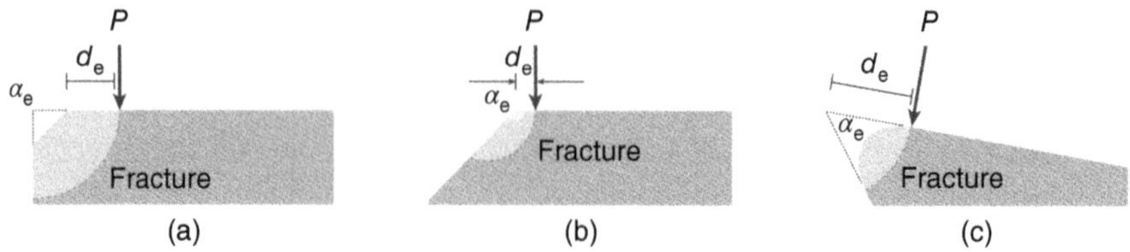

Figure 3.8 Definition of effective distance from edge (d_e) and included angle (α_e) for: (a) chamfered edge with a large flake; (b) chamfered edge with a small flake; and (c) arbitrary shaped edge. Source: Hangl 1996 [22]. Reproduced with permission from Springer.

angle (α_e) of >120° can result in relative edge toughness 5× that of the 90° edge. Figure 3.8 shows the effect of bevels and the adjustment in the definition of the edge distance (d_e) to be used with the edge toughness Equation (3.7) [22]. In general, the overall relative edge toughness increases with the addition of the bevels.

3.1.1.3 Sliding Indentation

So far, our description of indentation fractures has assumed a static indent – a nonmoving particle on a frictionless interface. When considering grinding processes, sliding and rolling must be factors, and friction between the particle and the glass surface is important. A sliding sphere with friction has been analyzed by Lawn [24]. Figure 3.9 compares the change in the calculated stress distributions for static and sliding indentation. Friction changes stress distribution such that peak tensile stress is at the trailing edge of the particle. The originating fracture (as viewed from the surface) will now have an arc shape, as opposed to the circular ring fracture commonly observed with frictionless Hertzian-cone contact [24, 25]. The geometry of trailing indent fractures is

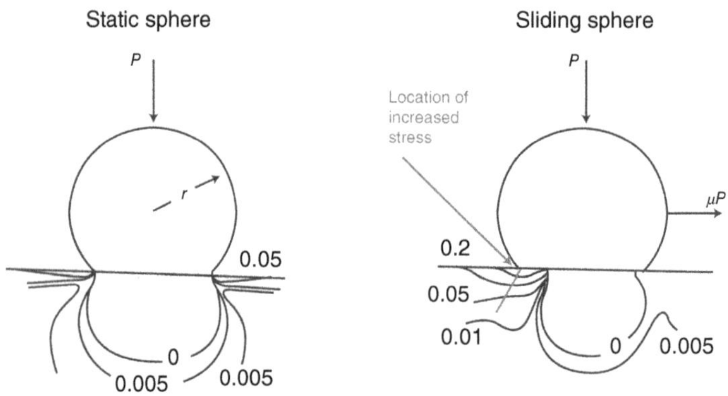

Figure 3.9 Comparison of the stress distributions for static (frictionless) and sliding (friction) Hertzian indentation. Source: Lawn 1993 [12]. Reproduced with permission from Cambridge University Press.

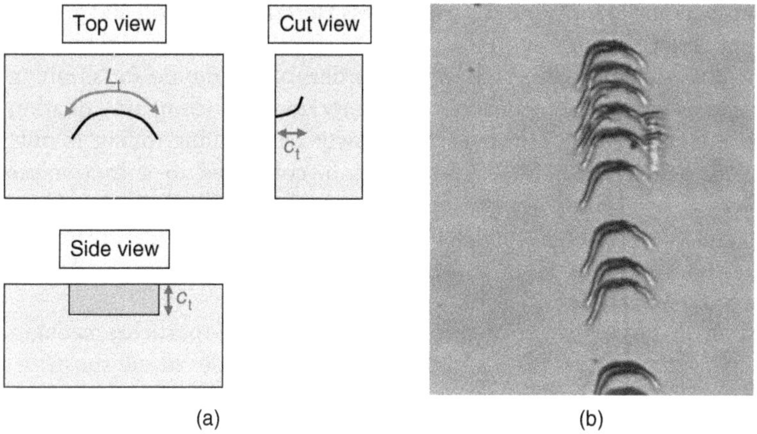

Figure 3.10 (a) Schematic of the geometry of an individual trailing indent fracture. (b) Optical micrograph of trailing indent fractures observed in a fine scratch that has been BOE etched (scratch width ~10 μm). Source: Suratwala et al. 2006 [4]. Reproduced with permission from Elsevier.

shown in Figure 3.10, both schematically and in an optical micrograph. A line of trailing indent fractures, also called chatter marks, is a commonly observed type of scratch. Chatter marks have also been shown to occur from a sharp sliding indenter [26]. The trailing indent initiation load (P_{ct}), crack depth (c_t), and surface-crack length (L_t) for a spherical sliding particle have a formalism similar to that of static indentation, given by [24]

$$P_{ct} = \frac{Ad}{2(1 + B\mu)^2}, \tag{3.9}$$

$$c_t = \left(\frac{\chi_h (1 + \mu^2)^2 P}{K_{Ic}} \right)^{2/3}, \tag{3.10}$$

$$L_t \cong \frac{1}{4}\pi(2a) \cong \frac{\pi}{2}\left(\frac{2}{3}\frac{k}{E_1}Pd \right)^{\frac{1}{3}}, \tag{3.11}$$

where B is a friction constant (unitless), μ is the friction coefficient (unitless), a is the contact radius for a Hertzian contact (m), d is the effective abrasive diameter, assuming a spherical particle (m), and k is a ratio of material constants given by

$$k = \frac{9}{16}\left((1 - \nu_1) + (1 - \nu_3)\frac{E_1}{E_3} \right), \tag{3.12}$$

where ν_1 and ν_3 are the Poisson's ratio for the workpiece and indenter abrasive particle, respectively. E_1 and E_3 are the Young's modulus of the workpiece and indenter particle, respectively. As the friction coefficient goes to zero, the expressions become identical to the frictionless Hertzian initiation and growth relationship described in Section 3.1.1.1. The frictional forces significantly reduce the initiation load (P_{ct}), sometime by orders of magnitude, for a trailing indent crack, as compared to a frictionless Hertzian crack (compare Equation (3.9) with Equation (3.1)). However, the growth of a trailing indent is only slightly increased with an increase in friction, compared to a frictionless Hertzian crack, as seen by comparing Equation (3.10) with (Equation (3.5)) [24]. The crack length of a trailing indent fracture is simply approximated as one quarter of the Hertzian contact circumference, as determined from Hertzian-contact analysis [12].

Depending on the local shape of the particle at contact, the particle's mechanical properties, the applied load, and the material properties of the substrate, other types of features besides trailing indent fractures may form a scratch [4, 26, 27]. The types of scratches that can be formed during the polishing of glass are shown in Figure 3.11. Swain described the general characteristics of a scratch for a sharp indenter at various load ranges (see Figure 3.12) [26]. At low loads ($P < 0.05$ N), a plastic deformation trench is generally formed without fractures. At intermediate loads ($0.1\ \text{N} < P < 5\ \text{N}$), a well-defined radial or trailing indent fractures occur, along with lateral cracks. At higher loads ($P > 5$ N), the plastically deformed track fractures convert to a rubble-like appearance, while lateral and trailing indent cracks are less pronounced.

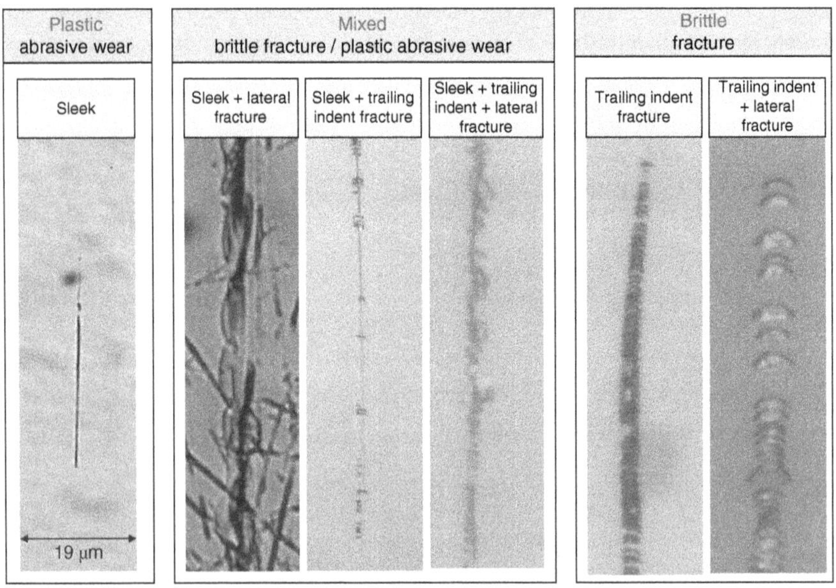

Figure 3.11 Categories of observed scratch types caused by rogue particles during polishing of fused-silica glass. Source: From Suratwala 2008 [28].

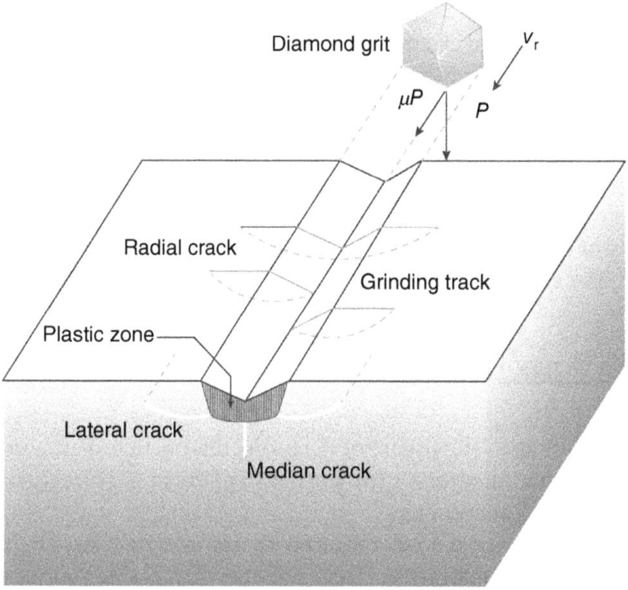

Figure 3.12 Types of fractures and deformation that may occur in a scratch. Source: Swain 1979 [26]. Reproduced with permission from Royal Society.

3.1.1.4 Impact Indentation Fracture

In certain process steps of optical fabrication, a moving particle or projectile may strike and fracture the workpiece. Examples include sandblasting and abrasive jet grinding, which intentionally fracture the surface for material removal; ultrasonic treatment of a workpiece in the presence of undesirable rogue particles may strike the surface; and during a workpiece's employment, for example, in aircraft, spacecraft, or laser-debris windows where exposed to the impact of foreign particles.

The principles used for static indentation generally apply to impact, where the impact load (P) is determined at contact from the velocity (v_3), particle diameter (d), and density (ρ_3) of the particle. Figure 3.2 illustrates particle impact fracture compared with static load fracture. The projectile may result in a cone fracture of length c_h (from a blunt projectile) or a radial fracture of radius c_r (from a sharp projectile), or both. The impact of hard particles on fused-silica surfaces has been previously examined [25, 29–31]. These studies have shown evidence of cone and radial fractures from both blunt and sharp indentation. By combining the existing models described by Lawn [12], Wiederhorn and Lawn [32], and Knight [33], two idealized cases of impact fracture (blunt and sharp) have been formulated. This impact model makes several basic assumptions, for example, that the projectile material is nondeforming and that the hardness of the projectile is greater than that of the target.

For blunt-projectile impact, the peak load (P_b) on the workpiece occurs when the projectile reaches its maximum elastic penetration into the substrate. Peak load can be determined by equating the kinetic energy of the projectile with that of elastic strain energy at maximum penetration, which leads to [33]

$$P_b = \left(\frac{5\pi\rho_3}{3}\right)^{3/5}\left(\frac{3k'}{4}\right)^{-2/5} v_3^{6/5}\left(\frac{d}{2}\right)^2, \tag{3.13}$$

where ρ_3 is the density of the projectile particle, d is the size of the particle, and v_3 is particle velocity. The subscript b designates blunt impact. k' is a material constant given by

$$k' = \frac{(1-v_1^2)}{E_1} + \frac{(1-v_3^2)}{E_3}. \tag{3.14}$$

Note that the peak load varies as the square of the projectile-particle diameter (d^2). Using the peak load from Equation (3.13), the depth of the impact fracture (c_h) can be determined using Equation (3.4).

The determination of the critical load required to initiate fracture (P_{ch}) depends on the quality of the substrate surface. For the surfaces containing flaws >1–10 μm, the initiation load varies linearly, as described by Equation (3.1). Substituting Equation (3.1) into Equation (3.13) and solving for v_3, the critical velocity for fracture initiation for a blunt projectile (v_{3bc}) without any plastic deformation is

$$v_{3bc}(d) = \frac{\left(\frac{3}{4}k'\right)^{1/3}}{\sqrt{\frac{5\pi\rho_3}{3}}}\left(\frac{2A}{d}\right)^{5/6}. \tag{3.15}$$

For sharp-projectile impact, contact area decreases, and therefore, stress within the contact zone rises dramatically. Hence, the applied stress usually exceeds the yield stress of the workpiece, causing irreversible plastic deformation of the workpiece. The generation of fractures during a sharp impact is analogous to that of sharp static indentation, as illustrated in Figure 3.2e. Radial and lateral cracks are typically observed during loading and unloading processes. An expression for the load at impact (P_s) can be determined by balancing the kinetic energy of the particle with the work to plastically deform the material [29, 32]. From this balance, the effective load is

$$P_s = \alpha_s \xi H_1^{1/3}\left(\frac{\pi d^3}{12}\rho_3 v_3^2\right)^{2/3}, \tag{3.16}$$

where α_s is the fraction of the kinetic energy that is used to plastically deform the surface (typically equal to 0.5), ξ is a geometric constant (equal to 4.8 for a Vickers-type indent), and H_1 is the hardness of the workpiece. The subscript s designates a sharp impact. This load expression assumes that the contact velocity (v_3) is slow relative to sonic velocities. In other words, quasi-static conditions apply [32].

As in the case of blunt projectiles, the size of radial cracks (c_r) from a sharp-projectile impact can be found using the peak load determined in Equation (3.16) and using its value in Equation (3.5). Both blunt and sharp projectiles will lead to fractures that scale as $P^{2/3}$. In contrast with blunt projectiles, the critical load to initiate fracture (P_{cr}) for a sharp impact is independent of projectile size. Rather, the fracture initiation load depends on the ratio $K_{Ic}{}^4/H_1{}^3$ of the substrate, as in Equation (3.2) [12]. For example, the critical load for fused silica has been determined as $P_{cr} = 0.02$ N. Substituting P_{cr} for P_s in Equation (3.16), the critical velocity for fracture initiation for a sharp projectile (v_{3sc}) is

$$v_{3sc}(d) = \left(\frac{P_{cr}}{\alpha_s \xi H_1{}^{1/3}}\right)^{3/4} \left(\frac{12}{\pi d^3 \rho_3}\right)^{1/2}. \tag{3.17}$$

Application of both the blunt and sharp particle-impact models is shown for fused-silica projectile particles onto a fused-silica workpiece to illustrate their utility. Figure 3.13a indicates when fracture initiation is expected, as predicted by Equation (3.13) for blunt indentation and by Equation (3.16) for sharp indentation. The critical velocity for damage initiation is considerably lower for a sharp impact than a blunt. Figure 3.13b shows the predicted Hertzian crack length (c_h) for an ideal blunt impact. Figure 3.13c shows the predicted radial-crack radius (c_r) for an ideal sharp impact.

Some experimental data is available to compare and validate the general behavior of this impact model [34, 35]. In a series of experiments, stainless-steel (SS) ball bearings ($d = 760$ μm) were shot at borosilicate glass (Borofloat) windows ranging in thickness from 0.4 to 3.0 mm and with a projectile velocity of 64–375 m s^{-1}. In many of the shots, the projectile is penetrated through the target glass. The data is compared to the impact model in Figure 3.14, which shows the cone-crack length in the workpiece as a function of projectile velocity. The points in the plot represent measured data concerning cone-crack length. The different symbols represent data measured on different thicknesses of plate. The solid line represents the model prediction for the Hertzian-crack length (c_h). The dashed lines represent the maximum crack length that can occur, based on the thickness of the workpiece. In most impacts, the cone fracture broke through the thickness of the workpiece; hence, the model cannot be used to predict the length of the cone fracture, though the data does not contradict the model in those circumstances. For the experiments carried out on targets of 3.0 mm thickness, the fracture did not break through (see triangle data points), and the calculated values matched up reasonably well with the measured data.

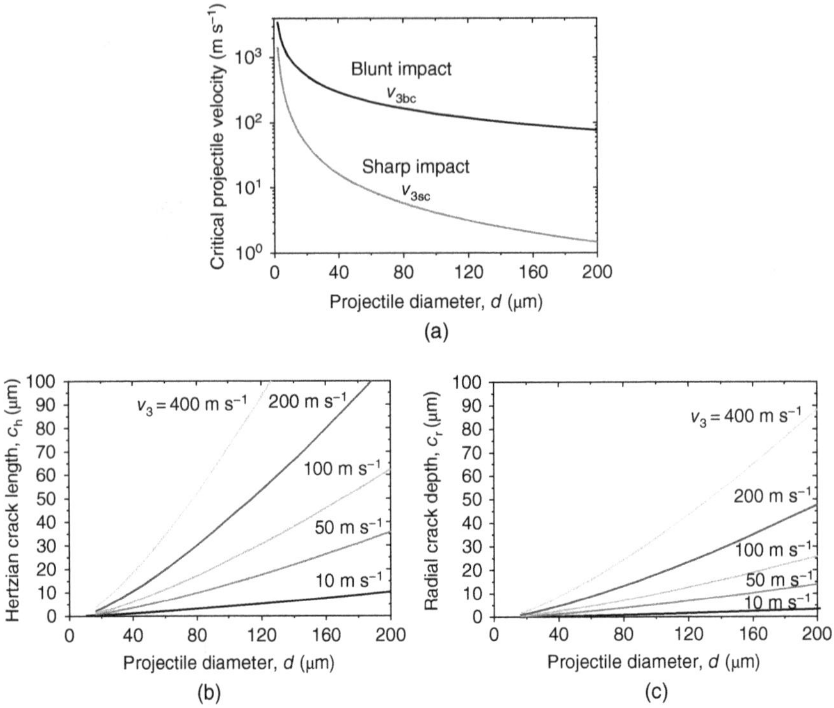

Figure 3.13 Calculated fracture characteristics on a fused-silica workpiece after impact by fused-silica particle projectiles using Equations (3.4), (3.5), (3.13)–(3.17): (a) critical velocity for fracture initiation by blunt and sharp projectiles as a function of projectile velocity; (b) Hertzian-crack length as a function of projectile velocity and diameter; and (c) radial-crack depth as a function of projectile velocity and diameter. Calculations use $\alpha_s = 0.5$, $\xi = 4.8$, $H_1 = 6$ GPa, $\rho_3 = 2.2$ gm cm^{-3}, $E_1 = E_3 = 73$ GPa, $\nu_1 = \nu_3 = 0.17$, and constants from Table 3.1.

The projectile indentation models above apply to impacting particles that are equal to or greater than the hardness of the workpiece. In the case of softer materials, such as many metals, the projectile will deform, reducing the effective load at impact. More sophisticated numerical simulations are likely required to examine the effect of deforming projectiles. In the case of brittle projectile materials with a lower hardness, fracture of the projectile itself must be accounted for. An analytical model that takes into account projectiles with hardnesses lower than the workpiece is not available; however, Shipway and Hutchings [36] have examined the effect of the H_3/H_1 ratio (where H_3 is the hardness of the projectile particle) on the erosion rate (which should be proportional to indentation damage) using a series of brittle projectile materials, as in see Figure 3.15. In general, for $H_3/H_1 > 1$ the erosion rate or indentation amount is essentially constant and the model described above should apply. For $H_3/H_1 < 1$, the erosion rate, and hence indentation depth,

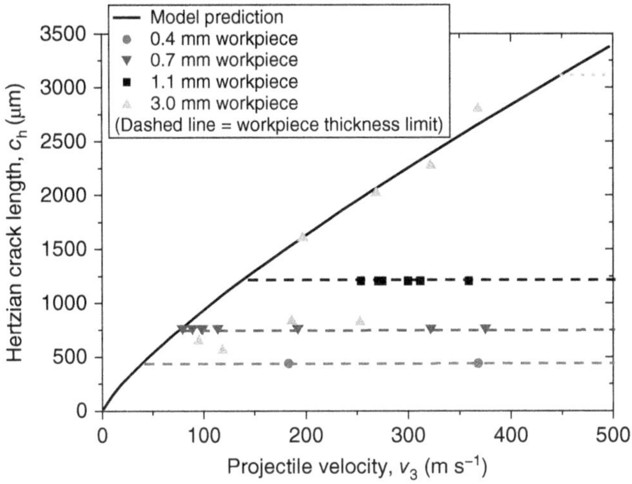

Figure 3.14 Comparison of measured cone-fracture lengths with that of the impact model for 760-μm-diameter SS projectiles at various velocities on borosilicate glass workpieces of different thicknesses. The points represent measured data, the solid line is that predicted by the impact model in Equations (3.4) and (3.13). The dashed lines indicate the length of the cone fracture at the point where it breaks through the thickness of the workpiece.

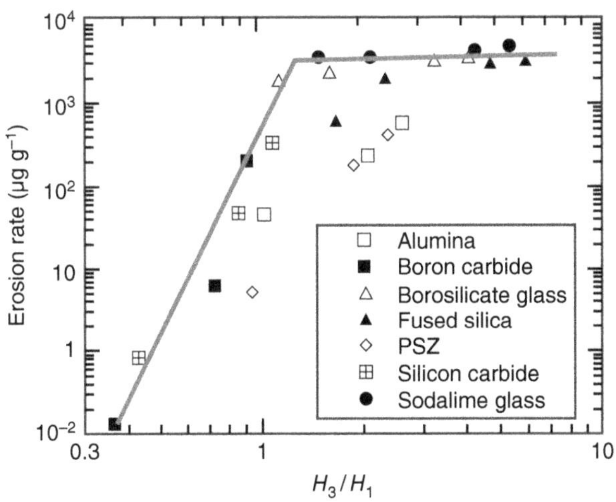

Figure 3.15 Erosion rate of various brittle materials as a function of the ratio of the hardness of the projectile to the hardness of the workpiece (after Shipway) [36]. The line represents an empirical upper-bound fit to the data, using Equations (3.18) and (3.19).

decreases dramatically. Using the data in Figure 3.15, the reduction in the erosion rate with projectile hardness can be used as an estimate of the drop in effective peak load at contact. This scaling factor is quantitatively defined as follows:

$$z_s = \frac{10^{6.3\frac{H_3}{H_1}-2.8}}{3160} \quad \text{for } H_3/H_1 < 1, \tag{3.18}$$

$$z_s = 1 \quad \text{for } H_3/H_1 \geq 1. \tag{3.19}$$

The z_s scaling factor can now be multiplied by the calculated peak load determined from Equations (3.13) and (3.16). For the specific case in Figure 3.13, using a fused-silica projectile on a fused-silica workpiece, $H_3/H_1 = 1$, and hence $z_s = 1$.

3.1.2 SSD During Grinding

3.1.2.1 Subsurface Mechanical Depth Distributions

As discussed at the beginning of this chapter, SSD consists of surface microcracks created on a processed workpiece during grinding. The generation of microfractures results from the same mechanism that removes material in the typical processes for grinding brittle materials (see Figure 1.9). Figure 3.16 provides a microscopic image of the ground surface of a glass workpiece after exposure to brief chemical etching. The surface has a rubbly structure revealing an ensemble of intersecting cracks, including lateral and trailing indent cracks.

Of practical importance is the depth distribution of cracks, which ultimately defines how much material removal is required at the next process step (see Figure 1.2). A wide variety of destructive and nondestructive techniques for measuring the amount and depth of SSD have been explored [38–43], many of

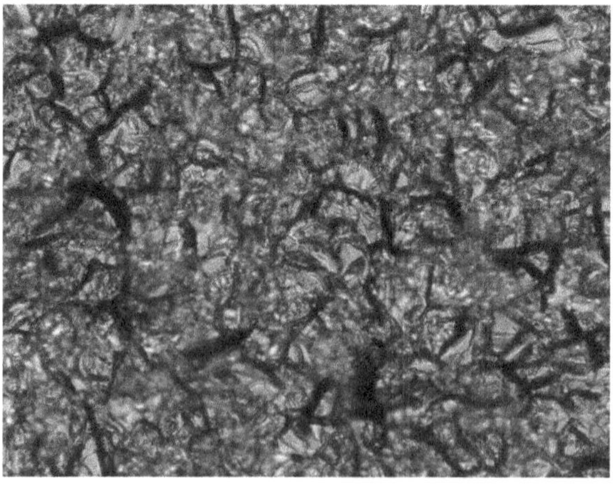

Figure 3.16 Optical micrograph of the ground surface of fused-silica glass, using a 150-grit generator grind. Full scale is 2.37 mm; the surface was etched ~1 μm to show the presence of surface cracks. Source: Suratwala et al. 2006 [4] and Wong et al. 2009 [37]. Reproduced with permission from Elsevier.

which are described in a review by Lee et al. [43]. Destructive methods include the following:

- ball dimpling [44]
- the magnetorheological finishing (MRF) spot method [45]
- the taper polishing method [46]
- the MRF [4, 47] wedge technique
- the Mulhearn method
- 3D cross-sectioning
- chemical etching
- dye impregnation

Nondestructive methods include the following:

- roughness–SSD correlations
- laser scattering
- confocal microscopy
- total internal-reflection microscopy (TIRM)
- optical coherent tomography (OCT)

Generally, nondestructive techniques require some inference or complex correlation, leading to greater uncertainty in the determination of depth. More precise are the destructive methods, which obviously require a sacrificial workpiece. It is desirable to choose a method that can be executed over an area large enough to get the desired statistics on crack-depth distribution and performed with a material removal process that does not create additional, interfering SSD.

The MRF wedge technique allows crack depth and length distribution to be determined statistically and directly over a large area [4, 48, 49]. Since the level of SSD is small or nonexistent with this polishing method, no interfering SSD is created to confuse the results (see Section 8.1). This SSD diagnostic method is particularly valuable when the final workpiece requires the removal of the deepest surface cracks and provides insights into how process parameters affect the shape of the SSD depth distribution.

The basic steps in the MRF wedge technique are illustrated in Figure 3.17. First, the grinding process of interest is performed on a sample workpiece. Then using MRF, a polished tapered wedge on the ground workpiece is created. Finally, after a slight chemical etch to enhance the visibility of cracks, crack statistics are characterized at various depths from the original ground surface. Details of this diagnostic are described in Section 7.2.2.

Figure 3.18 shows images of cracks observed as a function of depth for various grinding processes on fused-silica glass [4]. The number density of cracks on a typical ground surface is very high, such that the individual cracks intersect, giving a rubbly appearance (Figure 3.16). However, a few microns below the surface (after polishing through it), distinct individual cracks of a common morphology are visible, which decrease in number density with depth. Most

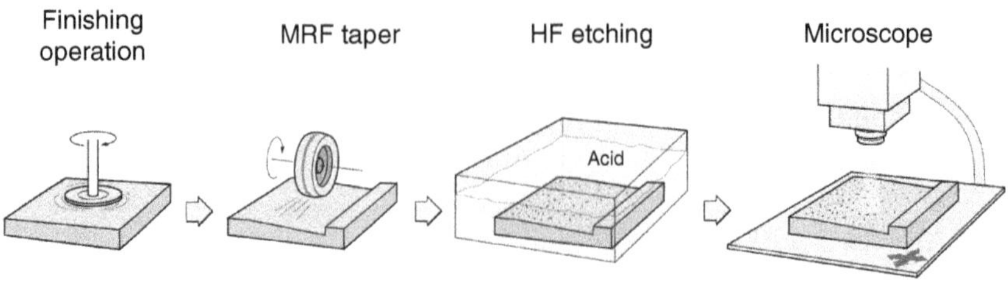

Figure 3.17 Schematic illustration of a wedge or taper polishing technique that can be used to directly measure SSD distribution. Source: Suratwala et al. 2006 [4]. Reproduced with permission from Elsevier.

of these cracks have a trailing indent character [4], also referred to as chatter marks [50] or stick-dig fractures.

Figure 3.19 compares the optical micrographs of the various ground surfaces after the rubbly surface has been removed by polishing, illustrating the typical individual cracks observed on a ground surface. The length or size of these trailing indent cracks (L_t) decreases with the size of the abrasive particle used for grinding. Specifically, L_t is ~0.15 to 0.30 times the diameter of the abrasive used for grinding [4, 48]. Note that both the width and length of the original cracks have increased in size by ~1 μm, owing to the opening up of the cracks by etching.

The statistical crack depth (Figure 3.20a) and length (Figure 3.20b) distributions determined from these images provide greater detail of the characteristics of the SSD. The crack-depth distributions are reported in terms of cumulative crack obscuration (crack-area fraction) as a function of depth into the original treated glass workpiece. Obscuration was used instead of crack-number-density as the unit to describe SSD density because the intersection of cracks near the surface causes significant uncertainty in the number density count of the cracking. The shape of the depth distributions follows a single-exponential dependence for the majority of the distribution, except near the end, where crack density drops very rapidly, appearing like an asymptotic cutoff. In some cases, the exponential dependence spans four or five orders of magnitude of obscuration. Also, the overall depth of SSD generally increases with an increase in the abrasive size of the grinding process. Crack-length distributions are plotted in terms of cumulative number distribution, as a function of crack length. For these distributions, only discrete cracks (nonintersecting cracks) were counted. Again, the mean crack length for each grinding process generally increases with an increase in the abrasive size (see Figure 3.20b).

In an interesting study by Menapace et al. [47], SSD was three-dimensionally reconstructed by iterative MRF polishing slices from a ground piece of fused-silica glass and subsequent microscopic imaging (see Figure 3.21a).

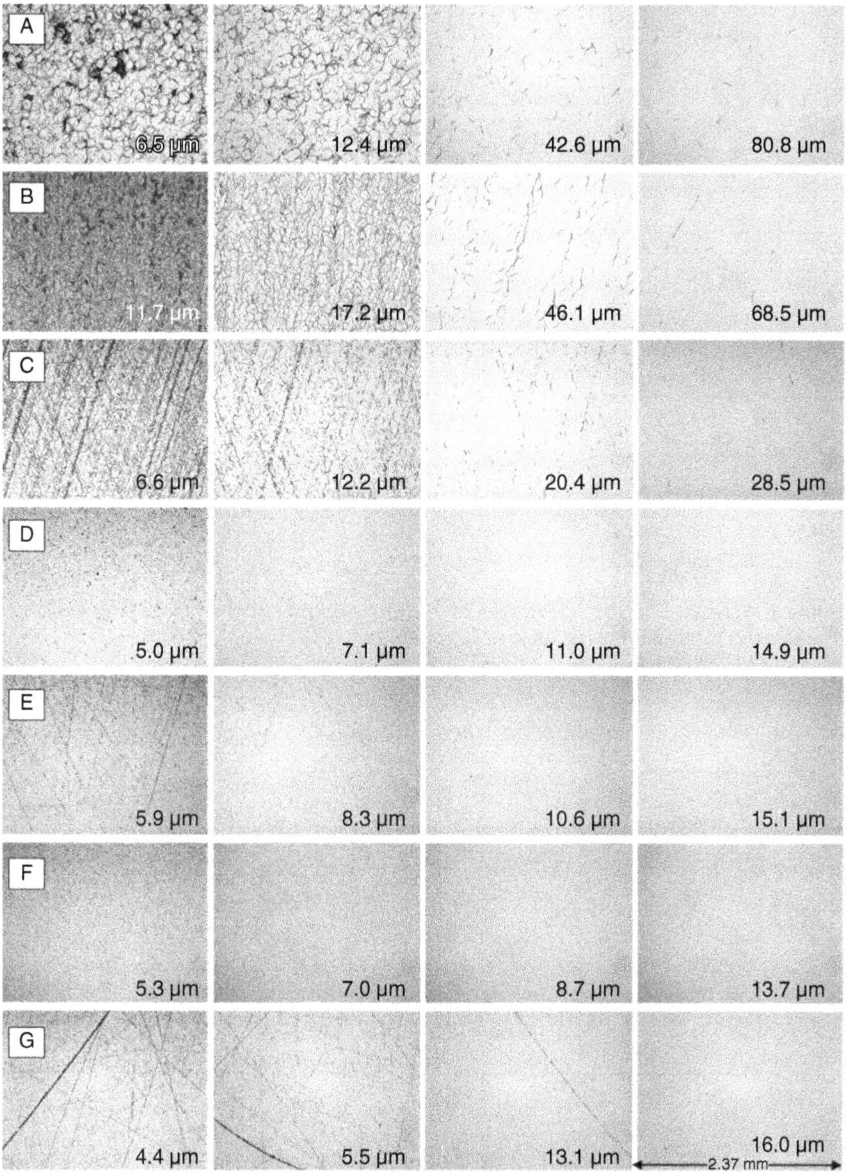

Figure 3.18 Optical-microscopic images of fused-silica workpieces treated with a variety of grinding processes. The samples are at various depths of removal using the MRF wedge technique. The value at bottom right is the depth below the original surface at which the image was taken. Sample A: sandblast; Sample B: 120-grit generator grind; Sample C: 150-grit generator grind; Sample D: 15-μm loose abrasive; Sample E: 15-μm loose abrasive, Sample F: 9-μm loose abrasive, Sample G: 7-μm fixed abrasive. Source: Suratwala et al. 2006 [4]. Reproduced with permission from Elsevier.

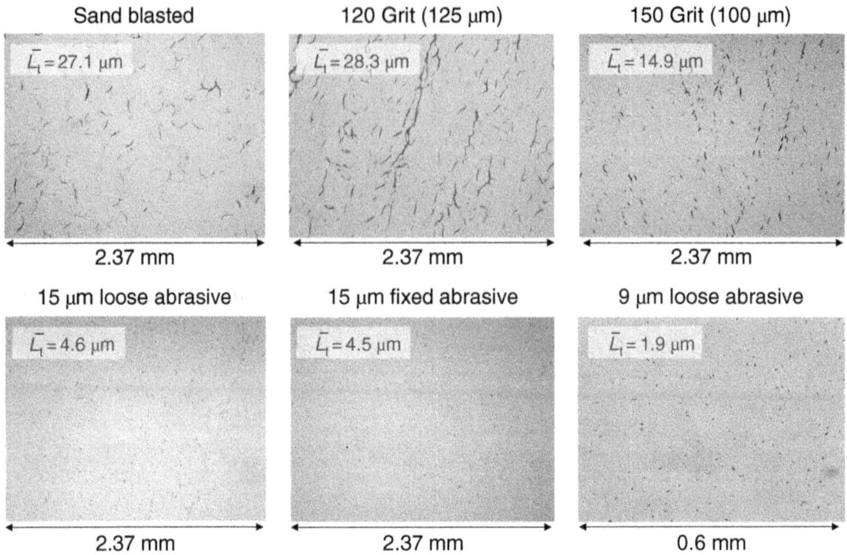

Figure 3.19 Optical-microscopic images of the fractures formed from various grinding processes on fused-silica glass. Source: Suratwala et al. 2006 [4]. Reproduced with permission from Elsevier.

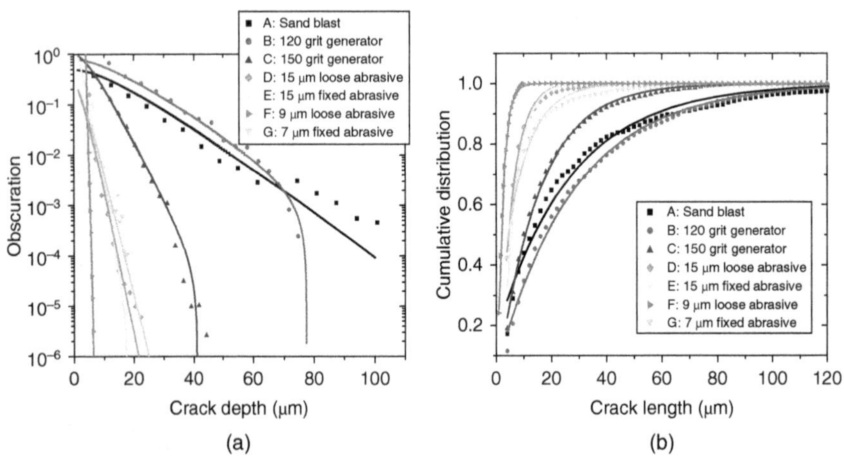

Figure 3.20 (a) Measured cumulative crack-depth distributions for fused-silica surfaces treated by a variety of grinding processes. The plots are semi log as crack obscuration versus crack depth. (b) Measured cumulative crack-length distributions for the same samples in (a). The points represent data points and the lines represent best fits using the grinding model (described in Section 3.1.2.5). Source: Suratwala et al. 2006 [4]. Reproduced with permission from Elsevier.

This technique allows observing the morphology and depth distribution more visually and affords greater insight into the morphology, because the data can be viewed from various angles, including the workpiece interior. For example, in Figure 3.21b, the intersections and various depths of the trailing indents cracks can be observed.

3.1.2.2 Relationship of Roughness and Average Crack Length to the Maximum SSD Depth

Another method to estimate SSD depth is to correlate it with the roughness of the ground workpiece using

$$c_{max} = k_{max}\delta, \tag{3.20}$$

where k_{max} is proportionality constant and δ is the roughness of the workpiece surface. This method is convenient because it is nondestructive, and roughness is a property of routine measurement in most optical-fabrication facilities. However, correlation is not as precise as with the MRF wedge technique.

k_{max} has been measured for fused silica, other glasses, and crystals. Figure 3.22 shows the data for one of these correlations. However, its reported values varies significantly for even the same material (such as fused silica): for example, Preston finds ~3 [50]; Aleinikov, ~4 [38]; Lambropoulos, <2 [42, 46, 52]; Randi, 1.4 [46]; P. Hed, 5–8 [46]; Neauport, ~3.3 [53]; and Suratwala and Miller, ~49 for deepest cracks and ~9 for average [4, 48]. The large variance in k_{max} values stem from variations in the method of measurement (e.g. profilometry versus white-light interferometry), spatial-scale range measured, roughness-metric definition (rms or PV), and length of the surface or area measured. Most have found that the PV roughness, δ_{PV} (the highest height minus the deepest valley) [42, 44, 52] – as opposed to average or

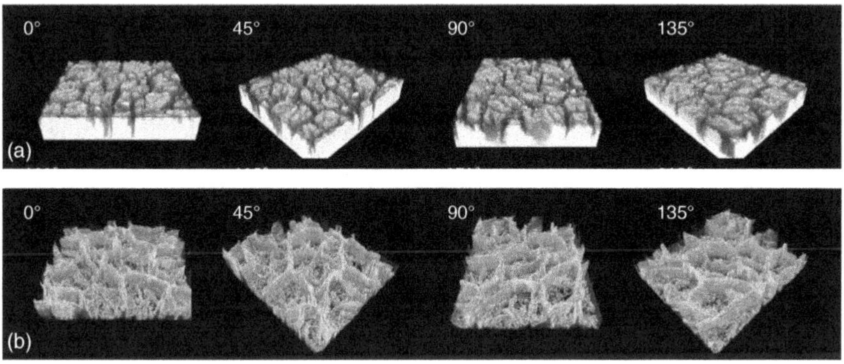

Figure 3.21 Measured SSD morphology of a 120-grit ground fused-silica surface viewed at different angles from the (a) top and (b) bottom. Data was collected from MRF slicing of the ground surface and imaging at various thickness depth, followed by 3D reconstruction. Source: Reproduced with permission of Menapace 2005 [51].

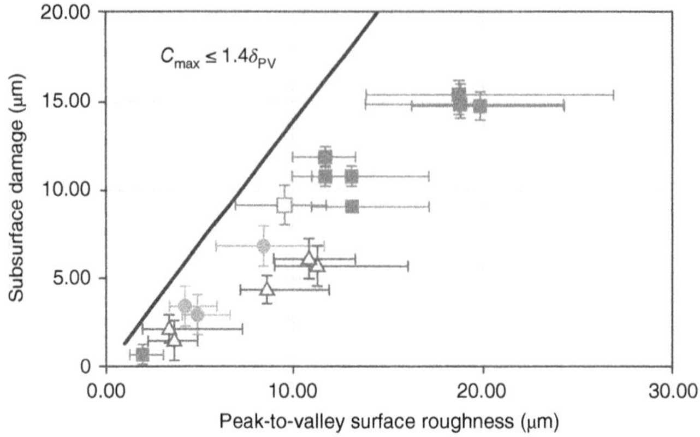

Figure 3.22 SSD depth versus PV surface roughness for various single crystals. Source: Randi et al. 2005 [46]. Reproduced with permission from The Optical Society.

RMS roughness – provides the better correlation to SSD depth [42, 44, 52]. However, the δ_{PV} measurement can be very sensitive to the length of the scan and position of the surface measured.

At a more fundamental level, the relationship between *surface roughness* and SSD depth can be studied in terms of the fracture mechanics of individual cracks. Lateral cracks can be considered as contributing largely to material removal and roughness, and radial cracks or trailing indent cracks as largely contributing to SSD depth. Lambropoulos et al. use this idea to derive a roughness–SSD correlation [52]. Assuming the PV *surface roughness* (δ_{PV}) is equal to the lateral-crack depth (c_ℓ), the proportionality constant (k_{max}) using Equation (3.20) is then

$$k_{max} = \frac{c_{max}}{\delta_{PV}} = \frac{c_{max}}{c_\ell} = \frac{c_{max}}{\frac{\chi_\ell (E_1/H_1)^{2/5}(P/N_L)^{1/2}}{H_1{}^{1/2}}}, \tag{3.21}$$

where c_{max} is the maximum subsurface-damage depth measured statistically over the surface and N_L is the number of abrasive particles loaded. Using the material properties for fused silica, k_{max} values ranging from 25 to 39 are determined. This compares reasonably well with k_{max} equal to 49, determined using the measured PV *surface roughness* [4].

3.1.2.3 Fraction of Abrasive Particles Mechanically Loaded

The indentation–fracture relationships described in Section 3.1.1 can be used to illustrate that the fraction of abrasive particles being loaded and causing fracture is quite small [4]. Take, for example, a 9-μm loose-abrasive grinding of fused silica and assume that all the abrasive particles between the lap and workpiece are evenly, mechanically loaded and the fill fraction of particles below the work piece is ~0.3 [54]. Since the friction coefficient is not known and its effect on crack depth is small, Equation (3.4) is used to estimate the depth of damage for a trailing indent crack. For a 10-cm diameter workpiece and 9-μm diameter abrasive, the total number of particles below the workpiece (N_T) is 10^8. Given a total load of 25 N, the load per particle is only 2×10^{-7} N. Substituting into Equation (3.4), a crack depth of only 3.5 Å is calculated. This prediction is $\sim 10^4$ smaller than the measured value of c_{max} (6 μm), as shown in Figure 3.20. A reasonable explanation for this large discrepancy is that only a small fraction of the particles is loaded.

The fraction of particles loaded for a given fracture depth ($f_{load}(c)$) can be described as follows:

$$f_{load}(c) = \frac{N_L}{N_T} = \frac{\chi_h P}{K_{Ic} c^{3/2}} \frac{\bar{d}^2}{4 f r_o^2}, \tag{3.22}$$

where N_L is the number of particles loaded, N_T is the total number of particles between workpiece and lap, c is crack depth, P is total applied load, $\bar{d}$ is the mean abrasive-particle size, f is the fill fraction, and r_o is the radius of the optic [4]. Using Equation (3.22) on the process parameters and data shown in Figure 3.20, the $f_{load}(c)$ is calculated. At shallow SSD depths (say, 1 μm) the f_{load} ranges from 10^{-3} to 10^{-5}. At SSD depths approaching c_{max}, f_{load} is 10^{-5}–10^{-7}. In other words, for the deepest SSD, only 1 of 100 000 to 10 million particles is loaded to cause the fracture. Others have come to similar conclusions regarding load-bearing particles [55, 56].

The larger particles in the tail end of the abrasive-particle distribution are the load-bearing particles. Figure 3.23 shows the abrasive-size distribution of a

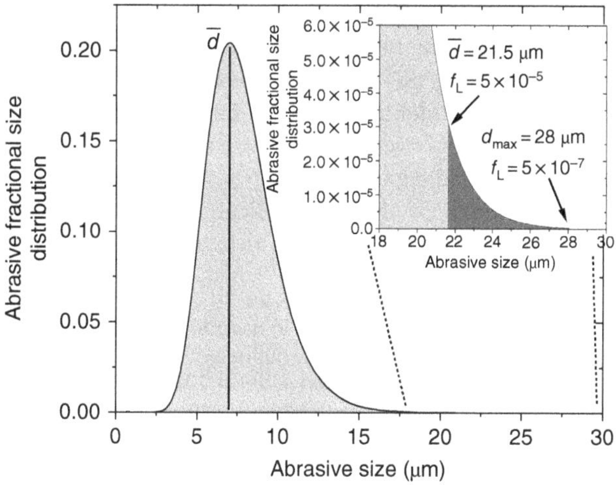

Figure 3.23 Abrasive fractional size distribution and the estimated range of loaded particles that cause fracture. Source: Suratwala et al. 2006 [4]. Reproduced with permission from Elsevier.

15-μm loose abrasive (Microgrit 15T) provided by the manufacturer, fit to a log-normal distribution [4]. By extrapolating the size-distribution data with a log-normal fit and using Equation (3.22), an estimate of the size of particles being loaded is determined as 22–28 μm, which is much larger than the average particle size of ~7 μm.

3.1.2.4 Relationship Between the Crack Length and Depth

Because cracks are not all the same depth for a given grinding process (Figure 3.20), it is clear that the load for a given particle varies. The load variation per particle is largely driven by the size of the particle, where larger abrasive particles bear a larger load (see Figure 3.24). Hence, the particle size distribution (PSD) of the abrasive is related to the fundamental crack distribution ($f_o(c)$). This relationship, however, is complicated by many possible factors, such as the following:

- Participating in the fracture event by tail-end particles only (the largest particles of the distribution)
- Agglomeration
- Comminution
- Rotation of the particles
- Foreign particles entering the interface
- Initiation of fractures by loaded particles, which may change according to preexisting fractures present on the surface (such as changes in stress profile or intensity at the crack tip)
- The roughness of the lap and workpiece
- The overall interface gap, which can alter the load or the particle

Most models of the grinding of brittle materials have focused on determining *material removal rate* versus predicting SSD depth distribution (see Chapter 5). To date, a comprehensive grinding model has yet to be developed that starts from workpiece- and lap-material properties, abrasive size and geometric distribution, kinematics, and loading- and feed-rate conditions.

Even without knowing the effects of these complicating factors, useful engineering correlations relating the crack-length distribution of surface

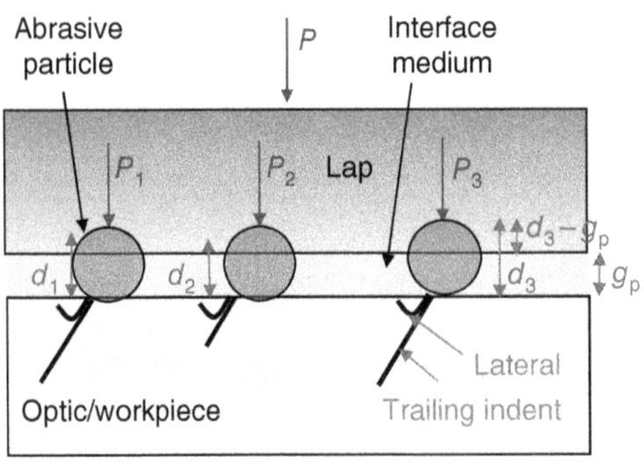

Figure 3.24 Schematic of a working model to describe the development of SSD during grinding. Source: Suratwala et al. 2006 [4]. Reproduced with permission from Elsevier.

cracks (which are easily observable) to crack-depth distribution has been formulated [4]. A schematic of such an engineering model is shown in Figure 3.24, where spherical abrasive particles are loaded at the interface between the lap and workpiece.

The sliding abrasive particles generate trailing indent cracks, contributing to SSD, and lateral cracks, contributing to material removal. The trailing indent fractures have a geometry outlined in Figure 3.10, and their crack growth is approximated using Hertzian-crack growth relationships (see Section 3.1.1.1) [4]. For simplicity, the surface of the lap and the workpiece are smooth and flat, the gap (g_p) is fixed, and not time varying. The load per particle is assumed to be linearly proportional to the depth of particle penetration in the lap or workpiece. In other words, for a fixed gap (g_p), it is the vertical dimension of the particle, which for spherical particles is the effective particle size (d) that is directly proportional to load. The linear dependence of load with particle size has also been used in other grinding and polishing models [55, 56]. The load for a given effective particle size (d) is then defined as follows:

$$P(d) = \frac{P}{N_L}\left(\frac{d}{\bar{d}}\right), \tag{3.23}$$

where P is the total applied load on the part, N_L is the number of particles being loaded on the workpiece, and $\bar{d}$ is the mean particle size of the abrasive. Using the various relationships described in Section 3.1.1.1, the relationship between crack depth and length simplifies to [4]

$$c = \frac{L_t}{\Omega}, \tag{3.24}$$

where

$$\Omega = \frac{\pi}{2}\left(\frac{K_{Ic}}{\chi_h}\right)^{2/3}\left(\frac{2kN_L\bar{d}}{3E_1P}\right)^{1/3}. \tag{3.25}$$

The value of Ω largely depends on workpiece material and crack-growth properties, abrasive size, and applied load. For a fused-silica workpiece, most grinding cases lead to values of Ω between 3 and 5 (i.e. not changing significantly) [4]. Hence Equation (3.24) suggests that the crack length is linearly related to crack depth. The experimental data also show a linear relationship between the mean crack length ($\overline{L_t}$) with both the average crack depth ($\bar{c}$) and the maximum SSD depth (c_{max}), which is consistent with the engineering model (see Figure 3.25a).

This relationship between crack length and depth has important practical implications (see also Section 6.2). By measuring the length of a crack on a workpiece, one can estimate its depth. According to Figure 3.25a, the crack depth on fused-silica glass will be, on average, 35% of the length. This ratio was also observed by Catrin et al. [57]. However, the crack may be as deep as 2.8 times the crack length. In other words, if a microfracture has a crack length of 10 μm (the same as a scratch width of 10 μm) on the surface of a workpiece, then

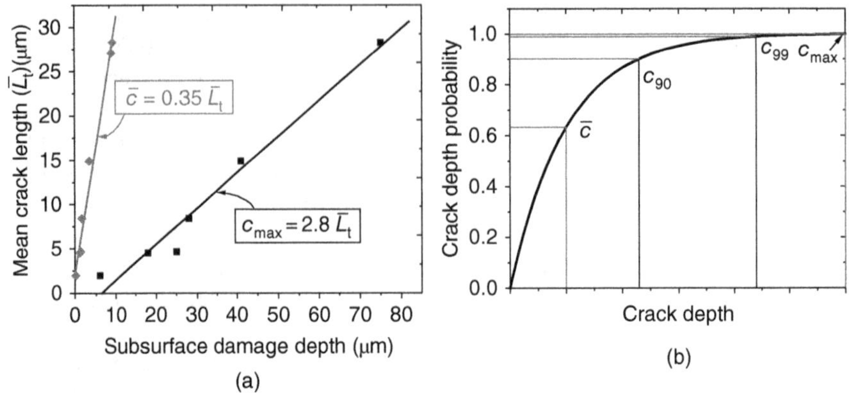

Figure 3.25 (a) Correlation of maximum SSD depth (c_{max}) and average SSD depth ($\bar{c}$) with the measured mean crack length ($\bar{L}$); (b) crack-depth removal probability as a function of amount removed from workpiece. Source: Suratwala et al. 2006 [4]. Reproduced with permission from Elsevier.

removing $0.35 \times 10\ \mu m = 3.5\ \mu m$ of material gives a 50% probability of removing that individual fracture. By the same logic, removing $2.8 \times 10\ \mu m = 28\ \mu m$ gives a 100% probability of removing the fracture (see Figure 3.25b).

3.1.2.5 SSD Depth-distribution Shape

The SSD depth distributions shown in Figure 3.20a in a semilog form follow a general exponential dependence. Two major factors that determine the shape of SSD depth distribution are the fundamental instantaneous distribution of cracks, $f_o(c)$, generated at some given time and the summation of $f_o(c)$ due to the continual shortening of cracks from material removal and new cracks generated at each time step. A simple model has been used to evaluate relationships between $f_o(c)$ and the final observed SSD depth distribution, $F_c(c)$ (see Figure 3.26) [4]. Knowing that the previously generated cracks are unlikely to grow (see Section 3.1.2.6), the cumulative distribution of crack depths is described as

$$F_c(c) = \int_c^{\infty} f_c(c)\,dc, \tag{3.26}$$

where $f_c(c)$ is the final incremental distribution of cracks, which depends on the fundamental instantaneous distribution of cracks ($f_o(c)$) as

$$f_c(c) = f_o(c) + f_o(c + \Delta) + f_o(c + 2\Delta) \ldots = \sum_i f_o(c + i\Delta), \tag{3.27}$$

where Δ is the removal increment amount and i is the number of removal increments. For each Δ, another set of cracks of the same fundamental distribution is generated, but starting at the new surface depth of $i\Delta$. Previously generated

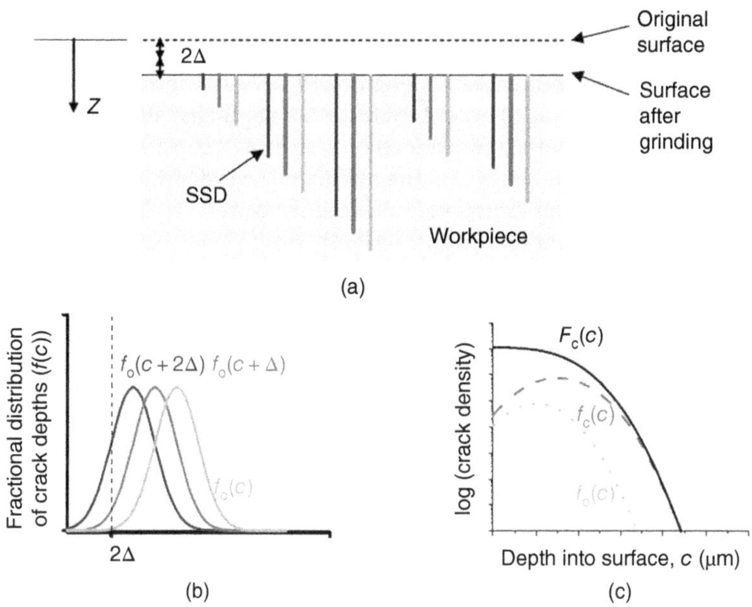

Figure 3.26 (a) Schematic illustration of the creation and removal of cracks as a function of surface removal; (b) instantaneous distribution of crack depths ($f_o(c)$) for different removal increments of Δ; (c) comparison of $f_o(c)$, $f_c(c)$, and $F_c(c)$ on a semi log plot. Source: Suratwala et al. 2006 [4]. Reproduced with permission from Elsevier.

cracks are continually made shorter and eventually removed. This simple model implies that the deepest cracks in the distribution are created in the last increments – that is, at the end of the grinding process. Another interesting result from this simple model is that the shape of the tail end of the calculated distribution appears as a single exponential, regardless of the functional form for $f_o(c)$ (e.g. Gaussian, log-normal, and Lorentzian), which matches the shape of the observed SSD distribution [4].

Using a single exponential fit to the measured crack-length distributions (see lines in Figure 3.20b) and using the established relationship between crack-length distribution and -depth distribution, the SSD crack-depth distribution (in terms of obscuration or crack area viewed from top) has been derived as [4]

$$O(c) = A_o w_c n_s \Omega \left(c + \frac{\overline{L}_t}{\Omega} \right) \exp\left(\frac{-c\Omega}{\overline{L}_t} \right) - A_o w_c n_s (L_{max} + \overline{L}_t) \exp\left(\frac{-L_{max}}{\overline{L}_t} \right), \tag{3.28}$$

where A_o is a pre-exponent constant, w_c is the width of the crack (dependent on how much the sample has been etched), n_s is areal number density of cracks at the surface, L_{max} is maximum crack length, and $\overline{L_t}$ is average crack length. The first term in Equation (3.28) describes the initial depth distribution, which is roughly a single exponential, and the second term is a constant related to the maximum depth of SSD (c_{max}) [4]. The calculated SSD depth distributions shown in Figure 3.20a, using Equation (3.28) and measured crack-length data characteristics, match the measured data well [4]. This formalism has the practical utility of enabling estimation of the SSD distribution after measuring the average and maximum crack lengths observed on the surface of a ground workpiece.

3.1.2.6 Effect of Various Grinding Parameters on SSD Depth Distributions

SSD depth distribution also correlates with some basic processing parameters such as load, abrasive size, and process sequence [4, 25, 38, 58]. Figure 3.27a shows the crack-depth distributions of a series of 9 μm loose-abrasive grinding processes performed on fused-silica glass at three applied loads (25, 220, and 580 N) or pressures (0.46, 4.1, and 10.7 psi) on a 100 mm diameter workpiece [4]. With an increase in load, crack depth increased, owing to the increase in load per particle (see Section 3.1.1.1). Also, at very high applied load, a second kink in the SSD depth distribution appears that was not observed at the lower load. This suggests that a second distribution of particles is loaded, causing fracture, or that some fraction of particles is loading differently, fracturing by another mechanism. Li et al. [59] showed a trend of increasing SSD depth with depth of cut (which is analogous to applied load or pressure) using fixed-abrasive grinding on fused silica.

Based on the model outlined in Figure 3.24, an increase in average abrasive size leads to a decrease in the areal density of load particles (N_L), which results in a larger load per particle. The larger load per particle increases fracture depths, and hence removal rate (Figure 1.3) and SSD depth. Figure 3.20a shows this trend with abrasive size for various loose and fixed abrasives on fused-silica glass. Dong and Cheng [60] showed that the dominant contributor to SSD depth using fixed abrasives is abrasive size and, more weakly, velocity during the grinding of SiC. Wang et al. [49] also showed the effect of abrasive size on BK7 glass.

To determine whether cracks from a previous grinding step will grow and alter the SSD depth distribution caused by the next grinding step, SSD distribution was measured after various grinding-step sequences. Figure 3.27b shows the SSD depth distributions measured on fused-silica glass using four grinding processes with four grits: 120, 150, 120 followed by 150, and 120 followed by etching and a 150-grit grind. Note that grit is a common descriptor for abrasive size, where the larger the value, the smaller the abrasive. Here, 120 grit has ~125 μm particles and 150 grit has ~100 μm particles. For each of these

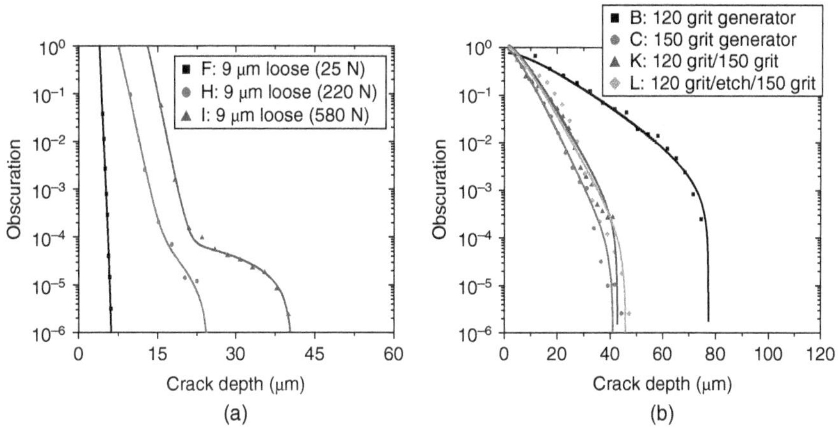

Figure 3.27 Measured cumulative crack-depth distributions for fused-silica surfaces (a) as a function of load, (b) with particle contamination, and (c) with various grinding steps performed in series. The plots are semi log plots of crack obscuration vs crack depth. Source: Suratwala et al. 2006 [4]. Reproduced with permission from Elsevier.

samples, ~100 μm of material was removed during the final 150 grit grinding to ensure that the removal was greater than the SSD depth caused by the 120 grit grinding alone, which was measured as 80 μm. Little difference in the SSD depth distribution was observed with each of the 150 grit final grindings, suggesting that cracks that preexisted in the 120-grit process did not grow much deeper than the cracks caused by the 150 grit process alone.

Finally, the various studies have evaluated the impact of nonconventional grinding processes and different workpiece or grinding materials on SSD. All appear to be qualitatively consistent with basic indentation-fracture mechanics and the grinding model outlined in Figure 3.24. Lv et al. [61] compared the SSD depth of ultrasonic grinding with conventional grinding on BK7 glass, concluding that ultrasonic grinding was more susceptible to deeper SSD depth. This is likely due to the greater load per particle imparted by ultrasonic energy. Neauport et al. [53] used a comprehensive set of loose abrasives to grind fused silica, concluding that alumina abrasives lead to lower SSD depth, as compared with higher-hardness, higher-modulus abrasives such as B_4C or SiC. Lower-hardness abrasives are shown to have a lower wear rate, as indicated in Figure 3.28 [14, 62].

SSD depth distributions have also been measured on various workpiece materials, including optical single crystals Randi et al. [46], hard ceramics, including AlON, SiC, and Al_2O_3 Shafrir et al. [63], and other glasses Esmaeilzare et al. [64]. However, to date, a comprehensive quantitative model to predict SSD depth as a function of workpiece material properties has not been developed and tested.

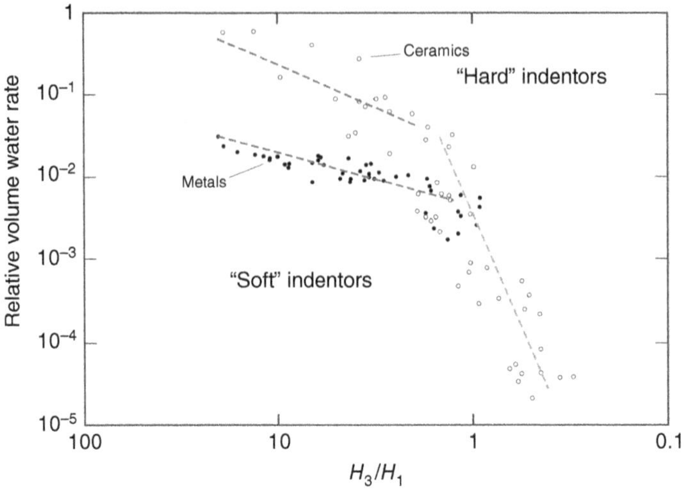

Figure 3.28 Volume wear rate for various metals and ceramics as a function of the relative hardness between the abrasive and substrate. Source: Hutchings 1992 [14]. Reproduced with permission from Taylor & Francis.

3.1.2.7 Rogue Particles During Grinding

Rogue particles are particles in the PSD that are significantly larger than the base distribution. Their presence can have a strong influence on SSD during grinding and polishing (see Section 3.1.3). Consider a series of alumina loose-abrasive particles with different PSDs, as presented in Figure 3.29. In Figure 3.29a, grinding was performed on fused-silica glass using various mixtures of these slurries with all other process conditions constant [65]. Figure 3.29b shows the resulting measured SSD depth distributions using various mixtures of 9 and 15 μm alumina loose abrasives. SSD depth increased significantly with the addition of a small amount of 15 μm slurry contaminant to the 9 μm slurry. Another example is illustrated in Figure 3.29c, which shows measured SSD depth distributions after grinding fused-silica glass by using various mixtures of 15 and 30 μm alumina loose abrasives. The minimum and maximum SSD depth in the plot are bound using the 15 μm abrasive alone and the 30 μm abrasive alone. With small additions of the 30 μm abrasive particles (that is, rogue particles) to the 15 μm abrasive slurry, SSD depth increases from the depth found with the base slurry at a fairly low fraction of rogue particles ($<10^{-4}$) or at an areal density of 200 cm^{-2}. Hence, it takes only a small number of rogue particles to increase SSD.

The results imply that only the largest particles in the distribution are active particles with a load that perform the work of creating cracks and removing material on the workpiece. The remaining particles are largely inactive. A similar conclusion was reached with the analysis described in Section 3.1.2.3. The

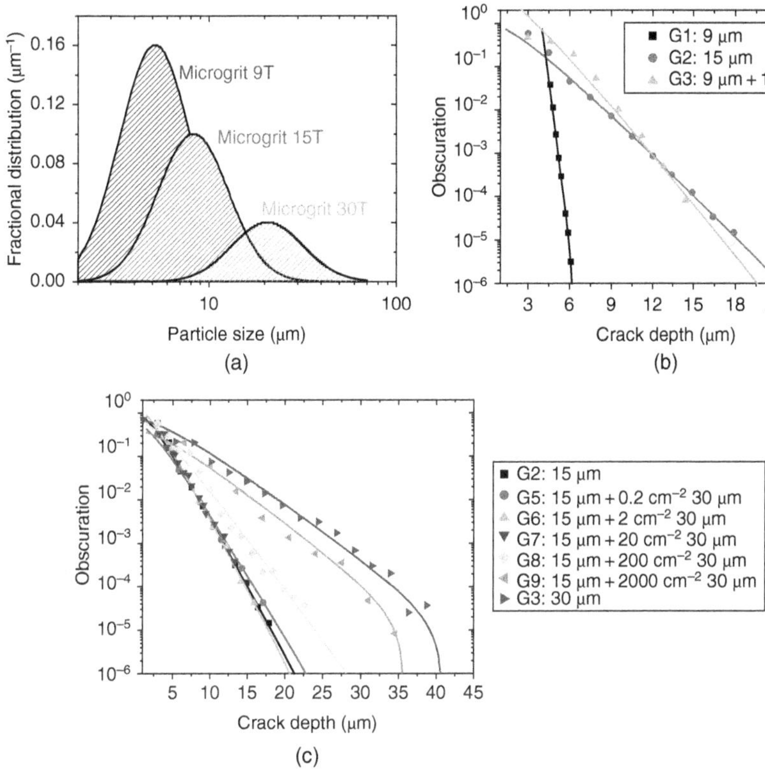

Figure 3.29 (a) PSDs of various alumina slurries used in grinding experiments; (b) SSD depth distributions on fused-silica workpieces using different alumina loose-abrasive PSDs; (c) SSD depth distributions of fused-silica workpieces after loose-abrasive grinding using 15, 30, and 15 μm contaminated with various amounts of 30-μm alumina particles. Source: Suratwala et al. 2008 [65]. Reproduced with permission from Elsevier.

practical implication is that the presence of rogue particles may strongly reduce the yield and efficiencies of the optical-fabrication process (see Figure 1.2). Hence, in managing SSD, significant controls are needed to minimize rogue particles, such as preventing contamination between processing steps.

The change in the SSD depth distribution with the addition of rogue particles during grinding can be quantified using the grinding model described in Sections 3.1.2.4 and 3.1.2.5. When two different PSDs are mixed and used for grinding, the shape of the SSD distribution will take the form [65]

$$O(c) = \int_{c}^{c_{\max}} ((1 - x_r)g_b(d) + x_r g_r(d))c^{1/2} Q_r \, dc, \tag{3.29}$$

where $g_b(d)$ is the PSD of the base slurry abrasive, $g_r(d)$ is the PSD of the rogue particles, x_r is the number fraction of particles from the rogue PSD, and Q_r is a constant given by

$$Q_r = \frac{3\pi K_{Ic} w_c n_s}{8\chi_h} \left(\frac{N_L \overline{d}^{\,2}}{P} \right)^{2/3} \left(\frac{2k}{3E_1} \right)^{1/3}, \tag{3.30}$$

where K_{Ic} is the fracture toughness of the substrate, N_L is the number of abrasive particles loaded, $\overline{d}$ is the mean abrasive-particle size, χ_h is the Hertzian-indent-crack growth constant, P is the applied load, w_c is the width of the cracks on the surface, n_s is the number density of cracks on the surface, and k is a material constant related to the modulus and Poisson's ratio of the substrate and indenter (see Equation (3.14)).

Using a log-normal distribution functional form for the abrasive PSD in Figure 3.29a, $n_s \sim 0.05\ \mu m^{-2}$, and $P/N_L \sim 0.25$ N, the SSD depth distribution was calculated using Equations (3.29) and (3.30) [65]. The calculated SSD depth distributions are reported in Figure 3.29b,c, and the calculated maximum SSD depths are reported in Figure 3.30 by the solid lines. This simple model predicts SSD depths that are fairly consistent with measured data as a function of rogue-particle concentration. Equations (3.29) and (3.30) can likely be used for other grinding processes to estimate SSD depth in the presence of rogue particles or as a method to estimate the size or concentration of rogue particles, based on observed changes in SSD depth distribution.

3.1.2.8 Conclusions on Grinding SSD

The SSD depth and length distributions for various grinding processes can be directly measured and statistically evaluated. Observed surface cracks are characterized as near-surface lateral or deeper trailing indent type (chatter-mark) fractures. The length of the trailing indent ($\overline{L_t}$) is correlated with a given grinding process, specifically the abrasive size ($\overline{d}$). The SSD depth distributions are typically described by a single exponential distribution, followed by an

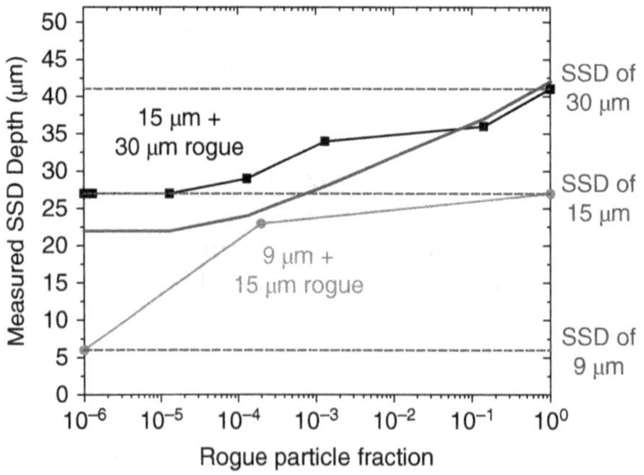

Figure 3.30 Maximum SSD depth as a function of rogue-particle concentration (of 15-μm or 30-μm alumina) relative to the SSD depth observed for 9-μm or 15-μm loose-abrasive grinding without the addition of rogue particles. Source: Suratwala et al. 2008 [65]. Reproduced with permission from Elsevier.

asymptotic cutoff in depth (c_{max}). The critical process parameters affecting SSD include abrasive size, PSD, and hardness, applied load, and workpiece mechanical properties.

Using fracture indentation relationships, it is shown that only a small fraction of abrasive particles are loaded and participating in the fracture, and the larger particles in the abrasive PSD bear the higher loads. Using a mechanical model to describe the grinding process, the measured crack-length distribution has been related to crack-depth distribution. The maximum SSD depth (c_{max}) correlates with both mean crack length ($\overline{L_t}$) and measured *surface roughness* (δ). The proportionality constant between roughness and crack depth can be reliable, but requires a standardization of the characterization. Also, the observed relationship between the mean crack length and the maximum SSD depth can be used as a rule of thumb to nondestructively estimate the depth of SSD by measuring the crack length of an individual SSD defect. For grinding performed in a sequence of steps, SSD depth distribution did not noticeably influence the SSD caused by the previous grinding step, provided that material removal exceeded the SSD depth of the previous step. Finally, a small amount of contaminant of larger abrasive particles (rogue particles) can greatly increase SSD depth. Hence, rogue particles must be carefully managed in an optical-fabrication process.

3.1.3 SSD During Polishing

SSD during polishing is defined as fractures generated by rogue particles, i.e. brittle scratches. As described in Section 1.3, material removal during polishing occurs below the fracture limit at lower load per particle, either by nanoplastic deformation or chemical reaction. During grinding, larger particles are used (typically >10 μm), resulting in fewer loaded particles per unit area between the workpiece and the lap, and thus resulting in loads per particle that exceed the initiation load for fracture (>0.01 N). During polishing, smaller particles are used (typically <3 μm), resulting in more loaded particles per unit area and loads per particle below the fracture-initiation load. For example, consider a glass workpiece polished with a typical ceria-based slurry (using 0.5 μm particles at an applied pressure of 0.3 psi, with a 0.3 fill fraction at the interface and a fraction of load-bearing particles of 10^{-4}). From this, one may estimate a load per particle of $\sim 10^{-6}$ N [65]. Clearly, the load on an average polishing particle is many orders of magnitude lower than needed to initiate fracture. Thus, any scratch formed on a polished surface implies a rogue particle or an asperity that is causing a much higher load than the average particle.

Preston was among the first to recognize the presence of SSD on finished surfaces, noting that etching exposes chatter-mark cracks (trailing indent fractures, or brittle scratches) [50]. The presence of rogue particles in the slurry during grinding or polishing is known to strongly influence the surface properties of a workpiece, either in terms of deeper damage or isolated scratching [12, 65–69]. Several studies on polishing integrated circuits have investigated the effects of rogue particles for wafers ground or polished using chemical–mechanical planarization (CMP). Basim et al. [66] spiked colloidal-silica slurries with rogue silica particles and found that scratch densities increased with rogue-particle size and concentration. By characterizing the slurry PSD using dynamic light scattering, Basim also illustrated that rogue particles present even at a very low concentration (<1 out of 100 000) can degrade *surface quality*. Ahn et al. [67] and Kallingal et al. [70] compared different slurries bases on colloidal silica and colloidal alumina of various pHs, filtering methods, or ultrasonic preparation method to reduce microscratches. The scratching results were explained in terms of how the process parameters affected the large particles in the distribution. Understanding the impact of rogue particles during polishing can ultimately aid in developing finishing processes yielding minimal or no SSD. This is critical in applications that require high-quality polished surfaces, including integrated circuits, high-strength windows, and optics for high-power laser or lithographic applications [3, 5].

The various types of scratches promoted by adding rogue particles are shown in Figure 3.11 [65]. These scratches are divided into three basic categories:

- Plastic, which show no brittle fracture, but only plastic modification to the surface (often referred to as sleeks)
- Brittle, which have only cracks (trailing indent or lateral)
- Mixed, which contain both plastic modification and cracks

The scratches are further categorized as to the type of crack, whether lateral or trailing indent. Plastic scratches would be expected from sharp indenters, where the local pressure exceeds the yield stress of the workpiece at the contact zone. Purely brittle scratches would be expected from blunter indenters.

In a systematic study of the impact of rogue particles, a series of rogue diamond particles of various sizes were doped into a 0.5 μm ceria polishing slurry, which was then used to polish fused-silica workpieces [65]. Figure 3.31 shows the PSD of the polishing slurry and rogue diamond particles. The resulting scratch characteristics (length, width, and number density) as a function of rogue particle characteristics (size and concentration) are summarized in Figure 3.32. The scratch number density increased exponentially with rogue-particle size at a fixed rogue-particle number concentration (Figure 3.32a). This suggests that the larger the rogue particle, the greater the probability of causing scratches, or the greater the efficiency. The magnitude of the difference in scratch number density according to particle size is also illustrated in the microscopic images shown in Figure 3.33. Interestingly, the scratch number density was much less sensitive to the rogue-particle concentration. For most of the rogue-particle concentrations examined (5–500 cm^{-2}), scratch concentration did not change by more than a factor of five. However, at

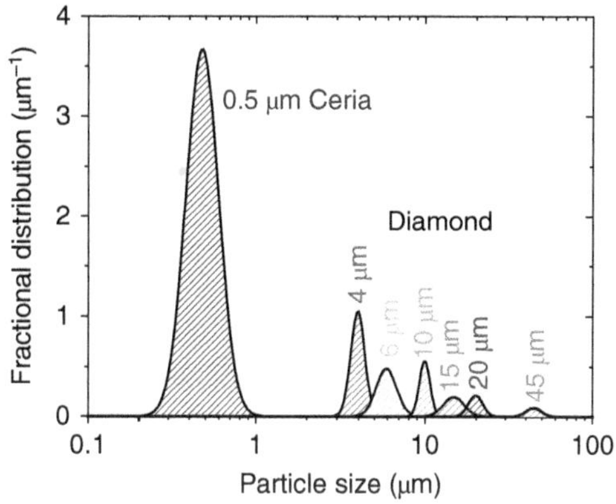

Figure 3.31 PSDs of the cerium-oxide slurry and rogue diamond particles added during polishing experiments. Source: Suratwala et al. 2008 [65]. Reproduced with permission from Elsevier.

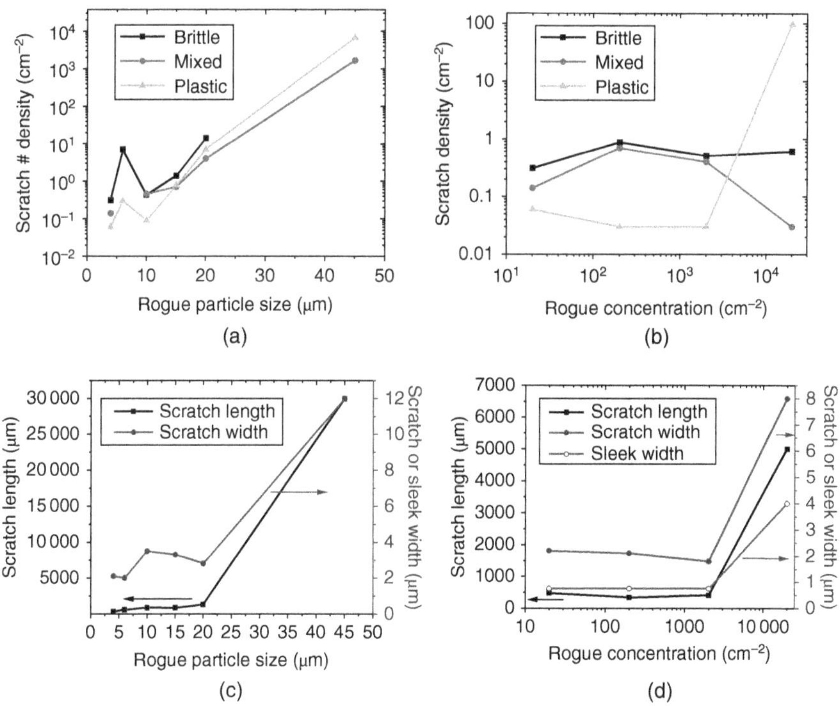

Figure 3.32 Scratch number density observed after ceria polishing of fused silica with rogue-diamond particle addition as a function of (a) rogue-particle size at a fixed rogue-particle concentration of 50 cm^{-2} and (b) rogue-particle concentration at a fixed rogue-particle size of 4 μm. Scratch dimensions (length, width) observed after ceria polishing of fused silica with rogue-diamond particle addition as a function of (c) rogue-particle size at a fixed rogue-particle concentration of 50 cm^{-2} and (d) rogue-particle concentration at a fixed rogue-particle size of 4 μm. Source: Suratwala et al. 2008 [65]. Reproduced with permission from Elsevier.

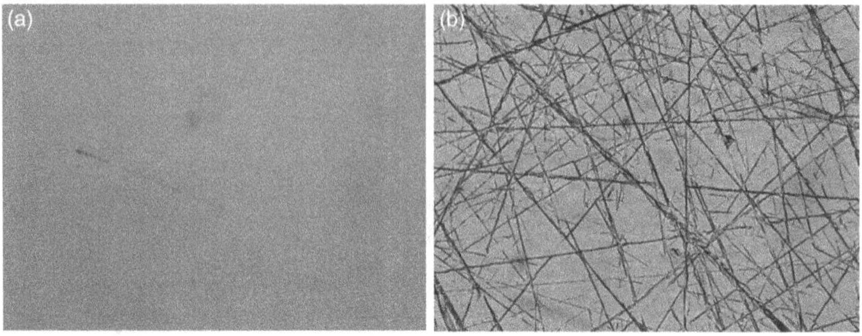

Figure 3.33 Optical micrographs of (a) a typical scratch observed on a polished surface upon addition of 10-μm diamond particles and (b) typical scratches observed on a polished surface upon the addition of 45-μm diamond particles. The horizontal full scale of each image is 237 μm. Source: Suratwala et al. 2008 [65]. Reproduced with permission from Elsevier.

a high rogue-particle concentration (5000 cm^{-2}), the plastic-scratch number density increased by a factor of 1000 (Figure 3.32b).

The width and the length of the scratches were observed to increase with rogue-particle size and were relatively insensitive to rogue-particle concentration, except at very high concentration (Figure 3.32c,d). The width increase is attributed to the larger contact zone expected with larger rogue particles; the width of the scratch was nominally 15–30% of the mean diamond diameter added to the slurry. Scratch length was also noted to increase with diamond-particle size in the range of 4–20 μm (Figure 3.32c). Figure 3.34 illustrates this trend more definitively by plotting the cumulative scratch-length distribution observed for various rogue diamond-particle sizes. The mean scratch lengths increased from ~330 to ~1300 μm with the rogue particle size. For these samples, the scratch number densities were ~2 to 25 cm^{-2} or nominally 150–2000 total scratches on a 100-mm-diameter round substrate. The average relative velocity of the particle relative to the optic was 80 cm s^{-1}. Hence, for the length of scratches observed, the time that each of the rogue particles was loaded ranged from 0.3 to 1.6 ms. Assuming that the scratches were created randomly during the time of polishing (1 h), a scratch is created every 1.8–24 s, on average. The conclusion is that scratching occurs sporadically.

A longer scratch suggests that the rogue particle loads the surface of the workpiece for a longer period, assuming a constant relative rogue-particle velocity. Figure 3.35 shows a schematic of the proposed mechanism of scratch creation by a rogue particle, where the time for a rogue particle to load is governed by the size of the particle and the viscoelastic properties of the lap. At some arbitrary time zero, a rogue particle finds its way to the interface between the workpiece and lap. Owing to its large size, it will bear a much higher load than the average

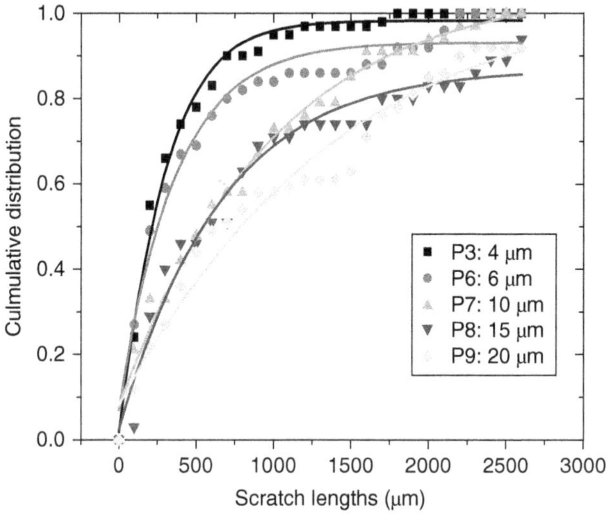

Figure 3.34 Cumulative scratch-length distribution measured for ceria-polished samples with rogue particles of different sizes and a constant rogue diamond concentration of 50 cm^{-2}. The average crack lengths determined from these distributions are reported in Figure 3.25. Source: Suratwala et al. 2008 [65]. Reproduced with permission from Elsevier.

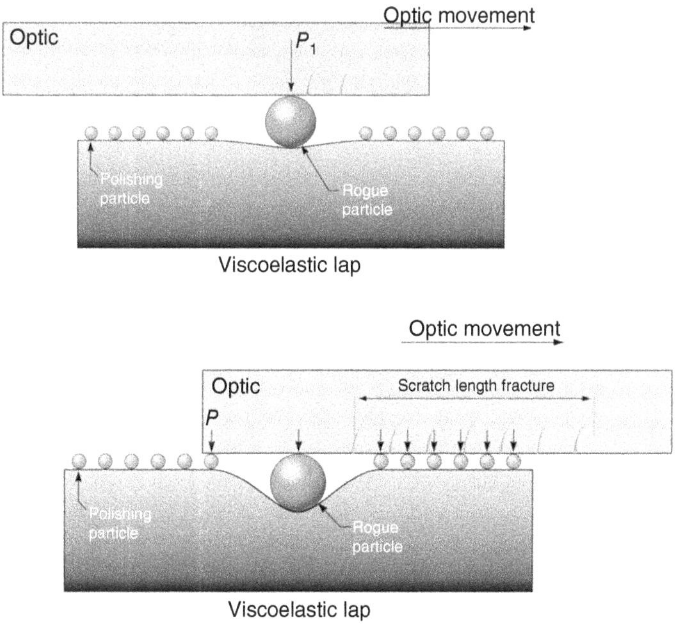

Figure 3.35 Schematic of proposed mechanism for the formation and resulting length of a scratch by a rogue particle on a viscoelastic lap. Source: Suratwala et al. 2008 [65]. Reproduced with permission from Elsevier.

particle on the lap. This load is sufficient to initiate and grow a brittle fracture or plastic deformation on the workpiece. While loaded and static relative to the lap, the rogue particle will penetrate into the viscoelastic lap until either

- The gap between the lap and optic is reduced to that of the polishing ceria particle, thus dropping the load on the rogue particle to match that of the average ceria particle and ending the scratch.
- The maximum elastic penetration has been reached (a function of the elastic modulus of the lap), resulting in incomplete rogue-particle penetration, little drop in load on the rogue particles, and indefinite scratch lengths across the length of the optic.

The validity of the viscoelastic mechanism for scratch length can be examined semiquantitatively by evaluating the measured scratch-length dependence not only as a function of rogue-particle size but also as a function of a number of parameters, such as lap viscosity (determined either by temperature or lap material) and applied pressure [65]. These results are summarized in Figure 3.36. In the case of lap material, scratch length was observed to increase

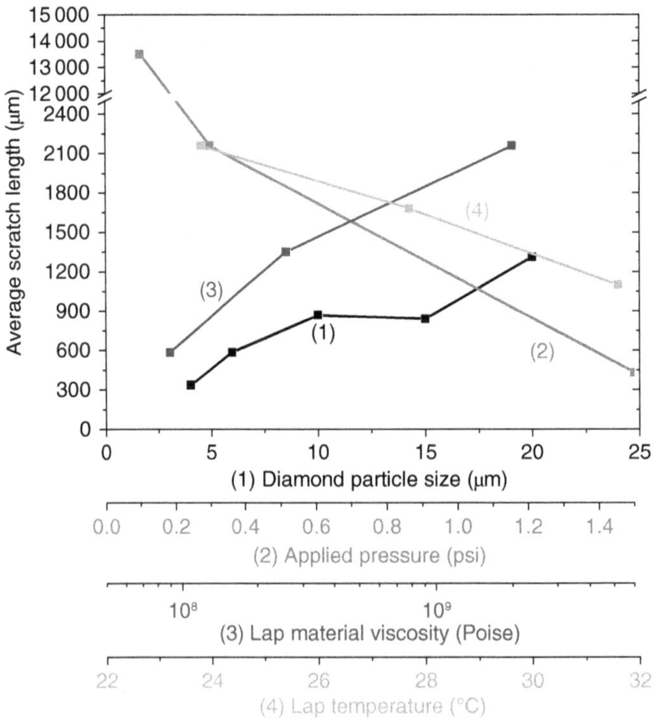

Figure 3.36 Average scratch length, measured as a function of diamond-particle size, applied pressure, lap material, and lap (pitch) temperature. Source: Suratwala et al. 2008 [65]. Reproduced with permission from Elsevier.

with increased lap viscosity. The temperature of the lap also influenced the scratch length; as the temperature increased, the lap viscosity decreased and scratch length decreased. Finally, as applied pressure was increased, scratch length was found to decrease. All these trends are consistent with the viscoelastic mechanism for explaining scratch lengths.

The viscoelastic penetration rate of the rogue particle has been quantified by modeling a hard spherical abrasive under a quasi-static load on viscoelastic lap [65]. This model uses an approach by Lee and Radok [71] that expands the original Hertzian elastic contact [72] on a simple linear viscoelastic substrate. The basis for this approach is the governing force-balance equation given by

$$E_2\varepsilon + \eta_2\frac{\mathrm{d}\varepsilon}{\mathrm{d}t} = \frac{P}{\pi a^2}, \tag{3.31}$$

where P is the applied load on the spherical particle, a is the Hertzian contact zone radius, and ε is resulting strain. The viscoelastic properties of the lap are described by the elastic modulus (E_2) and viscosity (η_2). For a given applied pressure (the term on the right side of Equation (3.31)), there is an elastic-limit response (the first term on the left side) and a time-dependent response (the second term on the left side). Accounting for a large strain with surface displacement of the viscoelastic lap outside the contact zone, Equation (3.31) can be rewritten in the following form (see Ref. [65] for details):

$$\frac{\partial \chi_v}{\partial t} + \frac{\chi_v}{\tau} = \frac{12P}{\pi d^2 \eta_2}(1 + {\chi_v}^{2/3}), \tag{3.32}$$

where

$$\chi_v = \varepsilon^3 = \frac{a^3}{\left(\frac{d^2}{4} - a^2\right)^{3/2}} \tag{3.33}$$

$$\tau_s = \frac{\eta_2}{3E_2}, \tag{3.34}$$

where d is the particle diameter, τ_s is the stress-relaxation time constant, and χ_v is the term related to strain. Equations (3.32)–(3.34) can be solved numerically to the determined contact zone as a function of time, $a(t)$.

The objective is to determine penetration as a function of time, i.e. to determine when the load will decrease to the point where scratching ends. Specifically, the particle needs to penetrate to the interface gap (which is approximately the diameter of the average polishing particle). Hence, using a typical polishing slurry of diameter 0.5 μm, a 4-μm rogue particle will need to penetrate ~3.5 μm before the scratch will end. In a more complex set of relationships, the contact zone (a) is related to the height of penetration of the particle into the lap due to two components: Hertzian penetration of a hard sphere and baseline surface displacement away from the contact zone. Details of these height relationships are described elsewhere [65].

Figure 3.37 shows the calculation results for the penetration of rogue particles as a function of time for the experimental results shown in Figure 3.32–3.34 [65]. These calculations use the known properties of the viscoelastic-polyurethane pad ($E_2 = 100$ MPa; $\eta_2 = 9 \times 10^7$ Poise) [73] and assume a load of 1 N, within the range needed to cause brittle scratches. When the line in the plot plateaus, the particle has penetrated to the interface gap, reducing the load on the particle and ending the scratch. The model predicts penetration times that are longer for larger rogue particles. This is consistent with the experimental results in Figure 3.34. For rogue particles in the range of 4–20 μm, penetration times of 1–5 ms are calculated. For a larger rogue particle (45 μm) at the same load of 1 N, as shown in Figure 3.37b, the particle does not penetrate the lap, even after 120 ms. For the relative velocity used in that study ($V_r = 80\ \mathrm{cm\ s^{-1}}$), calculated scratch lengths are significantly larger (>14 000 μm) than the measurement field of view (237 μm), leading to the appearance of essentially infinite-length scratches. This is again consistent with the experimental results in Figure 3.33b.

The abovementioned viscoelastic model can be simplified to the following useful relationship to determine the average scratch length $\overline{L}_s$:

$$\overline{L}_s \cong 2.2 \frac{V_r \eta_2 d^2}{P}, \tag{3.35}$$

where V_r is the average relative velocity of the particle, relative to the workpiece surface, d is the rogue-particle diameter, and P is applied load. Figure 3.38 shows the calculated crack lengths using Equation (3.35) at loads of 1 and 5 N (in the range needed to cause scratches) and a comparison with measured

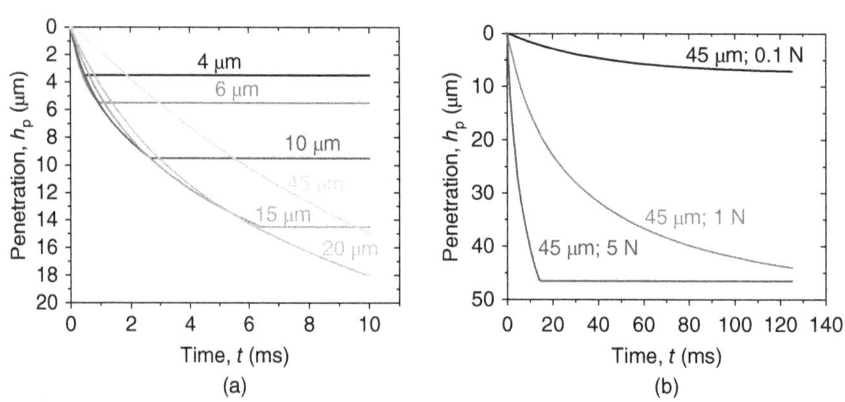

Figure 3.37 (a) Calculated depth of particle penetration as a function of time for different-size rogue diamond particles at a load of 1 N. (b) Calculated depth of particle penetration as a function of time for different loads. (Calculations use $E_2 = 100$ MPa, $\eta_2 = 9.7 \times 10^7$ P.) Source: Suratwala et al. 2008 [65]. Reproduced with permission from Elsevier.

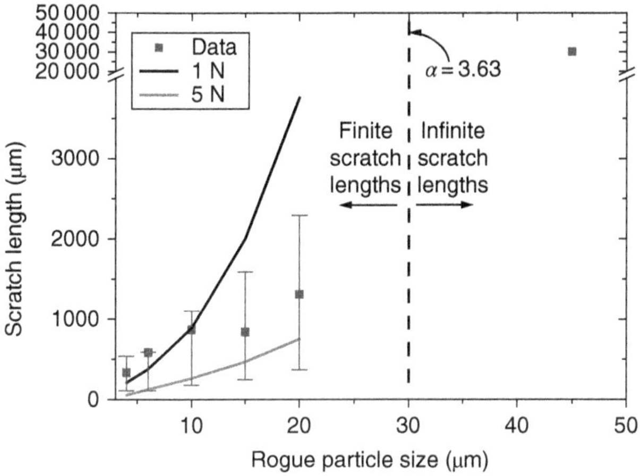

Figure 3.38 Average scratch length measured as a function of rogue-particle size, using data in Figure 3.24, compared with the simple viscoelastic model at two different loads of 1 and 5 N. The bars on the data point represent 80% of the distribution of scratch lengths observed. Source: Suratwala et al. 2008 [65]. Reproduced with permission from Elsevier.

scratch lengths as a function of rogue-particle size. The bars in the measured data represent 80% of the distribution of scratch lengths observed. Note that the data is bound within the load range of 1–5 N.

Another useful relationship from this model is the stress parameter (α), defined as follows:

$$\alpha = \frac{4P}{\pi d^2 E_2}, \tag{3.36}$$

where P is the applied load and E_2 is the lap modulus. This relationship is the ratio of the applied load to the equilibrium penetration. When this value is low, the load is insufficient to completely penetrate the rogue particle into the lap – hence, the scratch length will be infinite. When this value exceeds a critical value, in this case 3.64, the particle will penetrate and the scratch length will be finite. In Figure 3.38, the stress parameter value of 3.64 is reached at a nominal size of 30 μm for a load of 1 N. Above this particle size, the model predicts infinitely long scratches extending to the edge of the workpiece; this is again consistent with experimental data in which 45-μm particles led to infinitely long scratches.

To summarize, Equation (3.35) serves as a simple expression that can be used to determine the length of a scratch based on the size of the rogue particle, load, and kinematic and material properties of the lap. Equation (3.36) also serves as a useful relationship to determine whether scratches will be finite or infinite in length.

Aside from diamond particles, rogue particles of different material types have also been shown to cause scratches. Whether diamond, pitch, or dried-ceria agglomerates, scratches were created. The only exception was organic urban dust, which did not lead to any measurable SSD [65]. In a review of scratching during CMP by Kwon et al. [69], various sources of rogue particles were described. These included pad pores, which can house rogue particles, diamond conditioners, which may dislodge and act as rogue particles, and pad debris. Other rogue-particle sources are cross contamination from other processing steps, such as grinding particles, and particles dislodged from the edge of the workpiece. These results confirm that cleanliness and the elimination of *all* rogue particles sources during polishing are essential in obtaining near-scratch-free surfaces.

3.1.4 Effect of Etching on SSD

Chemical etching during optical fabrication has important functions with respect to SSD. First, it is a very useful way of exposing hidden surface microfractures, such as scratches. During polishing, a modified surface layer called the Beilby layer is formed on the workpiece and can hide surface microfractures (see Section 3.3). An example of how etching can reveal such fractures is shown in Figure 3.39 for a fused-silica workpiece [37]. A second important function of etching is as a treatment for ground surfaces to reduce the magnitude of SSD depth. This allows a reduction in subsequent polishing.

Chemical etching is further used to clean the workpiece surface (Section 3.2), to remove impurities from and within the polishing surface layer (Chapter 9), to remove highly compressive stress from ground surfaces (Section 2.5.1.4), and to dramatically increase the laser-damage threshold of laser optics (Chapter 9).

The method of etching used, whether with a liquid, gas, or ion etchant, and the composition of the etchant depends greatly on the composition of the workpiece and the type of etching desired (isotropic, anisotropic, congruent, or incongruent). The discussions in the following sections are mostly limited to hydrofluoric (HF)-acid-based aqueous etching of optical glass, specifically fused-silica glass, which has largely isotropic, congruent etching behavior.

Fluoride-based wet-chemical etching of silicate glasses is accomplished with a variety of reagents including (HF) or, under suitably acidic conditions, fluoride or bifluoride salts. The dissolution of silicate glasses results in the formation of the stable hexafluorosilicate, (SiF_6^{2-}) anion [4, 45, 50, 74–80]. When hydrofluoric acid is used, the overall reaction can be summarized as follows:

$$SiO_2 + 6HF \rightarrow SiF_6^{2-} + 2H_2O + 2H^+. \quad (3.37)$$

The underlying mechanism of such reactions involves a number of steps and intermediate species. For example, in an aqueous solution, HF acid acts as a

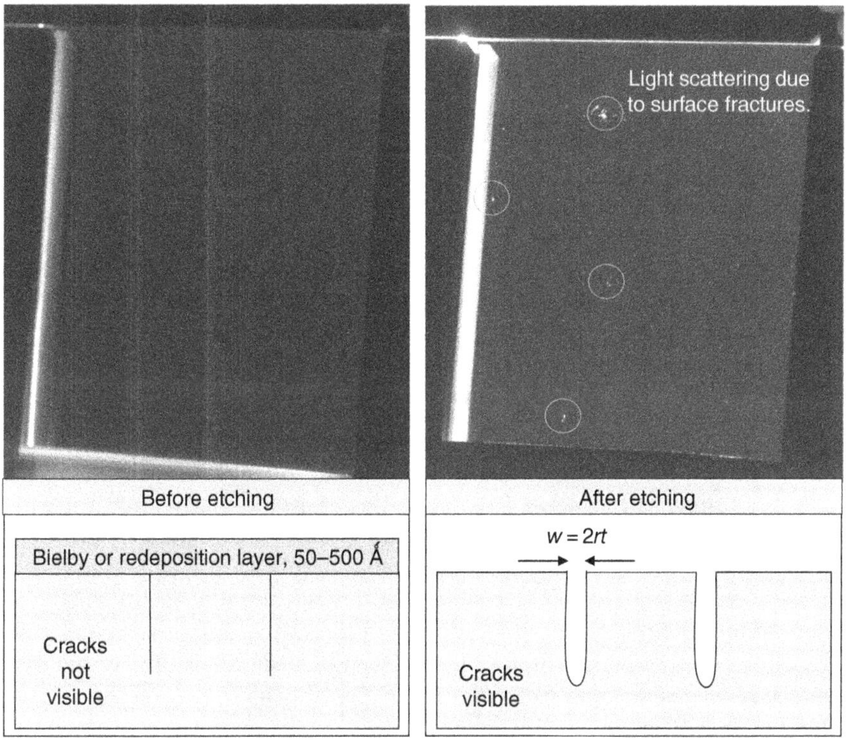

Figure 3.39 Top row: edge or side-lit images of a polished fused-silica surface, before etching (left) and after etching (right). Bottom row: a cross-section of the glass surface shows cracks hidden before etching (left), which open up after etching (right). Source: Wong et al. 2009 [37]. Reproduced with permission from Elsevier.

weak acid with an equilibrium between the undissociated acid, H^+ ions, and F^- anions,

$$HF \rightarrow H^+ + F^- \quad {}^{25}K_a = 6.7 \times 10^{-4}\ \text{mol}\,\text{L}^{-\text{o}}, \tag{3.38}$$

where ${}^{25}K_a$ is the equilibrium constant at 25 °C [81]. In the presence of undissociated HF, the fluoride ion reacts to form the bifluoride anion (HF_2^-), which is thought to be the primary species responsible for the attack on the silica glass matrix: [75, 79, 81]

$$HF + F^- \rightarrow HF_2{}^{-25}K = 3.96\,\text{mol}\,\text{L}^{-\text{o}}. \tag{3.39}$$

A variety of reagents, so long as they produce both fluoride (F^-) and hydrogen (H^+) ions, can form the bifluoride ions in situ and thus etch silicate glasses.

Buffered-oxide etch (BOE), which is a mixture of HF and ammonium fluoride (NH_4F), is often used as the etchant instead of HF, because of its stable etch rate and lower vapor pressure, reducing some of safety precautions required. In such solutions, NH_4F dissociates completely, providing an abundant source of F^- ions, which are free to react with the undissociated HF to form HF_2^- ions as seen in Reaction (3.39). In addition, the small fraction of dissociated HF in Reaction (3.38) provides the H^+ ions that catalyze the overall reaction [82].

BOE was originally developed for use in the semiconductor industry [74–76, 80]. For this reason, most studies of fluoride etching on glass surfaces focused on masking and dielectric film applications, describing the evolution of masked silica surfaces with etch time [74, 75, 83, 84]. Preston [50] was among the first to examine how HF etching influences optical fabrication, both on the exposure of scratches as an inspection aid and on surface topography changes on ground-glass surfaces. Spierings [74] later combined the results from the literature on glass science and integrated circuit (IC) fabrication in a comprehensive review that described the etching mechanism, morphological changes of glass surfaces with etching, and effects of etchant and glass composition.

3.1.4.1 Topographical Changes of SSD During Etching

The effect of BOE etching on a series of trailing indent cracks (i.e. a typical scratch) is shown in Figure 3.40 [37]. The top image displays optical micrographs at various etch times; the corresponding surface profilometry line-outs are at the bottom. The morphology of the scratch may change dramatically during etching. Under short etch times, the individual fractures remained isolated. At longer times, these cracks open to form cusps, described as rounded pits on the surface with depths equivalent to the crack depth. The cusps coalesce at longer etch times to form a nearly continuous trench on the surface of the optic. Because this coalescence occurs only in a single direction on the workpiece, which corresponds to the minimum distance between cracks, the scratch remains prominent even after 120 h of etching (or removal of ~200 μm of effective glass thickness). A schematic illustration of the etching morphology changes of neighboring cracks is shown in Figure 3.41. Note that BOE etching of fused silica occurs largely isotropically, etching at a roughly equal rate from all penetrable surfaces. With this etching mechanism, the actual scratch never goes away, because it etches at the planar workpiece surface and the bottom of each individual scratch at the same rate. Another way to describe this phenomenon is that for an isolated scratch with an original scratch depth of 10 μm, the PV height will always be 10 μm, regardless of the how much is etched.

Figure 3.42 shows the effect of BOE etching on two ground fused-silica surfaces, one processed with 150 μm grit-fixed abrasive and the other with 30 μm grit-loose abrasive [37]. As observed in the isolated scratch in Figure 3.40, etching of the ground surfaces forms a series of individual cusps as the silica glass erodes from the sides and bottom of each microfracture. Here, individual cusps coalesce in all directions as etching progresses, because the ground surfaces consist of an ensemble of closely spaced trailing indent fractures distributed randomly on the surface. Their coalescence increases the cusp size and yields an increase in the spatial period of *surface roughness* and a decrease in SSD depth.

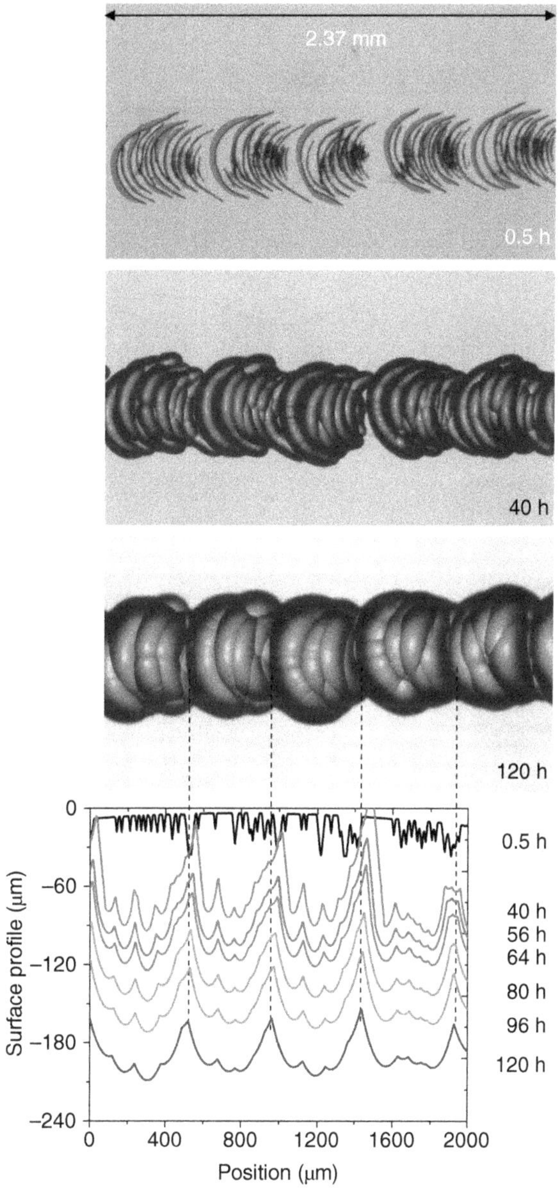

Figure 3.40 Optical micrographs and surface profilometry of a scratch (trailing indent cracks) after etching with BOE at various times. The optical micrographs are shown at the same scale as the line-out of the surface profile. Source: Wong et al. 2009 [37]. Reproduced with permission from Elsevier.

As described by Spierings [74], each cusp originates with the etching of an individual surface fracture. Cusp diameter (D_c) increases with the square root of etch time ($t^{1/2}$), as shown in Figure 3.43 [37]. Although their mechanisms differ significantly, the formalisms describing the growth rate of such cusps and

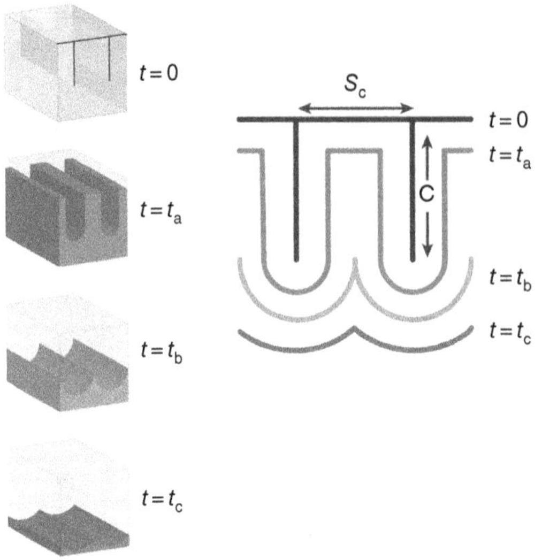

Figure 3.41 Simple, 2D, geometric model describing the evolution of the surface profile of cracks with etch time. Source: Wong et al. 2009 [37]. Reproduced with permission from Elsevier.

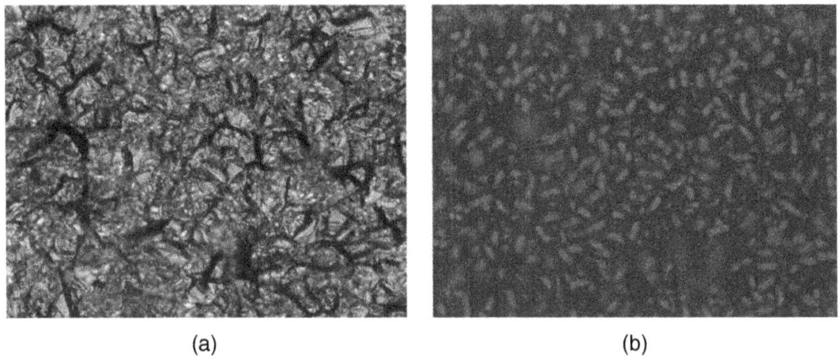

Figure 3.42 Optical micrographs and surface profilometry of two ground fused-silica surfaces after etching with BOE for various durations. Surfaces were processed with (a) 150-grit fixed abrasive grind and (b) 30-μm loose abrasive grind. Source: Wong et al. 2009 [37]. Reproduced with permission from Elsevier.

grain growth in polycrystalline ceramics are similar [85]. Specifically, the diameter of each cusp (D_c) grows at a rate inversely proportional to its size, following:

$$\frac{dD_c}{dt} = \frac{k_c}{D_c}, \tag{3.40}$$

where k_c is the cusp-growth constant. Integration of Equation (3.40) gives

$$D_c = \sqrt{2k_c t + D_{co}{}^2}, \tag{3.41}$$

where D_{co} is the initial cusp or crack size [37]. When $D_{co} \ll D_c$, then $D_c = \sqrt{2k_c t}$, which is consistent with the observed square root of etch time dependence observed in Figure 3.43. The value of k_c is 0.010 μm^2 s for the 150 μm grit-fixed abrasive workpiece and 0.006 μm^2 s for the 30 μm, loose-abrasive workpiece. The rate of cusp growth depends on crack depth, length, and spacing, and orientation distribution of the ground surface (see Section 3.1.4.2).

3.1.4.2 Influence of SDD Distribution on Etch Rate and Roughness

The number density of cracks on a ground surface is typically so high that they overlap, obscuring the identification or depth of individual fractures. As discussed in Section 3.1.2.1, after several micrometers of material are polished from a ground surface, the number density drops exponentially until a maximum depth of subsurface damage is reached.

For an isotropic etchant, the etch rate of the workpiece is expected to scale with the surface area exposed. Figure 3.36 shows a large drop in the etch rate with etch time on two ground fused-silica surfaces that plateaus to a nominally flat surface etch rate [37]. Just as SSD depth distribution decreases exponentially with depth, etch rate decreases exponentially with depth removed. The decrease in surface area with etch time is likely driven by the coalescence of the cusps or annihilative neighboring cracks, as shown in Figure 3.41 at time t_b. The behavior illustrated in Figure 3.44 suggests that the surface area initially available to the etchant on the ground faces is about 3.5 times larger than that of a similarly sized polished, fracture-free surface. As the cusps associated with

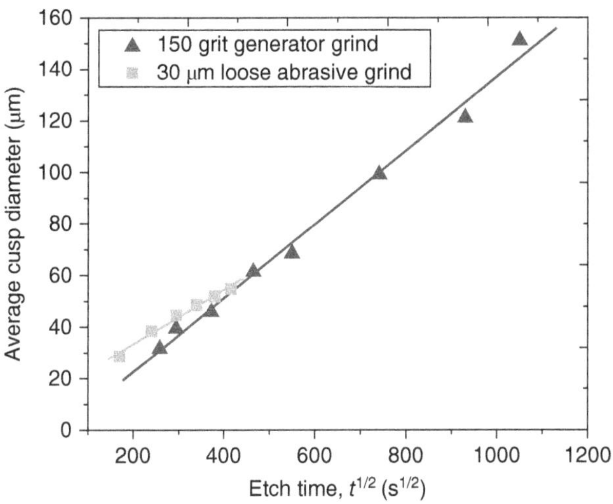

Figure 3.43 Measured cusp diameter vs etch time$^{1/2}$ for two ground fused-silica surfaces etched in 20 : 1 BOE. Source: Wong et al. 2009 [37]. Reproduced with permission from Elsevier.

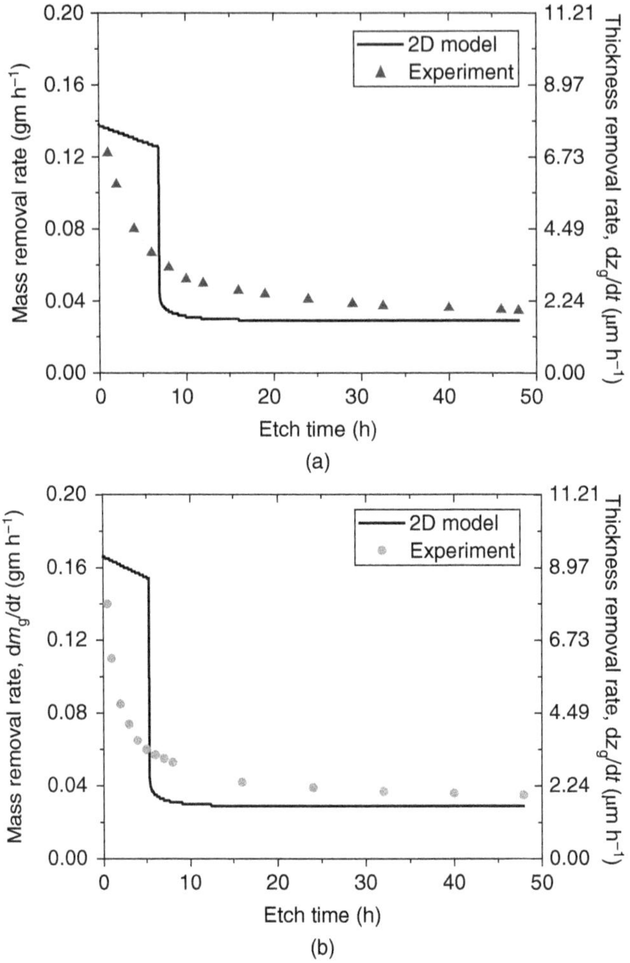

Figure 3.44 Measured and calculated mass and thickness removal rates using a 2D etch model for the (a) 150-grit fixed-abrasive ground surface and (b) 30-μm loose-abrasive ground surface. Source: Wong et al. 2009 [37]. Reproduced with permission from Elsevier.

each fracture coalesce during etching, the true surface area – that is, the area available to the etchant – diminishes until equivalent to the nominal geometric surface area, at which time the rate of material etched from each face of the disk is equivalent to the bulk etch rate.

The growth and coalescence of the cusps during the etching of a ground surface also influences surface roughness, specifically PV roughness. Figure 3.45 depicts PV roughness as measured by needle profilometry, as a function of etch time for the ground surfaces shown in Figure 3.42 [37]. The initial rise in measured roughness is an artifact of the finite width (~10 μm) of the profilometer

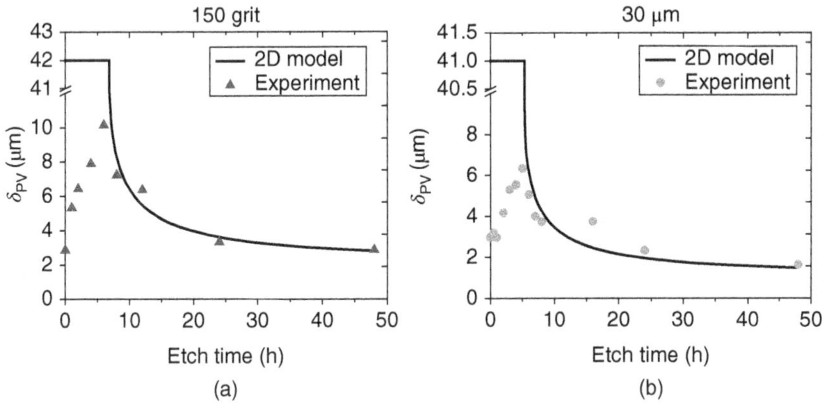

Figure 3.45 Measured and calculated roughness using 2D etch model for the (a) 150-grit fixed-abrasive ground surface and (b) 30-μm loose-abrasive ground surface. Source: Wong et al. 2009 [37]. Reproduced with permission from Elsevier.

stylus, which cannot fit in the narrow, high-aspect-ratio (~1 μm × 40 μm) cracks present before etching. Actual PV roughness should correspond to SSD depth, which is ~40 μm. At an etch rate of 1.6 μm h^{-1}, full penetration of the stylus occurs after ~4 h of etching, consistent with the peak in roughness observed in Figure 3.45. More importantly, PV roughness is observed to decrease with longer etch time.

A simple geometric 2D model can describe the key features of cusp coalescence and illustrate how etch rate and surface topography or roughness should evolve as a function of etch time. Consider a surface consisting of an infinite series of equally spaced, parallel cracks of uniform depth (c), separated by a distance (s_c), as in Figure 3.41. Fractures on the real surfaces are characterized by a distribution of crack depths, spacings, and orientations, which affect etching characteristics. As shown in Figure 3.41, at $t = 0$, etching has not started and the cracks remain closed. At time t_a, etching begins, resulting in a series of isolated cusps. At time, t_b, the cusps intersect and coalesce. The time required for intersection (t_b), and thus coalescence, is given by

$$t_b = \frac{s_c}{2r_b}, \tag{3.42}$$

where r_b is the is bulk etch rate. The effective thickness etch rate before the coalescence of adjacent surface fractures ($t < t_b$) is given by the bulk etch rate multiplied by the ratio of the surface area (at time t) to the flat surface area, given by [37]

$$\frac{dz_g}{dt} = r_b \frac{(2\pi - 8)r_b t + 4c + 2s_c}{2s_c} \quad \text{for } t < t_b. \tag{3.43}$$

After coalescence ($t \geq t_b$), using the geometry of the cusp outlined in Figure 3.41, the thickness etch rate is given by [37]

$$\frac{dz_g}{dt} = \frac{2r_b^2 t}{s_c}\sin^{-1}\left(\frac{s_c}{2r_b t}\right) \quad \text{for } t \geq t_b, \tag{3.44}$$

Using the determined values of the bulk etch rate ($r_b = 1.6$ μm h), average crack spacing ($s_c = 23$ μm for 150 grit-fixed abrasive and $s_c = 17$ μm for 30 μm grit-loose abrasive), and maximum SSD depth ($c_{max} = 42$ μm for 150 grit-fixed abrasive and $c_{max} = 41$ μm for 30-μm grit-loose abrasive) for c, the calculated etch rates are compared with measured values in Figure 3.44 using Equations (3.43) and (3.44) [4, 37]. Note that the use of single values for c and s_c in the 2D model is an oversimplification, because a distribution of crack depths and spacings exists on each real ground surface [4, 86]. However, this approach still captures some salient characteristics of etch-rate behavior with etch time. The calculated etch rate has highest value at the onset of the etching process with values similar to those measured, suggesting the initial calculated surface area is close to that present on the actual surface. The etch rate is predicted to decrease slightly until the cusps begin to coalesce, at which time surface area decreases significantly and, as a consequence, etch rate decreases until it converges with the etch rate of the bulk material (r_b). Using a distribution of crack depths and spacings in this model would smooth out calculated values with etch time, coming closer to the measured data presented in Figure 3.44.

The 2D etch model shown in Figure 3.41 can also be used to calculate roughness evolution with etch time as follows [37]:

$$\delta_{PV} = c \quad \text{for } t < t_b, \tag{3.45}$$

$$\delta_{PV} = r_b t - \sqrt{(r_b t)^2 - \frac{1}{4}s_c^{\,2}} + \delta_b \quad \text{for } t \geq t_b, \tag{3.46}$$

where δ_{PV} is PV roughness and δ_b is baseline roughness. Before cusp coalescence occurs, that is, $t < t_b$, PV roughness is defined by crack depth (c). At time t_b, δ_{PV} decreases instantaneously, because the cracks are assumed to have vertical sides in the present model. After coalescence, that is, $t \geq t_b$, *surface roughness* becomes independent of crack depth and follows Equation (3.46).

Figure 3.45 compares the calculated PV *surface roughness* for the two ground workpieces with the measured data. As indicated by the solid lines, the calculated cusp intersection and annihilation times are 6.9 h for the 150 grit-fixed abrasive ground surface and 5.3 h for the 30 μm grit-loose abrasive ground surface. Once the time required to etch beyond the physical dimensions of the profilometer probe is exceeded, the present model provides a reasonable prediction of the time dependent evolution of *surface roughness*. The observed decrease in PV *surface roughness* illustrates how etching can be used after final grinding to reduce the polishing needed to remove SSD.

The efficacy of etching in reducing SSD, and thus polishing time, is increased by grinding processes that yield a high density of fractures with relatively uniform dimensions that can readily coalesce in etching. This is illustrated by isotropic etching simulations performed on various distributions of cracks, using a 3D finite-difference etch model as shown in Figure 3.46 [37]. Three scenarios were simulated:

1. Randomly spaced and oriented cracks of the same length and depth
2. Ordered cracks of the same length and depth
3. Randomly spaced and oriented cracks of different lengths and depths.

In each case, the same number density of cracks was used. In the first scenario, the etch simulation reveals that in areas where the cracks are far apart, crack intersection and coalescence do not take place and PV *surface roughness* remains high. In the second scenario, the reduction in PV roughness is most efficient when cracks are equally spaced. This case confirms that those etching surfaces with the greatest uniformity of crack distribution (in terms of depth and spacing) provide the greatest benefit in reducing the subsequent polishing necessary to remove SSD. Because real ground surfaces typically contain a distribution of crack lengths, depths, orientations, and spacings, however, the third scenario is most representative. For simplicity, PV roughness was used as the metric for reduction in SSD with etching; the volume of SSD remaining would be a more quantitative, representative metric for reduction in polishing time after etching.

3.1.5 Strategies to Minimize SSD

A typical optical-fabrication process involves a series of process-step transitions, from shaping and grinding to polishing, as seen in Figure 1.1. Because removal rate decreases and process times increases by orders of magnitude in transitioning from grinding to polishing – at the same time as *surface figure* and roughness control increase – an economical finishing process requires multiple process steps, as illustrated in Figure 1.2.

Developing an efficient, high-yield, inexpensive optical-fabrication process that requires low SSD or defectivity on the final workpiece can be challenging. To compound the difficulty, an economical finishing process requires high yield with respect to *surface quality* at each process step. For example, a poor yield due to scratching after a given step may require more removal than planned at the next step or returning the workpiece to a previous step to remove the scratch.

Based on the fracture mechanics, tribology, chemical concepts, and the novel characterization methods described in this chapter, some basic guidelines and strategies are offered, both for developing an economical finishing process and achieving high *surface quality* yield with respect to SSD:

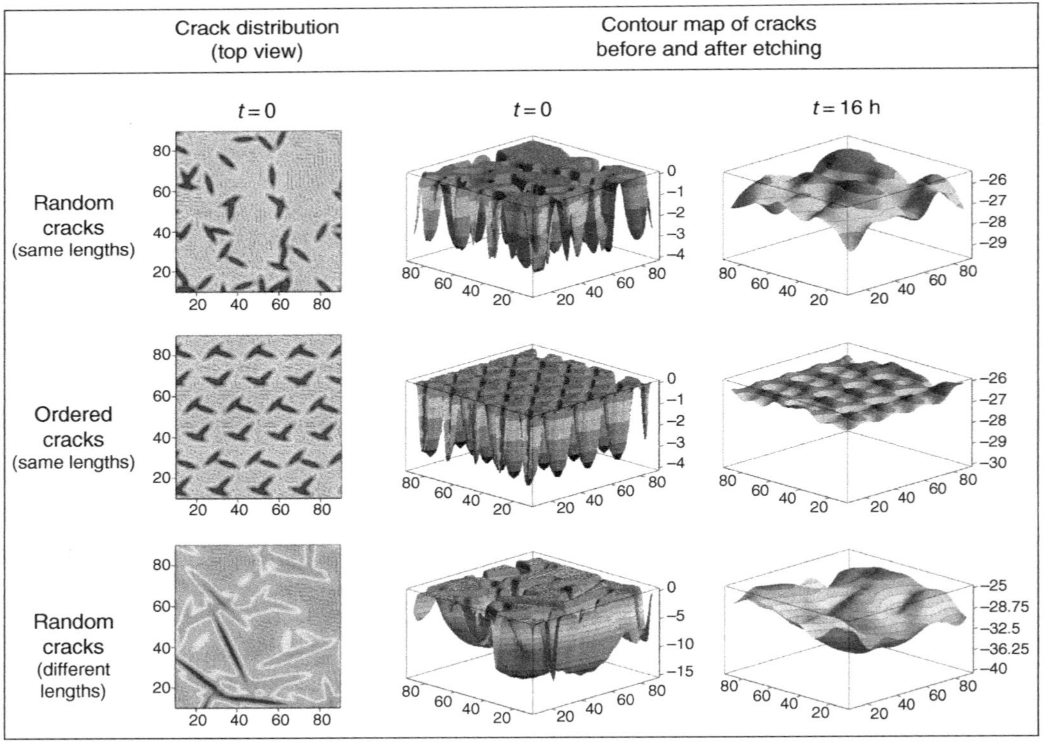

Figure 3.46 Isotropic 3D finite-difference etch simulations for three distributions of cracks on the surface. Source: Wong et al. 2009 [37]. Reproduced with permission from Elsevier.

1. *Measure and minimize SSD at each process step.* For each step, quantitatively measure the amount of SSD. For grinding SSD, use an SSD measuring technique (such as the taper wedge technique), and for polishing use the poststrip-etch scratch/dig measurement. Where SSD distributions are deeper than expected or desired, modify the process to achieve lower SSD by modifying slurry or tool particle-size distribution, area cleanliness, feed rate, etc.
2. *Define proper removal rate at each step such that all SSD from the previous step is removed.* Define a general process route that minimizes the process time at each step, but ensures that all damage from the previous step is removed at the subsequent step.
3. *Use etching to remove SSD just after grinding.* Evaluate whether chemical etching offers a cost-saving way to reduce SSD after grinding, potentially reducing a slower, costlier polishing time.
4. *Reduce or eliminate rogue particles in polishers or use polishing methods more immune to rogue particles.* A rogue particle or asperity that experiences a load greater than the critical load is required to initiate the formation of a scratch. Hence, scratches can be reduced by excluding rogue particles or making sure that rogue particles do not exhibit loads sufficient to initiate fracture. **THE BEST WAY TO PREVENT SCRATCHING OR DEEP SSD IS TO AVOID ROGUE PARTICLES.** Eliminating all sources of rogue particles can be very challenging, as sources include
   - large particles in the slurry particle-size distribution
   - wearing components that generate particles in the polishers
   - particles in the environment
   - cross contamination of abrasive or slurry particles from previous process steps
   - wet or dry agglomerates of slurry particles in the polishing system
5. *Ensure handling and cleaning, include safeguards that prevent rogue particles from touching the workpiece surface.* For example, avoid contact with any tooling surface containing environmental particles or asperities.
6. *Use scratch forensics to determine sources and estimate how much material to remove.* (see Section 6.2).

3.2 Debris Particles and Residue

In addition to SSD, the *surface quality* of the workpiece is affected by deposits of particles or residue (see Figure 3.1, middle). During optical fabrication, this type of *surface quality* relates to how effectively particles and chemical impurities used in a finishing process were removed. Some common examples of impurities are residual slurry particles, pitch, or wax residue. Inadequate cleaning of a workpiece surface can affect its roughness, wetting, absorption, scatter, and even resistance to laser damage. Depending on the workpiece's application, the effects of incomplete cleaning may range from cosmetic issues, to degraded optical performance, to catastrophic damage to the optical component. For high-quality laser optics requiring high laser-damage resistance, water rinse with low water-impurity levels down to part per trillion, are vital in reducing molecular impurities on the surface (see Chapter 9) [6, 87, 88].

3.2.1 Particles

Sources and types of particles include polishing slurries, fibers, pollen, dust, organics, metals, oxides, and other miscellaneous sources. Particle removal from the workpiece is driven by the adhesion forces (F_{13}) between the particle and the workpiece surface and forces applied (F_a) via a given cleaning process (see Figure 3.47). When the particle has *not* chemically reacted with the workpiece, total adhesion force is given by

$$F_{13} = F_{\text{vdw}} + F_{\text{electrostatic}} + F_{\text{double}} + F_{\text{capillary}} + F_{\text{gravity}} \tag{3.47}$$

$$F_{\text{vdw}} = \frac{h_{\text{vdw}} d}{16\pi z^2} \tag{3.48}$$

$$F_{\text{electrostatic}} = \frac{q_{\text{d}}^2 (\pi d^2)^2}{4\pi \varepsilon_{\text{r}} \varepsilon_{\text{o}} d^2} \tag{3.49}$$

$$F_{\text{double}} = \frac{\pi \varepsilon_{\text{o}} U^2 d}{2z} \tag{3.50}$$

$$F_{\text{capillary}} = 2\pi d\gamma \tag{3.51}$$

$$F_{\text{gravity}} = \frac{\pi d^3 \rho_3 g}{6} \tag{3.52}$$

where the major interaction forces are van der Waals forces (F_{vdw}), electrostatic image force ($F_{\text{electrostatic}}$), electrical double-layer force (F_{double}), capillary

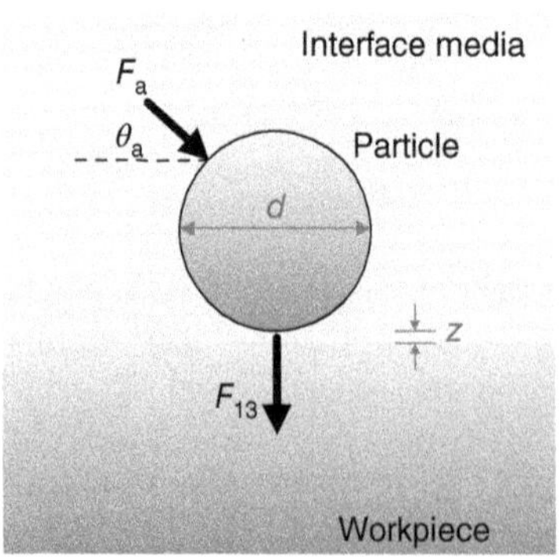

Figure 3.47 Illustration of adhesive force between a particle and the nature of applied force needed to dislodge the particle.

force ($F_{\text{capillary}}$), and gravitational force (F_{gravity}) [89]. d is the effective particle diameter, z is the separation distance between the particle and workpiece surface, γ is liquid-media surface tension, U is the work-function difference, h_{vdw} is the van der Waals constant, ε_r is a dielectric constant of interface media, q_d is the surface-charge density of the particle, ρ_3 is the particle density, g is gravitational acceleration (9.8 m s^{-2}), and ε_o is permittivity in vacuum (8.854×10^{-12} F m^{-2}). Van der Waals forces are generally attractive forces between atoms or molecules owing to permanent or induced dipoles, or both. Electrostatic image force (i.e. classical coulombic attraction force) describes the force between electrically charged particles. Electrostatic double-layer forces account for the charged species in the interface media, such as water (see discussion Section 4.5.1). Capillary forces are forces caused by liquid formation, typically water, at higher vapor pressures in small crevices between the particle and workpiece. Each of these forces is plotted as a function of particle size for the specific case of ceria particles on a fused-silica workpiece in Figure 3.48. Generally, either capillary forces in high relative humidity or van der Waals forces tend to dominate, and adhesion force increases with particle size.

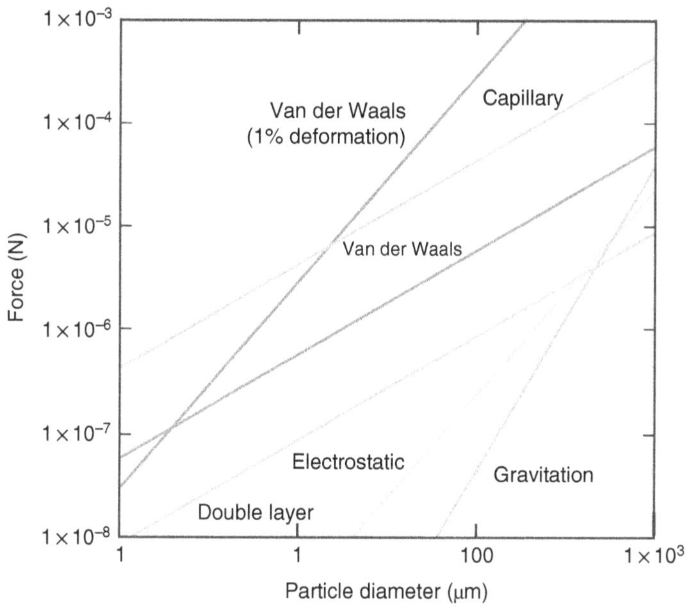

Figure 3.48 Various forces acting between a ceria particle and fused-silica workpiece surface as a function of particle size, using Equations (3.47)–(3.52), where $\rho_3 = 7.6$ gm cm^3, $z = 4$ Å, $h_{\text{vdw}} = 3$ eV, $q_d = 100$ electrons cm^{-2}, $\varepsilon_3 = 1$, $\gamma = 72$ mN m^{-1}, $U = 0.5$ V.

Consider a force applied by some means to remove the particle at some angle θ_a (Figure 3.47). Using a simple force balance and accounting for friction at the interface, the force needed to dislodge the particle from the surface is

$$F_a \geq \frac{\mu(F_{13} + F_a \sin\theta_a)}{\cos\theta_a}. \tag{3.53}$$

Note that increasing the application angle makes particle removal more difficult, because the applied forces add to the adhesion force. It is best to keep the application angle small or the force tangential as possible to the surface.

3.2.2 Residue

Residue is defined here as the formation of tiny particles via precipitation or the collection of molecular species (e.g. organic residue) as a workpiece dries after exposure to a liquid environment, usually aqueous. The aqueous solution may contain dissolved species such as metallic cations, ions, anions, or residual organic species, which may preferentially collect during drying at a pinned drying front or the end of a drying front. As the water evaporates, species with lower vapor pressures than the water will concentrate and eventually come out of solution, because the solubility limit has been reached or the water removed. The resulting residue may form crystalline salts, such as NaCl or KCl, or amorphous solids. Depending on the concentration of impurities, they may manifest on the surface macroscopically as stains or adhered particles.

There is really no such thing as a final rinse with no impurities – however wetting characteristics and surface topography may influence how impurities collect or migrate on the surface. Wetting is a description of a how a fluid will collect on a surface and is a function of the surface energy of the solid surface and the liquid surface energy. A droplet on the surface will form a characteristic contact angle θ_c (see Figure 3.49) on the surface given by

$$\cos\theta_c = \frac{\gamma_{SG} - \gamma_{SL}}{\gamma_{LG}} \tag{3.54}$$

where $\gamma_{SG}, \gamma_{SL}, \gamma_{LG}$ are the solid–gas, solid–liquid, and solid–gas surface energies, respectively.

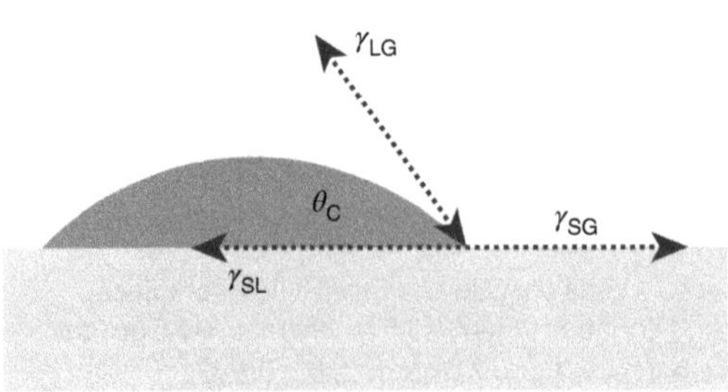

Figure 3.49 Illustration of the formation of a sessile liquid droplet on a surface. Source: https://en.wikipedia.org/wiki/Sessile_drop_technique. Licenced under CC BY 3.0.

The nonwetting of a rinsing solution (i.e. a large contact angle) or isolated droplets on a workpiece may detrimentally lead to the "coffee ring" effect. Take, for example, the final rinsing and drying of a workpiece after an aqueous rinse of the surface. If the water does not wet the surface well, droplets will form, and a drying droplet will concentrate any remaining impurities in a ring-like deposit along the perimeter of the pinned water droplet. This effect is well described by Deegan et al. [90, 91]. In this phenomenon, the liquid droplet starts to evaporate from the top surface, which causes the droplet to shrink in all dimensions. At some droplet size, the outer perimeter will be pinned – that is, the diameter of the droplet will not change further, as shown in Figure 3.50. At this point, further evaporation causes liquid to flow from the center of the droplet to the perimeter, carrying particles and dissolved species, because evaporation from the edge is replenished by the liquid in the interior. This flow is called advective current and results in greater solute or containment accumulation at the perimeter. This also results in greater evaporative flux ($J_s(r)$) at the perimeter, which enhances the collection of dissolved species and particles at the perimeter, as depicted in Figure 3.51. The evaporative flux is radially symmetrical and the amount of edge enhancement is dependent on the contact angle as [90, 91]

$$J_s(r) \propto (R_p - r)^{-\left(\frac{\pi - 2\theta_c}{2\pi - 2\theta_c}\right)} \tag{3.55}$$

where R_p is the pinned radius of the droplet.

In addition to wetting characteristics, the surface topography can affect the pinning of a liquid droplet, as illustrated in Figure 3.52. Consider a fused-silica workpiece with surface pits ~20 μm in size. The workpiece was rinsed with water containing a small concentration of dissolved food coloring and then allowed to dry. The main surface had good wetting, so a uniform, linear drying front occurred with no droplet pinning, and little or no residue was deposited

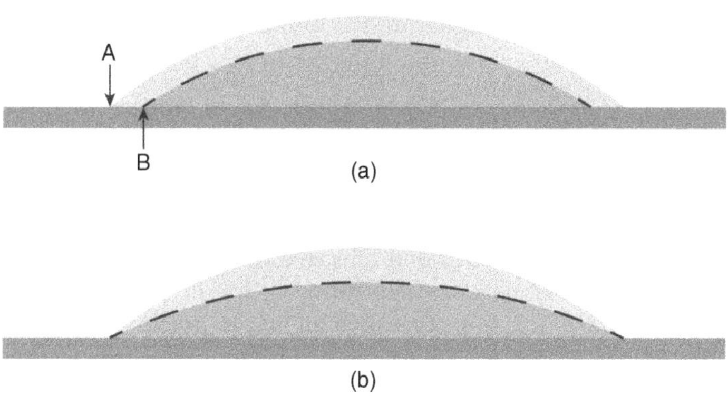

Figure 3.50 Evaporating liquid droplet on a surface that is (a) not pinned and (b) pinned on the edges. Source: Deegan et al. 2000 [91]. Reproduced with permission from American Physical Society

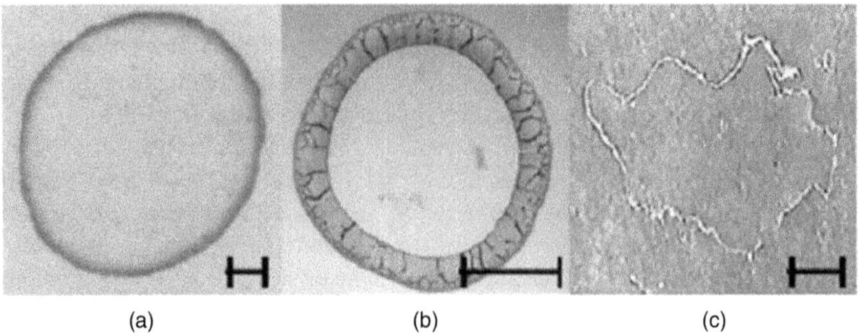

Figure 3.51 Preferential deposition of residue on a drying droplet of (a) coffee, (b) colloidal microsphere solution, and (c) saline. Source: Deegan et al. 2000 [91]. Reproduced with permission from American Physical Society.

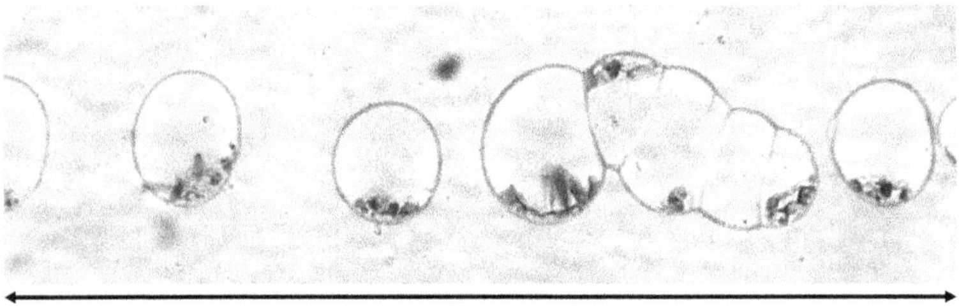

Figure 3.52 Fused-silica glass surface with pits (etch cusps) after drying with a dilute solution of food coloring, illustrating the preferential deposit of residue and precipitation in pits.

on the main surface. However, at the surface pits, the liquid water droplets were pinned and dried by a mechanism similar to that of the de-wetted water drop above. This results in a collection of residue at the edge of the pits. A schematic of this mechanism is shown in Figure 3.53.

3.2.3 Cleaning Strategies and Methods

The basic strategy for cleaning the workpiece surface involves applying enough applied force (F_a) to dislodge particles and effective rinsing to prevent particles and residue from depositing or re-depositing in the first place. While the force required to remove a larger particle is higher, in practice it is generally much more difficult to remove smaller particles from the surface. This is because it is more challenging to achieve a large enough applied force (drag force) approaching closer and closer to the surface. For example, using fluid or gas flow to dislodge a particle requires relatively high velocities near the surface. The challenge is that by definition, fluid velocity is zero right at the

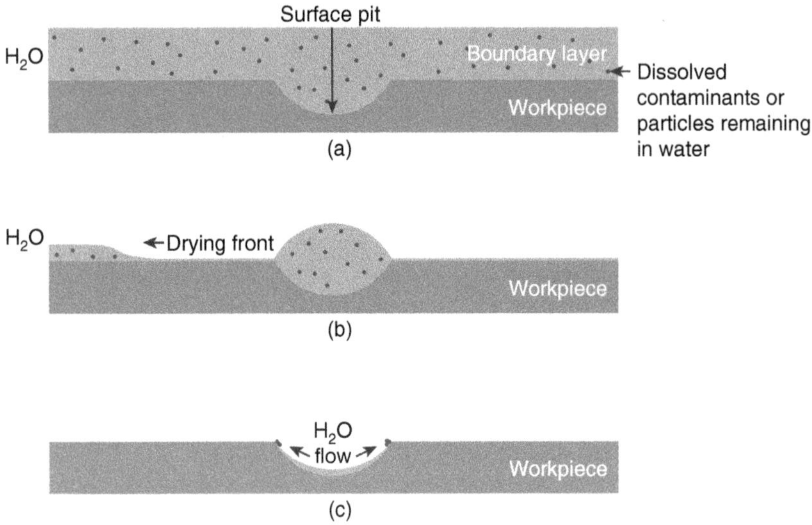

Figure 3.53 Schematic of mechanism of residue/precipitate collection in surface pits during drying of glass surface. (a) Final rinse, (b) drying front and pinned droplet at surface pit, (c) drying within surface pit and precipitation at edges.

surface – therefore, a very large velocity gradient near the surface is needed to remove the particle.

To produce a surface with the cleanest possible surface during rinsing, some key strategies should be implemented.

1. Attempt to achieve the highest level of dilution using solutions with high solubility for the targeted impurities. As the old chemists' saying goes, "dilution is the solution."
2. The purity of the cleaning solvent must be very high, especially during the final rinse and drying.
3. Third, use surfactants (i.e. soaps) to draw away impurities. The final rinse must sufficiently dilute the surfactant itself to ensure it is not left behind on the surface.

The methods for cleaning optical surfaces vary according to the nature of the polishing process and workpiece material. Izumitani [9], Karow [10], Bennett [92], Mittal [89], Kern [93, 94], and Reinhardt and Kern [95] discuss many cleaning methods for optical fabrication and semiconductor processing. One example is RCA clean, an established multistep cleaning process for semiconductor surfaces, with a step for the removal of organics by oxidation and of particles via zeta potential modification (using NH_4OH and H_2O_2). This is followed by HF etching to remove a few layers of oxide and then ionic cleaning

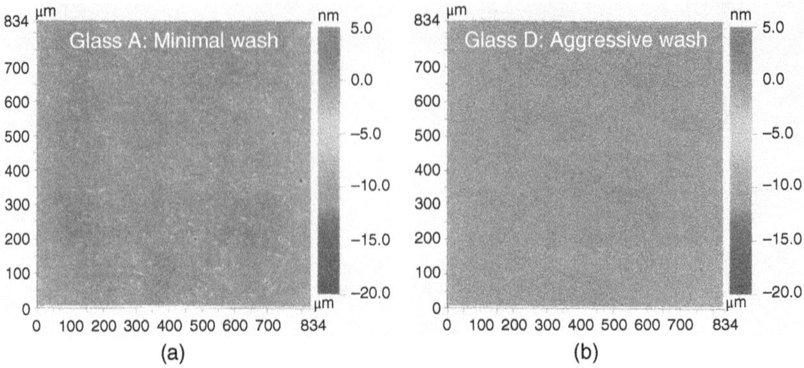

Figure 3.54 White-light interferometer image of laser phosphate-glass workpiece with different degrees of water rinsing after polishing using cerium-oxide slurry. Source: Suratwala et al. 2005 [86]. Reproduced with permission from Elsevier.

using HCl and H_2O_2 to remove trace metallic contaminants [93]. Other cleaning methods include high-pressure fluid washing, ultrasonic/megasonic treatment, drag wiping, CO_2 snow cleaning, vapor degreasing, chemical etching, strippable coatings, plasma cleaning, and ozone cleaning.

An important general rule outlined in Karow's classic, *Fabrication Methods for Precision Optics,* is that "the very large number of these tiny particles [slurry] is the source of haze, stains, and spots on the surface" and that allowing them to dry on the surface is detrimental. The best strategy is, therefore, not to let particles dry on the surface right after polishing. To accomplish this, effective *initial* rinsing is needed to removal all particles from the surface immediately after polishing, before the workpiece dries for the first time [86]. Cerium oxide is a common polishing compound for glasses that may chemically react with the glass surface when left on the surface after polishing. Subsequent washing of the particle at later steps may lead to surface pitting in the reacted-upon contact zone between the particle and workpiece. Pitting has been experimentally observed on fused silica glass (see Section 4.1) and on phosphate glass (Figure 3.54a), resulting a permanent haze scatter on the workpiece as a result [86, 96]. For phosphate glass, the effect of an aggressive, efficient initial rinse to reduce pitting is shown in Figure 3.54b. A proposed mechanism for the ceria reaction with laser phosphate glass is shown in Figure 3.55 [86].

3.3 The Beilby Layer

During the polishing of metals, crystals, and glasses, a modified amorphous surface layer is formed, ranging in thickness from a few nanometers to a µm. This layer is often referred to as the Beilby layer, though the name originally

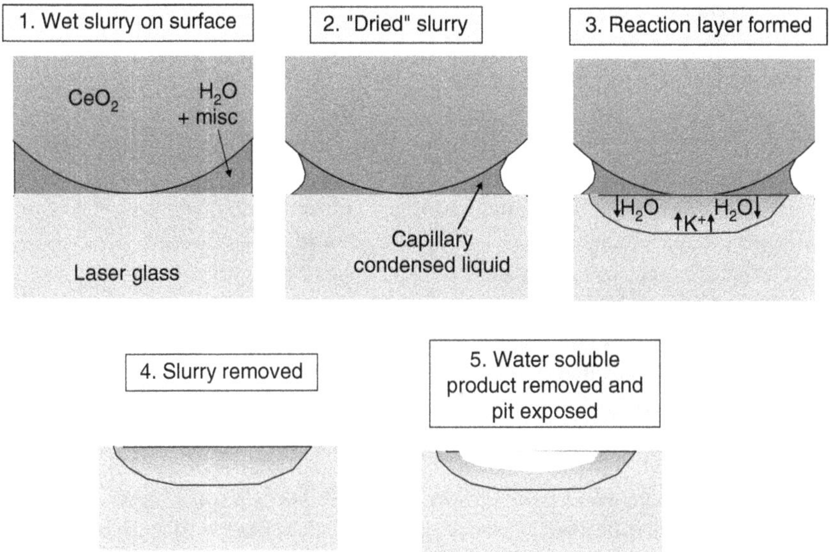

Figure 3.55 Illustration of how dried polishing-slurry particles on a workpiece surface may result in shallow pitting of the surface. Source: Suratwala et al. 2005 [86]. Reproduced with permission from Elsevier.

referred to metal polished layers only [2]. For polished glass surfaces, this layer is also called the polishing layer, the modified layer, or the hydrated layer. The Beilby layer is an important *surface quality* metric. This surface layer may have different properties from its bulk on the final workpiece, potentially changing workpiece behavior mechanically, chemically, or optically. Surface-property changes during polishing may also influence the resulting workpiece *surface roughness*, as discussed in Section 4.2.

Because the Beilby layer is very thin, it has been difficult to characterize this surface until the last few years. Nevot and Croce [97], and later Trogolo and Rajan [98], successfully imaged the Beilby layer of silica glass in cross-section using transmission electron microscopy, showing two distinct surface layers: one 3–4 nm thick and a deeper layer 15–20 nm thick. Liao et al. [99] recently performed similar measurements using field emission scanning electron microscopy (FE-SEM). Sensitive optical reflectivity, ellipsometry [100], and X-ray reflectivity [101] measurements indicate that the refractive index of this layer is typically 0.003–0.005 higher than the bulk material.

It has been convenient to characterize the Beilby layer based on the presence of impurities, using techniques such as secondary-ion mass spectroscopy (SIMS) to determine depth profiles. The Beilby layer on glasses contains elevated levels of impurities whose concentration typically decays exponentially to the bulk concentration within a few tens of nanometers of the surface

[7, 102–104]. The Beilby layer associated with silicate glasses can be removed by chemically etching the glass itself (e.g. using HF-based acids) or modified by chemically leaching the metal cations from the surface (e.g. using mineral acids such as nitric acid) [6, 50, 105].

Until recently, no direct measurement of water penetration in the polishing layer has been provided, though it is largely accepted that hydration is occurring. Wakabayashi and Tomozawa [106] gave indirect evidence that H_2O penetration is likely occurring, and Moon showed modification differences on a Si surface with respect to nanoplastic removal with and without chemical treatment [107]. Later H SIMS measurements of the surface confirmed the presence of H penetrating the polished surface layer of fused-silica glass (Figure 3.56) [108]. Since the instrumental noise floor is relatively high ($\sim 10^{19}$ atoms cm^{-3}) for measuring H, the actual penetration depth of H into a surface cannot be determined from this data. However, the results confirm a high surface concentration of H of $\sim 1–3 \times 10^{20}$ atom cm^{-3} (i.e. 10 000–30 000 ppm), because of H_2O penetration. H_2O penetration into glass surfaces has been shown to occur by diffusion, followed by reaction with the glass matrix to form hydroxyls [109–111].

A likely mechanism for the incorporation of other cation impurities in the surface is diffusion. If the Beilby layer formation is diffusion induced, there should be a steady-state relationship between *material removal rate* and the diffusion mass transfer of the impurity species. At steady-state, the diffusion flux of impurities into the workpiece is equal to the flux of impurities removed

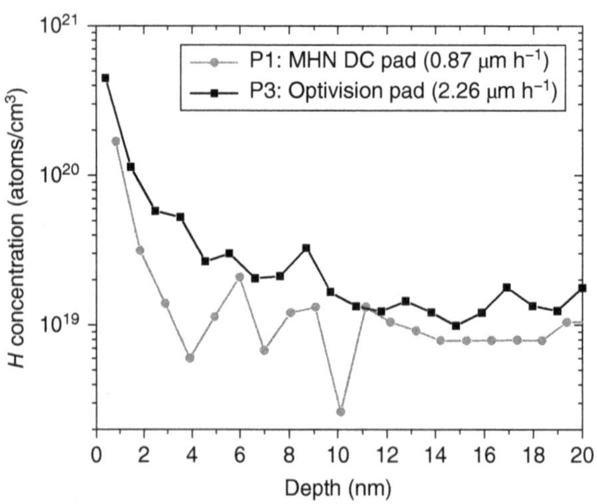

Figure 3.56 Depth profile of H surface, as measured by H SIMS, on selected polished fused-silica surfaces. Source: Suratwala et al. 2015 [108]. Reproduced with permission from John Wiley & Sons.

via removal of the surface during polishing. This balance is given by

$$C\frac{dh}{dt} = D\frac{dC}{dx}, \tag{3.56}$$

where C is the concentration of the diffusing impurity cation, dh/dt is the workpiece surface *material removal rate* due to polishing, D is impurity cation diffusivity, and x is the depth into the moving glass surface. Note that the right side of Equation (3.56) is simply Fick's first law. Using the Preston's material removal equation (Equation (1.3)) [50, 108] and solving Equation (3.56) gives

$$\frac{dh}{dt} = k_p\sigma_o V_r \cong \frac{D}{t_{\mathrm{Bielby}}}, \tag{3.57}$$

where k_p is the Preston coefficient, σ_o is the applied pressure, V_r is relative velocity during polishing, and t_{Beilby} is the effective thickness of the Beilby layer. This relationship suggests that increasing the *material removal rate* should result in a decrease in Beilby-layer thickness – that is, decrease the relative penetration of a given impurity.

The depth profiles of K and Ce for a series of fused-silica samples polished with cerium-oxide slurry – with additional KOH for pH balance, at different relative velocities (and hence different polishing removal rates) – are shown as a pair of semilog plots in Figure 3.57a,b [108]. Note the bulk-glass Si concentration was measured at a uniform value of 2×10^{22} atoms cm^{-3} (not shown in the plot), while the noise floor of each analyte is $\sim 10^{16}$ atoms cm^{-3}. K penetration into the glass surface was significantly greater (~500–900 nm) than

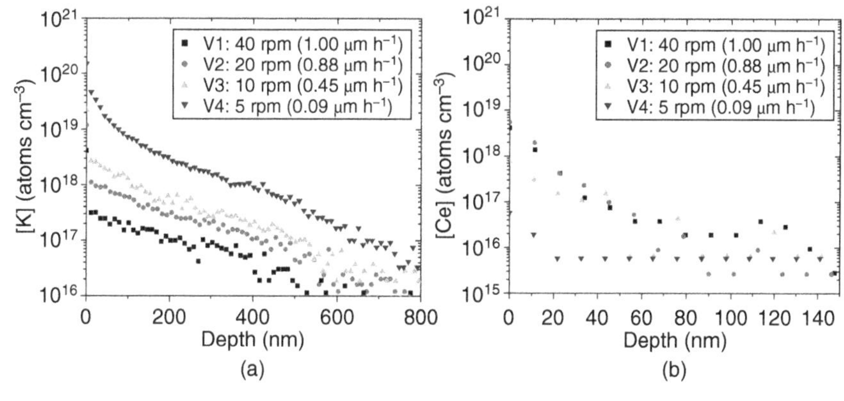

Figure 3.57 (a) SIMS-generated depth profile of K on fused-silica surfaces prepared at different material removal rates, measured 2 weeks after polishing. (b) SIMS-generated depth profile of Ce on the same fused-silica surfaces prepared at different material removal rates. The fused-silica workpieces were polished with stabilized Hastilite PO ceria slurry at pH 8.3 at different rotation rates to alter the material removal rate. Source: Suratwala et al. 2015 [108]. Reproduced with permission from John Wiley & Sons.

Ce penetration (~20–100 nm). The surface concentration for both impurities varied significantly, from 6×10^{16} to 2×10^{20} atoms cm^{-3}, with changes in polishing conditions. K penetration is consistent with the behavior expected for diffusion, where K penetration decreases with increase in *material removal rate* (Equation (3.57)). In contrast, the penetration of Ce into the near-surface layer is inconsistent with a diffusion mechanism, because Ce penetration increases with increase in *material removal rate* (Figure 3.57b). This suggests that another mechanism than diffusion, such as chemical reactivity, dominates the penetration of Ce into the glass surface. The discussion below describes penetration by both diffusion and chemical reactivity in greater detail to quantitatively explore this data.

3.3.1 K Penetration by Two-step Diffusion

Soaking fused-silica glass in KOH without polishing results in significant penetration of K (see Figure 3.58a), indicating that interface interactions during polishing are not required for K penetration. Also, diffusion of K continues to occur even after polishing when stored at room temperature (see Figure 3.58b). This is a most intriguing result, since it is often assumed that a polished workpiece is static and unchanging after polishing. Clearly, in the case of fused silica, the surface layer may continue to change during storage.

K penetration into the surface can be described by a two-step diffusion process. First, during polishing, diffusion occurs at a moving surface boundary and steady-state is reached, as described in the previous section [112]. Second, in storage, the source of K is now removed, the silica-glass surface boundary is fixed, and time-dependent K diffusion is determined by Fick's second law.

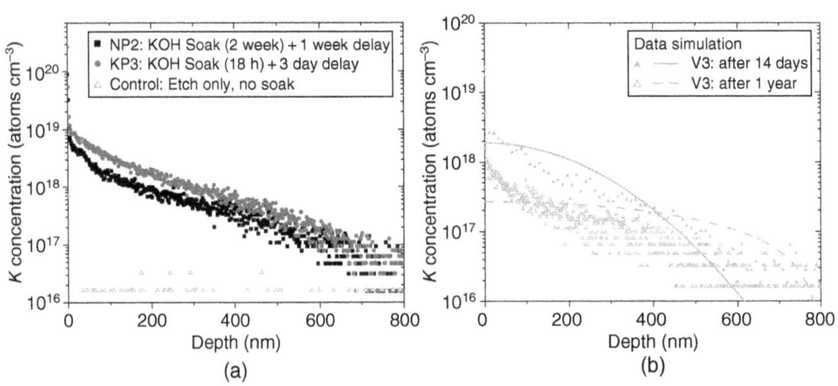

Figure 3.58 (a) SIMS-generated depth profile of K on etched fused-silica surfaces that were soaked in aqueous KOH (b) SIMS-generated depth profile of K 2 weeks after polishing and a year after polishing. The lines are calculated diffusion profiles using a simple one-step diffusion model. Source: Suratwala et al. 2015 [108]. Reproduced with permission from John Wiley & Sons.

Diffusion during polishing follows an advection diffusion model. Using a concentration-independent diffusion coefficient (D) for K in a semiinfinite solid in one dimension (x), the K concentration (C_K) will follow:

$$\frac{d}{dx}\left(D\frac{dC_K}{dx} + \frac{dh}{dt}C_K\right) = 0, \tag{3.58}$$

whose solution takes the form:

$$C_K(x) = \frac{A_o}{\left(\frac{dh}{dt}\right)}\exp\left(\frac{-\left(\frac{dh}{dt}\right)x}{D}\right), \tag{3.59}$$

where A_o is constant. Figure 3.59 shows the predicted profiles after the polishing step (dashed lines). Note the pre-exponential term in Equation (3.59) is the surface concentration, which decreases with dh/dt. Also, the exponential term in the equation will change with dh/dt – the slope of K depth profiles on a semilog plot should change with *material removal rate*. The measured data in Figure 3.57a did *not* show meaningful change in slope, despite the removal rate's changing by ~10×, which supports the conclusion that diffusion behavior follows a multistep process.

During storage, the source of K is removed and the silica-glass surface boundary is fixed. In this case, the time-dependent K diffusion is determined by Fick's

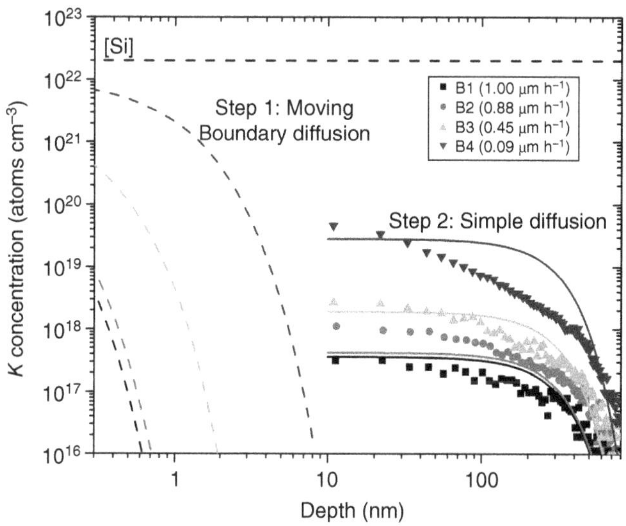

Figure 3.59 Comparison of the two-step diffusion model using Equations (3.58)–(3.60) with the measured depth profile of K. All measurements were performed 2 weeks after polishing. Source: Suratwala et al. 2015 [108]. Reproduced with permission from John Wiley & Sons.

second law in the form:

$$\frac{\mathrm{d}}{\mathrm{d}x}\left(D\frac{\mathrm{d}C_{\mathrm{K}}}{\mathrm{d}x}\right) = \frac{\mathrm{d}C_{\mathrm{K}}}{\mathrm{d}t}, \tag{3.60}$$

with the boundary conditions of zero concentration in the bulk glass and no removal of K from the surface of the glass. Using an estimated value of diffusivity (D) for both diffusion steps (10^{-16} cm^2 s^{-1}) and a best-fit value of A_o (10^{21} cm^{-3} μm^{-1} h), the calculated diffusion curves after a 2-week storage step (solid lines) are compared with the measured data for polishing at different removal rates (see Figure 3.59). The general shape, penetration depth, and *material removal rate* dependence of the two-step diffusion model is in good agreement with the experimental data. There is, however, a deviation in the detailed shape of final measured diffusion profiles, with the two-step model using a constant diffusion coefficient. The authors hypothesize that this deviation reflects changes in the diffusivity (D) of K with depth (x) possibly caused by changes in the structure of the glass in the Beilby layer [108].

3.3.2 Ce Penetration by Chemical Reactivity

In addition to the data shown in Figure 3.57, Figure 3.60a shows a series of Ce depth profiles for a number of other fused-silica workpieces that were polished using various Ce slurries and polishing pads [108]. The depth of Ce penetration varied significantly, depending on the polishing process used. Using the same data, Figure 3.60b shows surface Ce concentration as a function of the polishing-material removal rate. In contrast to the behavior of K penetration,

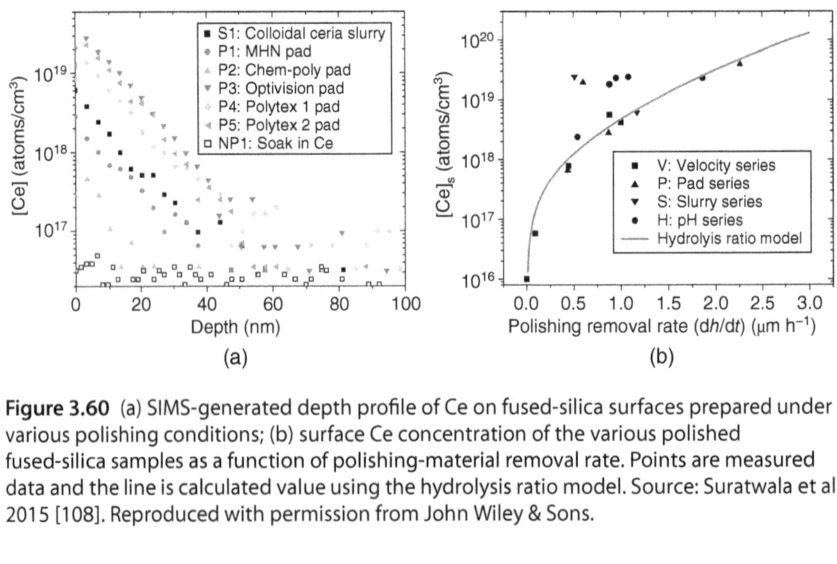

Figure 3.60 (a) SIMS-generated depth profile of Ce on fused-silica surfaces prepared under various polishing conditions; (b) surface Ce concentration of the various polished fused-silica samples as a function of polishing-material removal rate. Points are measured data and the line is calculated value using the hydrolysis ratio model. Source: Suratwala et al. 2015 [108]. Reproduced with permission from John Wiley & Sons.

Ce penetration into the fused-silica surface *increases* with polishing-material removal rate. In fact, a quantitative correlation exists between measured Ce surface concentration and measured *material removal rate,* despite large variations in the polishing processes. Also, in contrast to the behavior of K, simply soaking a fused-silica sample in ceria slurry did not result in Ce penetration into the surface of the polished sample (see open data points in Figure 3.60a). This suggests that the polishing process itself contributes to the transport of Ce into the near-surface layer of the glass.

It is known that ceria plays a key role in the removal of silica from the glass surface during polishing. The most widely accepted chemical mechanism is by condensation and subsequent hydrolysis reactions, also known as the chemical tooth model, given by [54].

$$\equiv \text{Si—OH} + \text{HO—Ce} \equiv \rightarrow \equiv \text{Si—O—Ce} \equiv + \text{H}_2\text{O} \tag{3.61}$$

$$\begin{aligned} &\equiv \text{Si—O—Si—O—Ce—O—Ce} \equiv + \text{H}_2\text{O} \rightarrow \\ &\equiv \text{Si—OH} + \text{HO—Si—O—Ce—O—Ce} \equiv \end{aligned} \tag{3.62}$$

where Reaction (3.61) is the condensation reaction between the Si—OH silica surface and Ce—OH ceria-particle surface and Reaction (3.62) is the subsequent hydrolysis of the Si—O—Si bond, leading to the removal of silica from the glass surface. The nature of these chemical reactions with respect to material removal is described in Chapter 5.

Another possible reaction has been proposed whereby hydrolysis occurs at the Ce—O—Ce bond, leaving Ce behind on the silica surface, given by [108].

$$\begin{aligned} &\equiv \text{Si—O—Si—O—Ce—O—Ce} \equiv + \text{H}_2\text{O} \rightarrow \\ &\equiv \text{Si—O—Si—O—Ce—OH} + \text{HO—Ce} \equiv \end{aligned} \tag{3.63}$$

For material removal to occur during polishing, the rate of Reaction (3.62) must be much greater than Reaction (3.63).

At steady-state during polishing, the rate of incorporation of Ce due to Reaction (3.63), and the rate of removal of Ce due to Reaction (3.62) from the moving silica surface will be equal. Using this rate balance, it has been shown that [108]

$$2av_\text{r}n_\text{p}S_\text{p}r_{\text{Ce:Si}} = 2av_\text{r}d_\text{m}n_\text{p}[\text{Ce}]_\text{s}, \tag{3.64}$$

where $2a$ is the contact diameter of a spherical polishing particle, v_r is the velocity of the polishing particle, n_p is the particle number density per unit area, S_p is the areal number density of Si atoms, d_m is removal depth or bond length, $r_{\text{Ce:Si}}$ is the hydrolysis reaction-rate ratio between Ce–O–Ce and Si–O–Si, and $[\text{Ce}]_\text{s}$ is the resulting steady-state Ce concentration at the surface. Note that $2av_\text{r}$ is the *areal* contact per slurry particle per unit of time and $2av_\text{r}d_\text{m}$ is the *volumetric* removal per slurry particle per unit of time. Simplifying Equation (3.64) gives

$$[\mathrm{Ce}]_\mathrm{s} = \frac{S_\mathrm{p} r_\mathrm{Ce:Si}}{d_\mathrm{m}}, \tag{3.65}$$

which suggests that the steady-state Ce surface concentration will be dominated by the ratio of the two hydrolysis reactions ($r_\mathrm{Ce:Si}$), because S_p (areal density of Si atoms) and d_m (removal depth) are largely constant [108].

To account for changes in the Ce surface concentration with changes in *material removal rate*, the hydrolysis reaction ratio ($r_\mathrm{Ce:Si}$) was proposed to change with interface temperature, which is a function of the *material removal rate* [108]. The interface temperature has been previously described for a single sliding particle, generating frictional heat at the moving contact interface given by [113–115]:

$$T = \frac{1.464aQ}{k_1\sqrt{\pi(0.874 + \mathrm{Pe})}}, \tag{3.66}$$

where

$$Q = \frac{\mu P v_\mathrm{r}}{\pi a^2}, \tag{3.67}$$

and

$$\mathrm{Pe} = \frac{v_\mathrm{r} a \rho_1 C_\mathrm{p1}}{2k_1} \tag{3.68}$$

where a is the contact radius, Q is frictional heat flux, μ is the friction coefficient, P is applied load, v_r is the particle velocity, ρ_1 is workpiece mass density, k_1 is workpiece thermal conductivity of silica, C_p1 is the workpiece heat capacity of silica, and Pe is the Peclet number. The *material removal rate* (dh/dt) is related to workpiece velocity by the Preston equation, as demonstrated in Equation (1.3), and hence Equation (3.67) can be rewritten as

$$Q = \frac{\mu P \frac{\mathrm{d}h}{\mathrm{d}t}}{\pi a^2 k_\mathrm{p} \sigma_\mathrm{o}}. \tag{3.69}$$

Using Equations (3.66)–(3.69) the calculated temperature rises as a function of *material removal rate, as* shown in Figure 3.61. For fused silica, $\rho_1 = 2.2\ \mathrm{gm\ cm^{-3}}$, $k_1 = 1.4\ \mathrm{W\ mK^{-1}}$, and $C_\mathrm{p1} = 740\ \mathrm{J\ kg\,K^{-1}}$. For the polishing experiments, $k_\mathrm{p} = 2.2 \times 10^{-13}\ \mathrm{m^2\ N^{-1}}$ and $\sigma_\mathrm{o} = 0.6$ psi. The average contact parameters were determined as $a = 55$ nm and $P = 10^{-4}$ N using polishing experiments and a contact-mechanics model called the Ensemble Hertzian Multi-Gap (EHMG) model [96], described in Section 4.7.1. The results show that particle–workpiece interface temperature increases linearly at a rate of 55 K for each 1 $\mu\mathrm{m\,h^{-1}}$ increase in removal rate. In a previous study [116], global temperature rise was measured during polishing under very

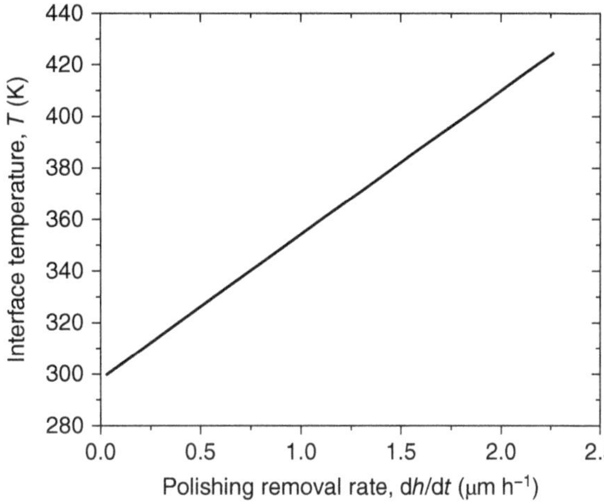

Figure 3.61 Calculated polishing particle–workpiece interface temperature as a function of material removal rate. Source: Suratwala et al. 2015 [108]. Reproduced with permission from John Wiley & Sons.

similar conditions, leading to a temperature rise of 1–2 K. Because the total particle–workpiece contact area is small relative to the whole workpiece surface area, it is expected that the local interface temperature will be much higher than the global temperature of the system.

Based on the above, the $r_{\mathrm{Ce:Si}}$ ratio should have an Arrhenius temperature dependence and Equation (3.65) can be rewritten as follows:

$$[\mathrm{Ce}]_{\mathrm{s}} = \frac{S_{\mathrm{p}}}{d_m} r_{\mathrm{A}} \exp\left(\frac{-E}{RT(\mathrm{d}h/\mathrm{d}t)}\right), \tag{3.70}$$

where r_{A} is the pre-exponential constant, E_{r} is activation energy, and $T(\mathrm{d}h/\mathrm{d}t)$ is the interface temperature as a function of removal rate. Figure 3.60b shows the comparison of the calculated Ce surface concentration designated by the solid line (using $r_{\mathrm{A}} = 80$ and $E_{\mathrm{r}} = 42$ kJ mol) with measured values from all the SIMS results for each of the polishing samples [108]. This model does a reasonable job in matching observed Ce surface concentration with removal rate. The best-fit activation-energy value of $E_{\mathrm{r}} = 42\ \mathrm{kJ\ mol^{-1}}$ appears to be a reasonable value, as hydrolysis activation of Si–O–Si has been reported to have similar values [117].

3.3.3 Chemical–Structural–Mechanical Model of the Beilby Layer and Polishing Process

Combining insights of the mechanism of Beilby layer formation with the chemical tooth model and determined polishing contact parameters [96], a detailed chemical–structural–mechanical picture of the Beilby layer and polishing process is described in Table 3.2 and Figure 3.62 [108]. Consider the specific case where a fused-silica workpiece is polished on a polyurethane pad, using a ceria slurry. As the workpiece is polished, a Beilby layer is formed with the incorporation of H_2O, Ce, and K impurities. The concentrations and depth of penetration

Table 3.2 Parameters used for developing chemical–mechanical–structural representation of the Beilby layer and polishing process.

Parameter	Symbol	Value	Si:X[a)]	Source
Beilby layer thickness	t_{Beilby}	50 nm	—	This study
H surface concentration	$[H]_s$	2×10^{20} atoms cm^{-3}	100 : 1	This study
Ce surface concentration	$[Ce]_s$	5×10^{19} atoms cm^{-3}	400 : 1	This study
K surface concentration	$[K]_s$	1×10^{18} atoms cm^{-3}	20 000 : 1	This study
Bulk SiO_2 structure	Q^4:Q^3:Q^2:Q^1	87 : 11 : 2 : 0	—	[26]
Surface SiO_2 structure	Q^4:Q^3:Q^2:Q^1	86 : 12 : 2 : 0	—	This study
Removal rate	dh/dt	0.80 μm h^{-1}	—	[16]
Particle size	r	400 nm	—	[16]
Average load/particle	P	10^{-4} N	—	[16]
Contact zone diameter	$2a$	110 nm	—	[16]
Elastic depth		3.8 nm	—	[16]
Removal depth	d_m	1 nm	—	[16]

a) Atomic ratio of Si (2×10^{22} atom cm^{-3}) to that of H, Ce or K.
Source: Suratwala et al. 2015 [108]. Reproduced with permission from John Wiley & Sons.

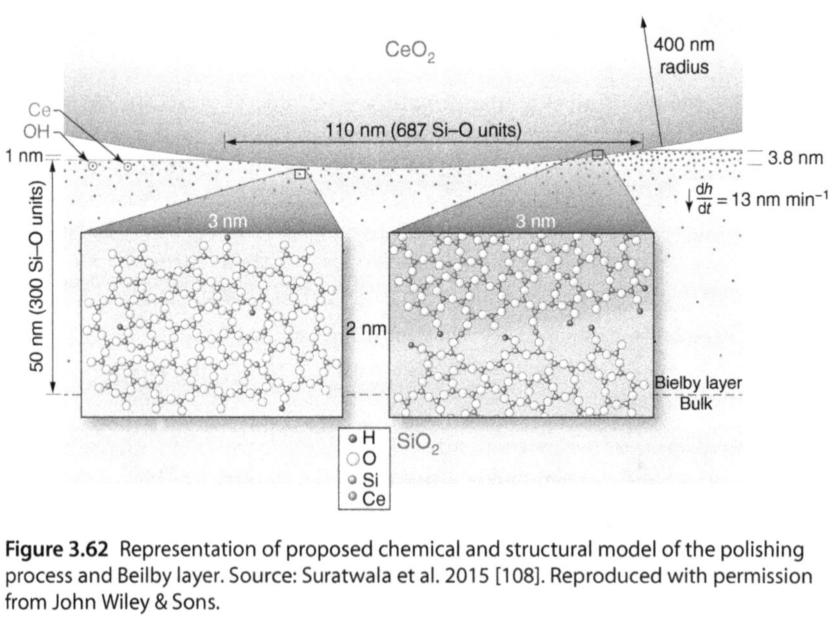

Figure 3.62 Representation of proposed chemical and structural model of the polishing process and Beilby layer. Source: Suratwala et al. 2015 [108]. Reproduced with permission from John Wiley & Sons.

of each of the impurities are quite different. Table 3.2 compares surface concentrations: H is the highest, with an H:Si ratio of 1 : 100, followed by Ce with a Ce:Si ratio of 1 : 400, and K with a K:Si ratio of 1 : 20 000. Given this discrepancy, the depth of the Beilby layer is difficult to define unambiguously, as these species penetrate to significantly different depths. The depth a given species penetrates may be time- or polishing-rate dependent. For convenience, the depth of the Beilby layer in Table 3.2 is defined in terms of Ce penetration depth, specifically where the Ce concentration is no longer detectable from a SIMS-measurement noise floor of $\sim 10^{16}$ atoms cm^3. Given the exponential decay in the concentration, the vast majority of the Ce is present within the first few nanometers of the surface [108].

The bonding structure of the fused silica near-surface is more difficult to determine experimentally. The bulk structural information for silica glass has been previously well characterized by solid-state nuclear magnetic resonance (NMR) by many researchers (e.g., Ref. [118]). The Si Q species describes the quaternary oxygen tetrahedron in the glass structure, where Q^4 represent four Si–O neighbors, Q^3 represents three Si–O neighbors, etc. The greater the fraction of Q^4, the more cross-linked the structure of the glass. It is reasonable that K and H behave as network modifiers, decreasing the cross-linking of the glass structure. Similarly, Ce would be expected to behave as a network former [85]. By assuming that all K and H species detected in the surface layer act as network modifiers, the distribution of Q species in the surface layer to that of bulk fused silica is calculated and summarized in Table 3.2. The calculated Q-species ratio at the surface shows only a minor reduction in cross-linking due to an overall small impurity concentration relative to bulk Si concentration.

The mechanical aspects of the contact between a polishing particle and glass surface are also shown in Table 3.2. A mechanics-based polishing model was used, based on multiple Hertzian contacts of slurry particles at the workpiece–pad interface, in which the effective interface gap is determined through an elastic-load balance using the slurry PSD (see Section 4.7.1) [96]. From this, the load, penetration, and contact-zone distributions were determined and the removal function for a single sliding particle was experimentally determined [96]. The average particle size (resulting in plastic-type removal) was $\sim$800 nm, with an average load of 10^{-4} N, resulting in an elastic contact zone of 110 nm, penetration depth of 3.8 nm, and removal depth of 1 nm.

Using the information in Table 3.2, a schematic of a single particle–workpiece contact is shown in Figure 3.62. Here a single CeO_2 800-nm particle slides from left to right under load on the glass surface. The particle size, contact zone, penetration depth, and Beilby layer thickness are shown to scale. The presence of chemical impurities, such as H and Ce, in the surface layer is indicated by colored dots, the density of which corresponds to the relative areal concentrations expected in the scale length of the model. The inserts provide a more detailed view of the glass structure at just below the particle–glass interface.

Because of the exponential decay in the concentration of impurities, the Beilby layer resembles the bulk glass within just a few nanometers of the surface. Also, because the K concentration is so low, it does not appear, on average, within the scale length of the model shown in Figure 3.62 [108].

References

1 Bennett, J.M. and Mattsson, L. (1999). *Introduction to Surface Roughness and Scattering*, 2e, viii, 130. Washington, DC: Optical Society of America.
2 Beilby, G.T. (1921). *Aggregation and Flow of Solids*. London: Macmillan and co., Ltd.
3 Spaeth, M.L., Wegner, P.J., Suratwala, T.I. et al. (2016). Optics recycle loop strategy for NIF operations above UV laser-induced damage threshold. *Fusion Sci. Technol.* 69 (1): 265–294.
4 Suratwala, T., Wong, L., Miller, P. et al. (2006). Sub-surface mechanical damage distributions during grinding of fused silica. *J. Non-Cryst. Solids* 352 (52–54): 5601–5617.
5 Feit, M.D. and Rubenchik, A.M. ed. (2004). Influence of subsurface cracks on laser-induced surface damage. Proceedings of SPIE, Volume 5273, pp. 264–272.
6 Suratwala, T.I., Miller, P.E., Bude, J.D. et al. (2011). HF-based etching processes for improving laser damage resistance of fused silica optical surfaces. *J. Am. Ceram. Soc.* 94 (2): 416–428.
7 Miller, P.E., Bude, J.D., Suratwala, T.I. et al. (2010). Fracture-induced subbandgap absorption as a precursor to optical damage on fused silica surfaces. *Opt. Lett.* 35 (16): 2702–2704.
8 Laurence, T.A., Bude, J.D., Shen, N. et al. (2009). Metallic-like photoluminescence and absorption in fused silica surface flaws. *Appl. Phys. Lett.* 94 (15): 151114.
9 Izumitani, T. (1986). *Optical Glass*, x, 197. New York: American Institute of Physics.
10 Karow, H.H. (1992). *Fabrication Methods for Precision Optics* (ed. J.W. Goodman), 1–751. New York: Wiley.
11 Brown, N.A. (1981). *Short Course on Optical Fabrication Technology*. Lawrence Livermore National Laboratory.
12 Lawn, B.R. (1993). *Fracture of Brittle Solids*, 2e, xix, 378. Cambridge, New York: Cambridge University Press.
13 Lawn, B. and Wilshaw, R. (1975). Indentation fracture: principles and applications. *J. Mater. Sci.* 10 (6): 1049–1081.
14 Hutchings, I.M. (1992). *Tribology: Friction and Wear of Engineering Materials*, viii, 273. Boca Raton, FL: CRC Press.

15 Anderson, T.L. (1995). *Fracture Mechanics: Fundamentals and Applications*, 2e, 688. New York: CRC Press.

16 Lawn, B. and Evans, A. (1977). A model for crack initiation in elastic/plastic indentation fields. *J. Mater. Sci.* 12 (11): 2195–2199.

17 Lawn, B. and Marshall, D. (1979). Hardness, toughness, and brittleness: an indentation analysis. *J. Am. Ceram. Soc.* 62 (7–8): 347–350.

18 Lambropoulos, J.C., Xu, S., and Fang, T. (1997). Loose abrasive lapping hardness of optical glasses and its interpretation. *Appl. Opt.* 36 (7): 1501–1516.

19 Fang, T. and Lambropoulos, J.C. (2002). Microhardness and indentation fracture of potassium dihydrogen phosphate (KDP). *J. Am. Ceram. Soc.* 85 (1): 174–178.

20 Danzer, R., Hangl, M., and Paar, R. (2000). Edge chipping of brittle materials. *Ceram. Trans.* 122: 43–55.

21 Morrell, R. and Gant, A. (2001). Edge chipping of hard materials. *Int. J. Refract. Met. Hard Mater.* 19 (4): 293–301.

22 Hangl, M., Danzer, R., and Paar, R. ed. (2013). Edge toughness of brittle materials. ECF11, Poitiers 1996.

23 Almond, E. (1990). N.J. McCormick, Edge flaking of brittle materials. *J. Hard Mater.* 1: 154–156.

24 Lawn, B.R. (1967). Partial cone crack formation in a brittle material loaded with a sliding spherical indenter. *Proc. R. Soc. London, Ser. A* 299 (1458): 307–316.

25 Chaudhri, M.M. and Brophy, P.A. (1980). Single particle impact damage of fused silica. *J. Mater. Sci.* 15 (2): 345–352.

26 Swain, M.V. ed. (1979). Microfracture about scratches in brittle solids. *Proc. R. Soc. London, Ser. A* 366 (1727): 575–597.

27 Li, K. and Warren Liao, T. (1996). Surface/subsurface damage and the fracture strength of ground ceramics. *J. Mater. Process. Technol.* 57 (3): 207–220.

28 Suratwala, T., Miller, P., Feit, M., and Menapace, J. (2008). Scratch forensics. *Opt. Photonics News* 20 (9): 12–15.

29 Ritter, J., Jakus, K., and Panat, R. ed. (1998). Impact Damage and Strength Degradation of Fused Silica. MRS Proceedings. Cambridge University Press.

30 Chaudhri, M. and Kurkjian, C. (1986). Impact of small steel spheres on the surfaces of "normal" and "anomalous" glasses. *J. Am. Ceram. Soc.* 69 (5): 404–410.

31 Hagan, J. (1979). Cone cracks around Vickers indentations in fused silica glass. *J. Mater. Sci.* 14 (2): 462–466.

32 Wiederhorn, S. and Lawn, B. (1979). Strength degradation of glass impacted with sharp particles: I, Annealed surfaces. *J. Am. Ceram. Soc.* 62 (1–2): 66–70.

33 Knight, C., Swain, M.V., and Chaudhri, M. (1977). Impact of small steel spheres on glass surfaces. *J. Mater. Sci.* 12 (8): 1573–1586.
34 Eder, D.C., Whitman, P.K., Koniges, A.E. et al. (2006). Optimization of experimental designs by incorporating NIF facility impacts. *J. Phys. IV* 133: 721–725.
35 Eder, D., Fisher, A., Koniges, A., and Masters, N. (2013). Modelling debris and shrapnel generation in inertial confinement fusion experiments. *Nucl. Fusion* 53 (11): 113037.
36 Shipway, P. and Hutchings, I. (1996). The role of particle properties in the erosion of brittle materials. *Wear* 193 (1): 105–113.
37 Wong, L., Suratwala, T., Feit, M.D. et al. (2009). The effect of HF/NH_4F etching on the morphology of surface fractures on fused silica. *J. Non-Cryst. Solids* 355 (13): 797–810.
38 Aleinikov, F. (1957). The effect of certain physical and mechanical properties on the grinding of brittle materials. *Sov. Phys. Tech. Phys.* 2 (12): 2529–2538.
39 Brinksmeier, E. (1989). State-of-the-art of non-destructive measurement of sub-surface material properties and damages. *Precis. Eng.* 11 (4): 211–224.
40 Lucca, D., Brinksmeier, E., and Goch, G. (1998). Progress in assessing surface and subsurface integrity. *CIRP Ann. Manuf. Technol.* 47 (2): 669–693.
41 Parks, R. (1990). Subsurface damage in optically worked glass. In: *Science of Optical Finishing*, Technical Digest Series, vol. 9. OSA.
42 Lambropoulos, J.C., Li, Y., Funkenbusch, P.D., and Ruckman, J.L. ed. (1999). Noncontact estimate of grinding-induced subsurface damage. SPIE's International Symposium on Optical Science, Engineering, and Instrumentation. International Society for Optics and Photonics.
43 Lee, Y. (2011). Evaluating subsurface damage in optical glasses. *J. Eur. Opt. Soc. Rapid Publ.* 6: 11001.
44 Hed, P.P. and Edwards, D.F. (1987). Optical glass fabrication technology. 2: Relationship between surface roughness and subsurface damage. *Appl. Opt.* 26 (21): 4677–4680.
45 Menapace, J., Davis, P., Wong, L. et al. ed. (2005). Measurement of process-dependent subsurface damage in optical materials using the MRF technique. Proceedings of SPIE.
46 Randi, J.A., Lambropoulos, J.C., and Jacobs, S.D. (2005). Subsurface damage in some single crystalline optical materials. *Appl. Opt.* 44 (12): 2241–2249.
47 Menapace, J.A., Davis, P.J., Steele, W.A. et al. (2005). MRF applications: measurement of process-dependent subsurface damage in optical materials using the MRF wedge technique. Proceedings of SPIE, Volume 5991.
48 Miller, P.E., Suratwala, T.I., Wong, L.L. et al. ed. (2005). The distribution of subsurface damage in fused silica.

49 Wang, Z., Wu, Y., Dai, Y., and Li, S. (2008). Subsurface damage distribution in the lapping process. *Appl. Opt.* 47 (10): 1417–1426.

50 Preston, F.W. (1922). The structure of abraded glass surfaces. *Trans. Opt. Soc.* 23 (3): 141.

51 Menapace, J.A., Davis, P.J., Steele, W.A. et al. ed. (2005). Utilization of magnetorheological finishing as a diagnostic tool for investigating the three-dimensional structure of fractures in fused silica. Boulder Damage Symposium XXXVII: Annual Symposium on Optical Materials for High Power Lasers. International Society for Optics and Photonics.

52 Lambropoulos, J., Jacobs, S.D., Gillman, B. et al. (1997). Subsurface damage in microgrinding optical glasses. *Ceram. Trans.* 82: 469–474.

53 Neauport, J., Destribats, J., Maunier, C. et al. (2010). Loose abrasive slurries for optical glass lapping. *Appl. Opt.* 49 (30): 5736–5745.

54 Cook, L. (1990). Chemical processes in glass polishing. *J. Non-Cryst. Solids* 120 (1–3): 152–171.

55 Bulsara, V.H., Ahn, Y., Chandrasekar, S., and Farris, T. (1998). Mechanics of polishing. *J. Appl. Mech.* 65 (2): 410–416.

56 Chauhan, R., Ahn, Y., Chandrasekar, S., and Farris, T. (1993). Role of indentation fracture in free abrasive machining of ceramics. *Wear* 162: 246–257.

57 Catrin, R., Neauport, J., Legros, P. et al. (2013). Using STED and ELSM confocal microscopy for a better knowledge of fused silica polished glass interface. *Opt. Express* 21 (24): 29769–29779.

58 Lambropoulos, J.C., Fang, T., Funkenbusch, P.D. et al. (1996). Surface microroughness of optical glasses under deterministic microgrinding. *Appl. Opt.* 35 (22): 4448–4462.

59 Li, Y., Zheng, N., Li, H. et al. (2011). Morphology and distribution of subsurface damage in optical fused silica parts: bound-abrasive grinding. *Appl. Surf. Sci.* 257 (6): 2066–2073.

60 Dong, Z. and Cheng, H. (2014). Study on removal mechanism and removal characters for SiC and fused silica by fixed abrasive diamond pellets. *Int. J. Mach. Tools Manuf.* 85: 1–13.

61 Lv, D., Wang, H., Zhang, W., and Yin, Z. (2016). Subsurface damage depth and distribution in rotary ultrasonic machining and conventional grinding of glass BK7. *Int. J. Adv. Manuf. Technol.* 86 (1–11): 2361–2371.

62 Moore, M.A. (1978). Abrasive wear. *Int. J. Mater. Eng. Appl.* 1 (2): 97–111.

63 Shafrir, S.N., Lambropoulos, J.C., and Jacobs, S.D. (2007). Subsurface damage and microstructure development in precision microground hard ceramics using magnetorheological finishing spots. *Appl. Opt.* 46 (22): 5500–5515.

64 Esmaeilzare, A., Rahimi, A., and Rezaei, S. (2014). Investigation of subsurface damages and surface roughness in grinding process of Zerodur® glass–ceramic. *Appl. Surf. Sci.* 313: 67–75.

65 Suratwala, T., Steele, R., Feit, M.D. et al. (2008). Effect of rogue particles on the sub-surface damage of fused silica during grinding/polishing. *J. Non-Cryst. Solids* 354 (18): 2023–2037.

66 Basim, G., Adler, J., Mahajan, U. et al. (2000). Effect of particle size of chemical mechanical polishing slurries for enhanced polishing with minimal defects. *J. Electrochem. Soc.* 147 (9): 3523–3528.

67 Ahn, Y., Yoon, J.-Y., Baek, C.-W., and Kim, Y.-K. (2004). Chemical mechanical polishing by colloidal silica-based slurry for micro-scratch reduction. *Wear* 257 (7): 785–789.

68 Blaineau, P., André, D., Laheurte, R. et al. (2015). Subsurface mechanical damage during bound abrasive grinding of fused silica glass. *Appl. Surf. Sci.* 353: 764–773.

69 Kwon, T.-Y., Ramachandran, M., and Park, J.-G. (2013). Scratch formation and its mechanism in chemical mechanical planarization (CMP). *Friction* 1 (4): 279–305.

70 Kallingal, C., Duquette, D., and Murarka, S. (1998). An investigation of slurry chemistry used in chemical mechanical planarization of aluminum. *J. Electrochem. Soc.* 145 (6): 2074–2081.

71 Lee, E. and Radok, J.R.M. (1960). The contact problem for viscoelastic bodies. *J. Appl. Mech.* 27 (3): 438–444.

72 Oyen, M.L. (2005). Spherical indentation creep following ramp loading. *J. Mater. Res.* 20 (08): 2094–2100.

73 Lu, H., Fookes, B., Obeng, Y. et al. (2002). Quantitative analysis of physical and chemical changes in CMP polyurethane pad surfaces. *Mater. Charact.* 49 (1): 35–44.

74 Spierings, G.A.C.M. (1993). Wet chemical etching of silicate-glasses in hydrofluoric-acid based solutions. *J. Mater. Sci.* 28 (23): 6261–6273.

75 Kikyuama, H., Miki, N., Saka, K. et al. (1991). Principles of wet chemical-processing in ULSI microfabrication. *IEEE Trans. Semicond. Manuf.* 4 (1): 26–35.

76 Kikuyama, H., Waki, M., Kawanabe, I. et al. (1992). Etching rate and mechanism of doped oxide in buffered hydrogen-fluoride solution. *J. Electrochem. Soc.* 139 (8): 2239–2243.

77 Zhou, Y.Y., Funkenbusch, P.D., Quesnel, D.J. et al. (1994). Effect of etching and imaging mode on the measurement of subsurface damage in microground optical-glasses. *J. Am. Ceram. Soc.* 77 (12): 3277–3280.

78 Kline, W.E. and Fogler, H.S. (1981). Dissolution of silicate minerals by hydrofluoric-acid. *Ind. Eng. Chem. Fund.* 20 (2): 155–161.

79 Judge, J.S. (1971). A study of the dissolution of SiO_2 in acidic fluoride solutions. *J. Electrochem. Soc.* 118 (11): 1772–1775.

80 Proksche, H., Nagorsen, G., and Ross, D. (1992). The influence of NH_4F on the etch rates of undoped SiO_2 in buffered oxide etch. *J. Electrochem. Soc.* 139 (2): 521–524.

81 Broene, H.H. and Devries, T. (1947). The thermodynamics of aqueous hydrofluoric acid solutions. *J. Am. Chem. Soc.* 69 (7): 1644–1646.

82 Kline, W. and Fogler, H. (1981). Dissolution kinetics: catalysis by strong acids. *J. Colloid Interface Sci.* 82 (1): 93–102.

83 Iliescu, C. and Tay, F.E. ed. (2005). Wet etching of glass. CAS 2005 Proceedings 2005 International Semiconductor Conference. IEEE.

84 Witvrouw, A., Du Bois, B., De Moor, P. et al. ed. (2000). Comparison between wet HF etching and vapor HF etching for sacrificial oxide removal. Micromachining and Microfabrication. International Society for Optics and Photonics.

85 Kingery, W.D., Bowen, H.K., and Uhlmann, D.R. (1976). *Introduction to Ceramics*, 2e, 1032. New York: Wiley.

86 Suratwala, T.I., Miller, P.E., Ehrmann, P.R., and Steele, R.A. (2005). Polishing slurry induced surface haze on phosphate laser glasses. *J. Non-Cryst. Solids* 351 (24–26): 2091–2101.

87 Bude, J., Miller, P.E., Shen, N. et al. (2014). Silica laser damage mechanisms, precursors and their mitigation. In: *Proceedings of SPIE, Volume 9237, Laser-Induced Damage in Optical Materials* (ed. G.J. Exarhos, V.E. Gruzdev, J.A. Menapace, et al.).

88 Bude, J., Miller, P., Baxamusa, S. et al. (2014). High fluence laser damage precursors and their mitigation in fused silica. *Opt. Express* 22 (5): 5839–5851.

89 Mittal, K.L. (1994). *Particles on Surfaces: Detection: Adhesion, and Removal*. CRC Press.

90 Deegan, R.D., Bakajin, O., Dupont, T.F. et al. (1997). Capillary flow as the cause of ring stains from dried liquid drops. *Nature* 389 (6653): 827–829.

91 Deegan, R.D., Bakajin, O., Dupont, T.F. et al. (2000). Contact line deposits in an evaporating drop. *Phys. Rev. E* 62 (1): 756.

92 Bennett, J.M. (1994). Contamination on optical surfaces-concerns, prevention, detection, and removal. In: *Particles on Surfaces: Detection: Adhesion, and Removal*, 101. Springer-Verlag.

93 Kern, W. (1990). The evolution of silicon wafer cleaning technology. *J. Electrochem. Soc.* 137 (6): 1887–1892.

94 Kern, W. (1993). *Handbook of Semiconductor Wafer Cleaning Technology*, 111–196. Park Ridge, NJ: Noyes Publication.

95 Reinhardt, K. and Kern, W. (2008). *Handbook of Silicon Wafer Cleaning Technology*. William Andrew.

96 Suratwala, T., Feit, M., Steele, W. et al. (2014). Microscopic removal function and the relationship between slurry particle size distribution and workpiece roughness during pad polishing. *J. Am. Ceram. Soc.* 97 (1): 81–91.

97 Nevot, L. and Croce, P. (1980). Characterization of surfaces by grazing X-ray reflection--application to the study of polishing of some silicate glasses. *Rev. Phys. Appl.* 15 (3): 761–780.

98 Trogolo, J. and Rajan, K. (1994). Near surface modification of silica structure induced by chemical/mechanical polishing. *J. Mater. Sci.* 29 (17): 4554–4558.

99 Liao, D., Chen, X., Tang, C. et al. (2014). Characteristics of hydrolyzed layer and contamination on fused silica induced during polishing. *Ceram. Int.* 40 (3): 4479–4483.

100 Yokota, H., Sakata, H., Nishibori, M., and Kinosita, K. (1969). Ellipsometric study of polished glass surfaces. *Surf. Sci.* 16: 265–274.

101 Wallace, W., Wu, W., and Carpio, R. (1996). Chemical-mechanical polishing of SiO_2 thin films studied by X-ray reflectivity. *Thin Solid Films* 280 (1): 37–42.

102 Kozlowski, M.R., Carr, J., Hutcheon, I.D. et al. ed. (1998). Depth profiling of polishing-induced contamination on fused silica surfaces. Laser-Induced Damage in Optical Materials: 1997. International Society for Optics and Photonics.

103 Miller, P. (1988). *Trace Impurities in High Average Power Optical Materials.* Livermore, CA: Lawrence Livermore National Laboratory.

104 Bach, H. ed. (1983). *Analysis of Surface Layers. Optical Surface Technology.* International Society for Optics and Photonics.

105 Miller, P.E., Suratwala, T.I., Bude, J.D. et al. (2012). Methods for globally treating silica optics to reduce optical damage. Google Patents.

106 Wakabayashi, H. and Tomozawa, M. (1989). Diffusion of water into silica glass at low temperature. *J. Am. Ceram. Soc.* 72 (10): 1850–1855.

107 Moon, Y. (1999). Mechanical aspects of the material removal mechanism in chemical mechanical polishing (CMP).

108 Suratwala, T., Steele, W., Wong, L. et al. (2015). Chemistry and formation of the Beilby layer during polishing of fused silica glass. *J. Am. Ceram. Soc.* 98 (8): 2395–2402.

109 Davis, K. and Tomozawa, M. (1995). Water diffusion into silica glass: structural changes in silica glass and their effect on water solubility and diffusivity. *J. Non-Cryst. Solids* 185 (3): 203–220.

110 Moulson, A. and Roberts, J. (1961). Water in silica glass. *Trans. Faraday Soc.* 57: 1208–1216.

111 Doremus, R. (1995). Diffusion of water in silica glass. *J. Mater. Res.* 10 (9): 2379–2389.

112 Crank, J. (1975). *The Mathematics of Diffusion.* Oxford: Clarendon press.

113 Chen, Y., Zhang, L., Arsecularatne, J., and Montross, C. (2006). Polishing of polycrystalline diamond by the technique of dynamic friction, Part 1: Prediction of the interface temperature rise. *Int. J. Mach. Tools Manuf.* 46 (6): 580–587.

114 Jaeger, J.C. ed. (1942). Moving sources of heat and the temperature of sliding contacts. *Proc. R. Soc. New South Wales* 76: 203.

115 Tian, X. and Kennedy, F.E. (1994). Maximum and average flash temperatures in sliding contacts. *J. Tribol.* 116 (1): 167–174.

116 Suratwala, T., Feit, M.D., Steele, W.A., and Wong, L.L. (2014). Influence of temperature and material deposit on material removal uniformity during optical pad polishing. *J. Am. Ceram. Soc.* 97 (6): 1720–1727.

117 Cypryk, M. and Apeloig, Y. (2002). Mechanism of the acid-catalyzed Si-O bond cleavage in siloxanes and siloxanols. A theoretical study. *Organometallics* 21 (11): 2165–2175.

118 Eckert, H. (1992). Structural characterization of noncrystalline solids and glasses using solid state NMR. *Prog. Nucl. Magn. Reson. Spectrosc.* 24 (3): 159–293.

4
Surface Roughness

Because it influences optical-scatter losses and downstream laser modulation [1], the *surface roughness* of an optical-glass component is an important parameter for many optical systems (including lasers and telescopes). During optical polishing, a number of interactions between the workpiece, polishing slurry, and pad may influence the resulting workpiece roughness at different spatial scale lengths. The phenomena affecting large spatial scale length (>1 mm) on the workpiece (i.e. *surface figure*) are described in Chapter 2 (see Figure 2.1). By contrast, in this chapter, the phenomena and process parameters affecting short spatial scale lengths (<1 mm) (i.e., *surface roughness* from AFM2 scale roughness to μ-roughness) are described (see Figure 1.8). Fine-scale roughness, typically measured by atomic force microscopy (AFM), is referred to as AFM1 (≤5 μm) and AFM2 roughness (≤50 μm), as represented on the right of the plot. Roughness at this scale is influenced by parameters such as the single-particle removal function, Beilby layer properties, slurry particle size distribution (PSD), pad topography, pad mechanical properties, and slurry particle redeposition. The next higher spatial scale length (the micrometer to millimeter range, known as micro- or μ-roughness), is usually measured by white-light interferometry (Figure 1.8). μ-Roughness is affected not only by the smaller spatial-scale length phenomena given above, but also by factors governing slurry-interface interactions, such as the spatial distribution of slurry particles present at the interface. These parameters and phenomena are listed Figure 4.1, consisting of a schematic of the workpiece–lap interface at the spatial scale length of interest that influences the final polished *surface roughness*. Figure 4.1 provides a valuable outline for this chapter.

4.1 Single-Particle Removal Function

The amount of material removed by each individual particle from the workpiece as a function of load (referred to as the removal function) serves as a

Materials Science and Technology of Optical Fabrication, First Edition. Tayyab I. Suratwala.

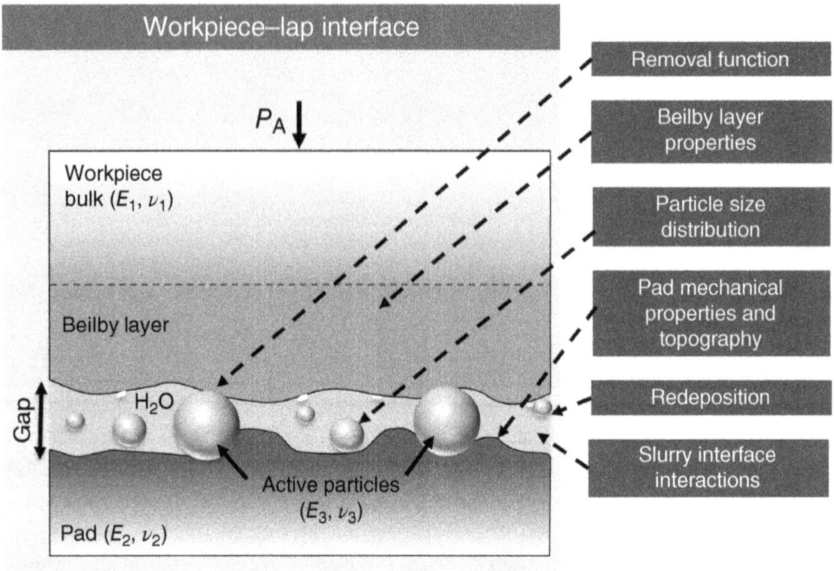

Figure 4.1 Schematic of workpiece–lap interface and key parameters affecting polished surface microroughness. Source: Suratwala et al. 2016 [2]. Reproduced with permission from John Wiley & Sons.

critical parameter that strongly influences both *surface roughness* and the global *material removal rate* (see Chapter 5). Key removal mechanisms include brittle fracture, chemical or physical dissolution, plastic deformation, and chemical reaction (see Figure 1.9). Except for the case of chemical dissolution, the amount removed is governed by the load on the acting particle. The load dependence on removal is illustrated in Figure 4.2, which shows the load and corresponding depth of interaction on fused-silica glass on a log–log plot for a range of loads and depths spanning many orders of magnitude [3]. The shaded regions corresponds with different material responses, ranging from elastic (in which no permanent mechanical modification occurs to the workpiece at very low load), to plastic (in which permanent plastic deformation or removal occurs at intermediate loads), to fracture (in which various types of cracks may form at high loads).

In the context of Figure 4.2, chemical removal will dominate in the elastic-load regime where the particle can make physical contact and react, but the loads are insufficient to mechanically alter the workpiece. The most widely accepted chemical removal mechanism for ceria particles on silicate glasses is that proposed by Cook [4], who found that chemical material removal occurs through condensation and subsequent hydrolysis reactions (see Reactions (3.61) and (3.62)). The surface of the cerium-oxide particle is cerium hydroxide,

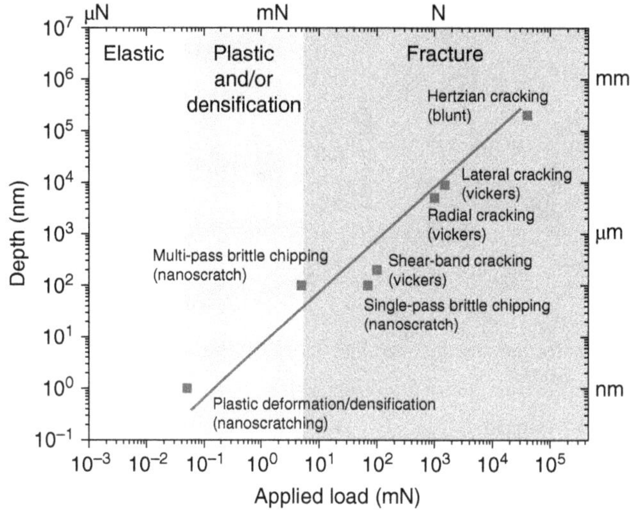

Figure 4.2 Relationship between applied load and resulting depth in fused silica during indentation and nanoscratching over a large range of loads. Source: Shen et al. 2016 [3]. Reproduced with permission from John Wiley & Sons.

which condenses with the glass (silanol) surface to form a Ce—O—Si bond. The strength of this new bond is greater than that of the Si—O—Si bond (i.e., the glass). Hence, hydrolysis of the Si—O—Si occurs, removing a "molecule" of silica away from the workpiece. Restated, polishing occurs as ceria particles repeatedly tear away the silica network at the molecular (i.e., Angstrom) scale.

In the intermediate load regime, removal may occur by plastic flow [3, 5–9]. Some have reported that polishing by chemical mechanical planarization (CMP) results from a chemically reactive ductile layer [10]. Interestingly, plastic flow also occurs at the nanometer level, even on very brittle materials, such as glass, in this load regime [3, 11–13]. This is illustrated by a simple experiment to reveal the impact of single-pass ceria particles in a slurry on the surface of a workpiece, as shown in Figure 4.3 [14]. Starting with a very smooth polished fused-silica workpiece (2.3 Å RMS), ceria slurry particles (~100 nm in size) were either statically loaded and dried on the surface (Figure 4.3A) or slid in a single pass along the surface (Figure 4.3B) at a common optical-polishing pressure (0.3 psi). The plot on the right of each surface image is a height lineout through the center of the image. In the static ceria-particle loading case, pits were observed, typically 10–50 nm wide and ~1 nm deep. The lateral size of observed pits indicates the size of the contact zone between the ceria particle and fused-silica surface. In the sliding-ceria-particle case, linear tracks were observed in the direction of the ceria sliding, with a typical depth of ~1 nm. These results define some important aspects of the material-removal function

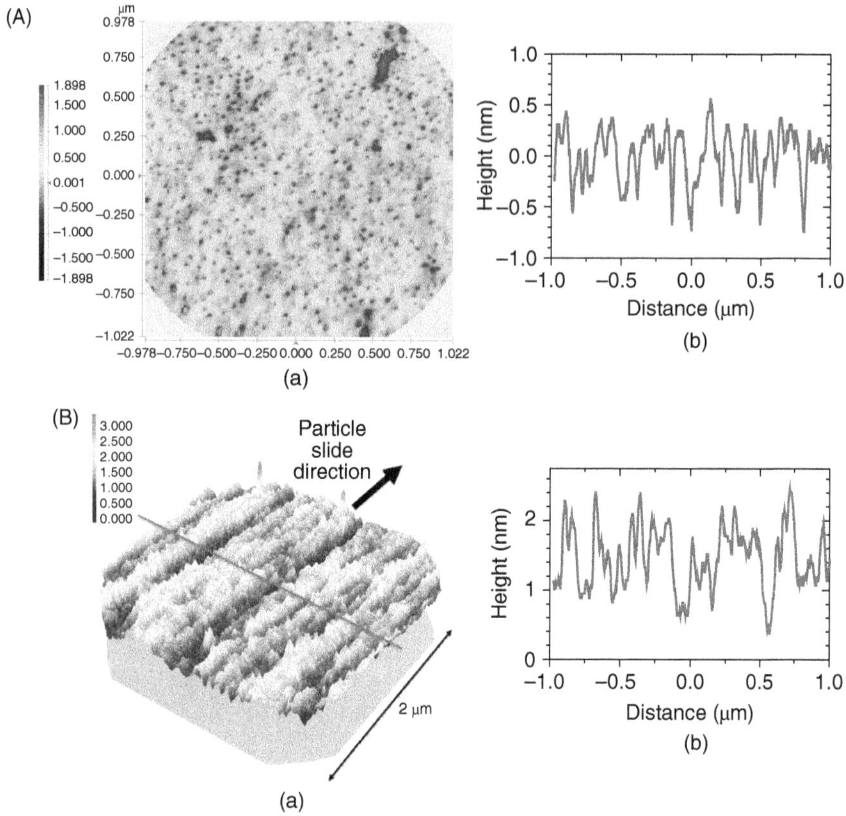

Figure 4.3 (A,a) AFM image ($2 \times 2\,\mu m^2$) of fused-silica surface after ceria slurry was statically loaded under pad at 0.3 psi, allowed to dry, and removed using nitric-peroxide etch; (A,b) lineout of surface. (B,a) AFM image ($2 \times 2\,\mu m^2$) of fused-silica surface after single-pass ceria-slurry slide ($10\ cm\ s^{-1}$ at 0.3 psi); (B,b) lineout of surface perpendicular to particle sliding direction. Source: Suratwala et al. 2014 [14]. Reproduced with permission from John Wiley & Sons.

of a single ceria particle during polishing. The depth of removal of ~1 nm is ~6 Si—O—Si units, suggesting that removal or modification can occur over depths beyond a single molecular level, through either nanoplastic flow deformation or densification. Hence, though it is known that ceria polishing can occur by chemical removal, at higher loads per particle, its removal can also occur by nanoplastic flow.

During optical-glass polishing, the loads experienced by slurry particles on a glass workpiece are typically quite small, ranging from 0.1 to 200 μN [4, 14, 15]. To understand and predict the *material removal rate* and resulting *surface roughness* obtained during polishing, the chemical and plastic removal

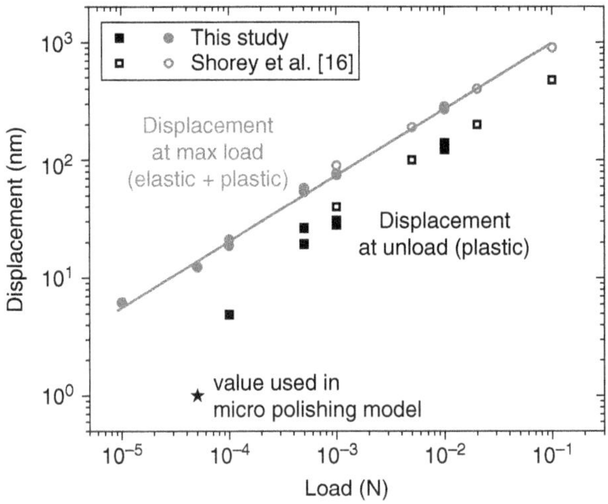

Figure 4.4 Measured displacement at maximum load, representing elastic and plastic deformation, and at unload, representing permanent plastic deformation on fused-silica surface after nanoindentation at various loads. Source: Suratwala et al. 2014 [14]. Reproduced with permission from John Wiley & Sons.

functions due to individual particles under relevant load conditions must be quantified.

The degree of plastic deformation can be quantified using mechanical nanoindentation. Figure 4.4 summarizes the static nanoindentation elastic and permanent deformation on fused-silica glass versus applied load [14, 16]. The indenter displacement into the fused-silica workpiece at maximum load describes the combined elastic, plastic, and densification deformation, while the indenter displacement after unload describes plastic and densification deformation only. The displacement at maximum load follows a power-law dependence, as does the displacement at unload, except at very low loads. The smallest permanent deformation displacement measurable was ~5 nm at an applied load of 10^{-4} N (or 100 μN). Hence, the critical load needed for plastic deformation could not be determined by this method. Evaluation at lower loads is needed to be more representative for polishing.

Later, a novel method was used to quantify the nanoplastic removal function using nanoscratching at more relevant loads [3, 17]. Shen et al. [3] used this technique to quantify the removal amount from the surface of various glasses as a function of load (see Figure 4.5). Using a "stiff" diamond AFM tip (150-nm radius) on an AFM, linear nanoscratches were formed on the surface of polished glasses in air and aqueous environments. AFM images of the nanoscratches generated in air and water at various loads from 20 to 170 μN on the polished fused-silica, borosilicate-glass, and phosphate-glass surfaces are shown in Figure 4.5a,b. The diamond tip slid top to bottom in the sample

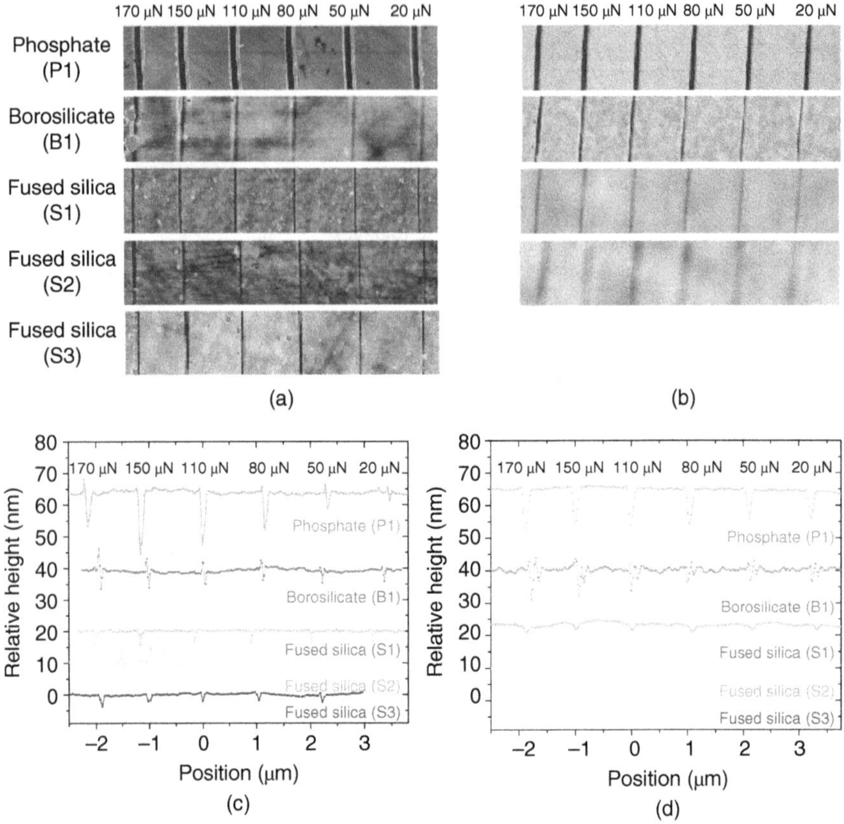

Figure 4.5 (a) AFM images of nanoscratching in air on various samples at various loads (170, 150, 110, 80, 50, and 20 μN). Scratch direction is top to bottom. Samples (from top): phosphate glass (P1), borosilicate glass (B1), fused silica (colloidal-silica slurry, 20 rpm) (S1), fused silica (ceria slurry, 10 rpm) (S2), and fused silica (ceria slurry, 40 rpm) (S3). (b) Images of nanoscratches performed in liquid water on the same samples at the same loads. (c) Average 1D lineouts of all nanoscratched samples shown in (a). (d) Average 1D lineouts of the nanoscratches shown in (b). Source: Shen et al. 2016 [3]. Reproduced with permission from John Wiley & Sons.

images. Linear tracks several hundred nanometers wide were clearly visible for each of the nanoscratches, regardless of the glass type, treatment, or load-range explored. No fracturing was observed. The average nanoscratch cross section for each load and workpiece material is shown in Figure 4.5c,d.

The nanoscratch depths on the fused-silica surfaces showed little or no load dependence over the load range explored here. The deformation depth for borosilicate glass depended weakly on load similar to fused silica, while the phosphate glass showed a systematic increase in deformation depth with increasing load. However, at much higher loads (700–6000 μN, as measured by Pergande et al. [18]), the plastic deformation depth increases on fused silica with load in a more expected manner, at a rate of 40 nm for every 1000 μN. The nanoscratches in water, which is more representative of the polishing environment, appear wider and shallower for fused-silica samples (Figure 4.5d). This effect was less significant for borosilicate and phosphate glass. The nanoscratch load dependence, however, was similar in the air and water environments for all three types of glass.

The trend in the removal depth with glass material, wherein fused silica < borosilicate < phosphate, is qualitatively consistent with the material properties of the glass that govern plasticity – that is, the inverse of the global hardness of the glass for a given radius of loading tip curvature [19]. The hardness (H_1) values for the glasses fused silica, borosilicate, and phosphate are 6.0, 5.1, 3.4 GPa [20], respectively. Their inverse compares well with the trend in removal depth observed.

Using the nanoscratch method to quantitatively determine the removed amount by a single particle is complicated by the fact that plastic flow and densification may both occur (see Section 7.2.5). It was determined using a sub-T_g thermal annealing technique that densification contributed significantly to deformation during nanoscratching [3, 21]. Evidence of significant relaxation of the nanoscratch is present after annealing, suggesting a significant fraction of the deformation on the first nanoscratch pass is due to densification on both the fused silica and the borosilicate glass. These results are consistent with previous reports of densification in fused silica during nanoindentation [22, 23].

Hence, to determine the removed amount due to plastic flow alone, a technique of repeated nanoscratching on the same location was used, where the delta removal amount with each pass represents the removal amount of the single particle [3]. Figure 4.6a shows representative cross-sectional depth profiles of repeated nanoscratching at the same location on fused silica, borosilicate glass, and phosphate glass, in air and water. The number of passes (1, 5, or 10) shown at the top of the figure represents the number of times the same diamond tip was used to nanoscratch at the same location. As expected, the volume removed during nanoscratching increased with the number of passes. During repeated nanoscratching, the diamond tip may not always align precisely with the previous scratch, which may lead to removal of material from the edge of the existing scratch and potential underestimation of the scratch-depth removed per pass. Therefore, rather than characterizing nanoscratches by their measured depth directly, an effective removal depth per pass was determined as a function of the number of nanoscratching passes, summarized in Figure 4.6b. The effective removal depth per pass (d_p) is defined as:

$$d_p = \frac{A_p}{n_{ps} w_p}, \tag{4.1}$$

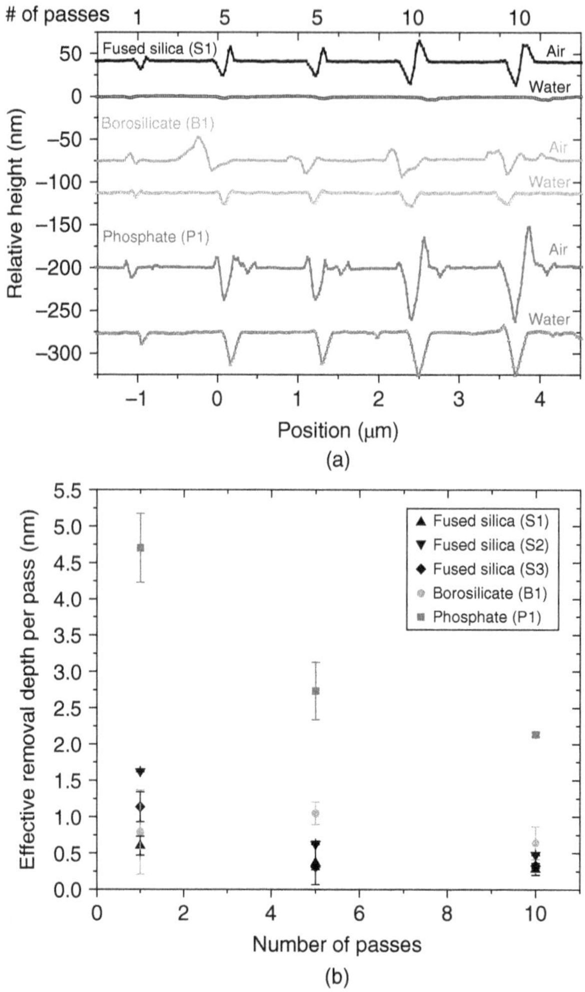

Figure 4.6 (a) Average lineouts of nanoscratches on fused silica (S1), borosilicate glass (B1), and phosphate glass (P1) at 110 μN for 1, 5, and 10 passes performed in air and in liquid water; (b) corresponding effective nanoscratch removal depth per pass (d_p) versus number of passes of nanoscratching for three glass types. Source: Shen et al. 2016 [3]. Reproduced with permission from John Wiley & Sons.

where A_p is the measured cross-sectional area of the plastic nanoscratch, n_{ps} is the number of passes, and w_p is the measured width of the resulting nanoscratch. The effective depth removal per pass decreases with the number of passes, stabilizing after ~10 passes (see Figure 4.6b). In this study, the effective removal depth per pass, determined in water, after 10 passes was found to best represent the repeated nanoscratching that occurs during optical polishing.

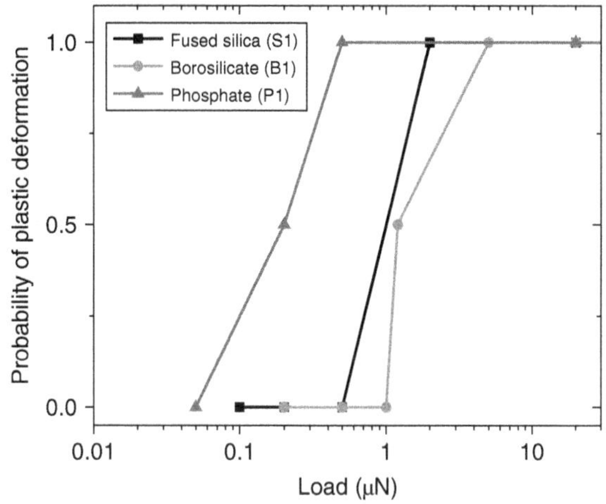

Figure 4.7 Probability of observing plastic deformation after nanoscratching at various applied load for fused silica (S1), borosilicate glass (B1), and phosphate glass (P1). The transition represents the critical load for permanent deformation. Source: Shen et al. 2016 [3]. Reproduced with permission from John Wiley & Sons.

The determined values at a load of 110 μN are 0.3–0.55 nm/pass for fused silica, 0.85 nm/pass for borosilicate glass, and 2.4 nm/pass for phosphate glass.

The transition between chemical and nanoplastic removal was also determined by nanoscratching at even lower loads (0.05–20 μN) (see Figure 4.7). Based on this, the critical load corresponding to the onset of any detectable permanent deformation was determined to be ~1.0 μN for fused silica glass, ~1.2 μN for the borosilicate glass, and ~0.2 μN for the phosphate glass. The variation in critical load is likely related to the hardness of the glass [3]. The implication of this critical load is that during polishing under identical conditions (i.e., same load/particle distribution) will lead to a greater fraction of particles removing material by plastic removal for the phosphate glass compared with fused silica and borosilicate glass due to the lower P_c value for the phosphate glass.

The information above can be tied together to describe quantitatively the removal function of a single polishing particle as a function of load for a given workpiece material (see Figure 4.8) [2, 3]. The shaded areas in the figure represent the effective depth of material removed. The removal function has two major regimes: chemical at low loads and nanoplastic at intermediate loads. The critical load for chemical reaction is approximated based on van der Waals interaction forces of ~10^{-9} to 10^{-8} N [24]. Also, from a previous study, the chemical removal depth for fused silica was inferred as 0.04 nm [14].

The removal function not only illuminates polishing mechanisms, it also aids in predicting the *surface roughness* of the polished workpiece. Combining the removal function with a determined load-per-particle-distribution in the interface, the removal for every particle is known, and hence roughness can be determined from the combined removal of all particles on the workpiece surface. This methodology is described in greater detail in Section 4.7.

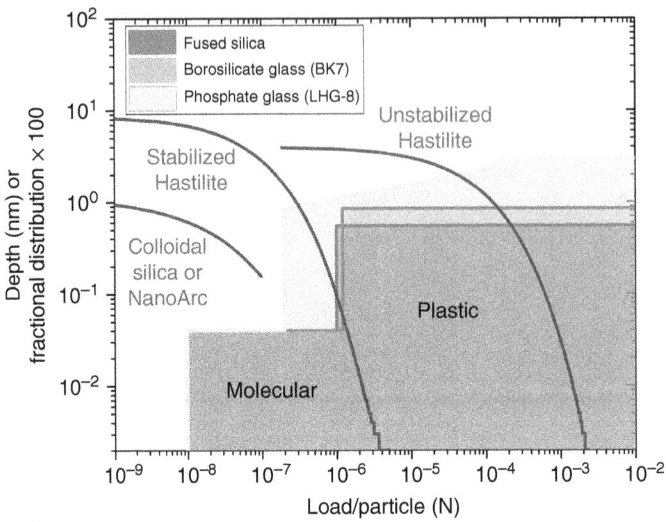

Figure 4.8 Measured removal function (shaded areas) as a function of load for individual slurry particles and calculated load per particle distribution (lines) for three slurries, using the EHMG model (described in Section 4.7). Source: Suratwala et al. 2016 [2]. Reproduced with permission from John Wiley & Sons.

4.2 Beilby Layer Properties

As discussed in Section 3.3, the chemical and structural properties of the Beilby surface layer formed during polishing are different relative to the bulk workpiece. Hence, these changes may affect the removal function of a single particle, and subsequently the final *surface roughness* on the workpiece.

Measuring the changes to the mechanical properties or chemical reactivity properties in the Beilby layer directly is more challenging, owing to its shallow depth (tens of nanometers) and to the exponentially changing properties within that surface layer. Even nanoindentation at ultralow loads (say, 100 μN) results in elastic penetration well exceeding the depth of the Beilby layer. Hence, to date, no convincing changes to the chemical reactivity of the Beilby layer glass surface have been reported. It is postulated that hydration of the surface layer increases the reactivity of the surface layer to Ce, likely increasing the effective chemical removal by cerium oxide (see Reactions (3.61)–(3.63)). Fu et al. [12] formulated a model based on where the polishing particle removes material via abrasion of the modified hydrated surface layer.

The effect of the Beilby layer on the single particle removal function can be explored, however, using nanoscratching, as discussed in Section 4.1. In Figure 4.5, three fused-silica surfaces – S1 using colloidal-silica slurry at 20 rpm, S2 using ceria slurry at 10 rpm, and S3 using ceria slurry at

40 rpm – showed noticeably different permanent deformation depths, ranging from 2 to 6 nm. The fused-silica workpiece (sample S3), polished at the higher rotation rate and hence higher *material removal rate*, showed less deformation at all loads. Considering that higher *material removal rate* in polished workpieces using ceria slurry results in greater Ce penetration in the Beilby layer, it is postulated that this difference in Beilby-layer character is responsible for the change in the removal function.

4.3 Slurry PSD

Another critical parameter influencing the resulting polished workpiece *surface roughness* is the PSD of polishing slurry (see Figure 4.1). Some studies have examined the effect of slurry PSD on scratches either by spiking the slurry with larger particles [25, 26] or by changing the overall PSD [27]. However, these studies focused on instances where the load per particle exceeded the initiation load for fracture (typically >0.1 N), leading to brittle fracture. Cumbo et al. [28] examined various aspects of the PSD, including the evolution of slurry PSD during polishing, leading to the break up of agglomerates. This work underscores the importance of slurry chemistry, particularly particle colloidal stability via particle-surface charge and slurry pH. Cumbo emphasized the importance of achieving a well-dispersed slurry to obtain the smoothest glass surfaces possible. The large presence of agglomerates in certain ceria slurries is illustrated in Figure 4.9, showing the reduction in agglomerates as the slurry is ultrasonically agitated.

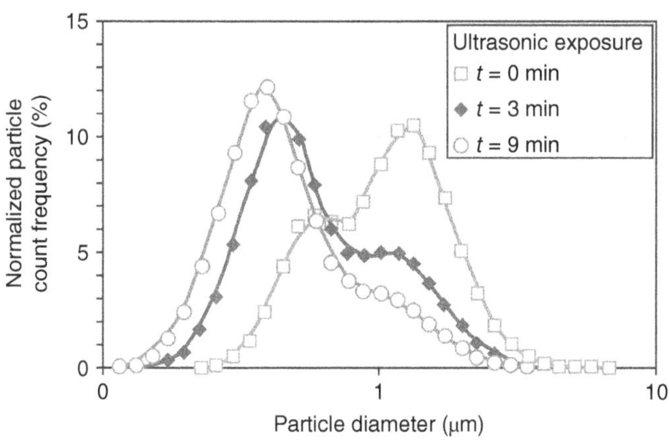

Figure 4.9 Evidence for the presence of slurry particle agglomerates: change in slurry-particle size distribution of ceria slurry via ultrasonic agitation. Source: Cumbo et al. 1995 [28]. Reproduced with permission from The Optical Society.

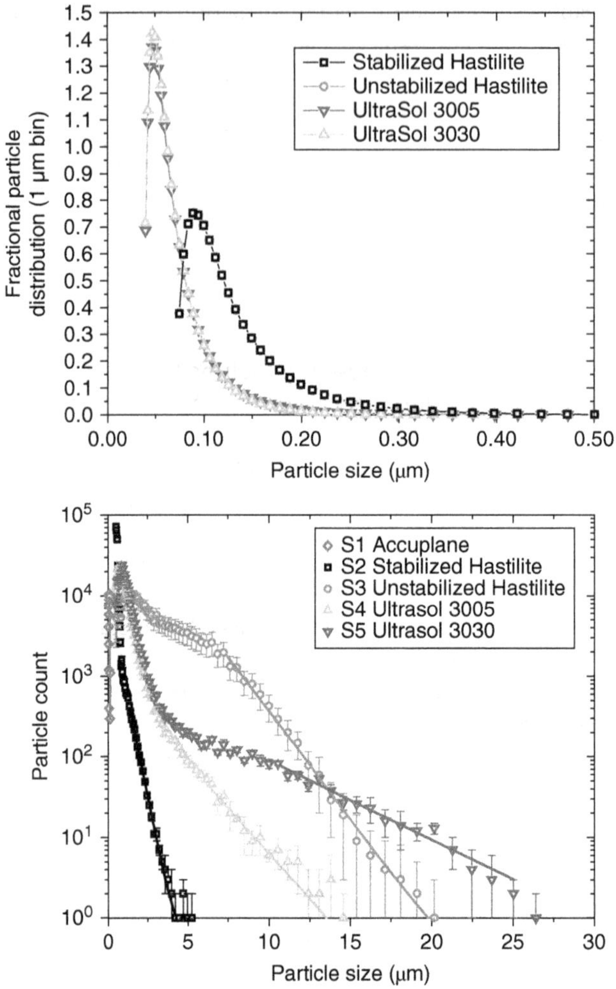

Figure 4.10 (a) Measured PSD on a linear scale using static light scattering for various ceria polishing slurries. (b) Measured PSD for the same ceria slurries on a log scale, using single-particle optical sensing. Source: Suratwala et al. 2014 [14]. Reproduced with permission from John Wiley & Sons.

A later study showed that final workpiece *surface roughness* is quantitatively correlated to slurry PSD (specifically, the logarithmic slope for the largest particles at the exponential tail end of the PSD) [14]. Figure 4.10a shows the fractional number PSD, $f(r)$, of some ceria slurries measured by static light scattering. Stabilized and unstabilized Hastilite ceria slurries have very similar distributions, with a mean size of ~100 nm; likewise, Ultra-Sol 3005 and Ultra-Sol 3030 have very similar distributions, with a mean size of 50 nm. However, the tail ends of the slurry PSDs, which involve a small fraction of total particles, are found to be very different. Figure 4.10b shows the fractional number PSD, $f(r)$, of the slurries as measured by single-particle optical sensing on a semi-log plot [14]. Though both sets of slurries have the same mean particle size, the tail of the distributions differ and can be described by a single exponential size dependence in the form:

$$f(r) = A_o e^{\frac{-2r}{d_{PSD}}}, \tag{4.2}$$

where r is the particle radius, A_o is a pre-exponential constant, and d_{PSD} is the inverse exponential slope in the PSD. Note the lines in Figure 4.10b.

Figure 4.11 shows the surface morphology and *surface roughness* (as measured by AFM) of fused-silica workpieces polished with the slurries described in Figure 4.10 [14]. Surface roughness varied significantly. The lowest roughness was achieved with the slurry with the fastest decaying tail of the PSD, and the highest with the slurry with the slowest decaying tail. This correlation is quantitatively illustrated in Figure 4.12 which shows a cross-plot between the exponential inverse slope of the volumetric PSD (d_{PSD}) and RMS roughness (δ),

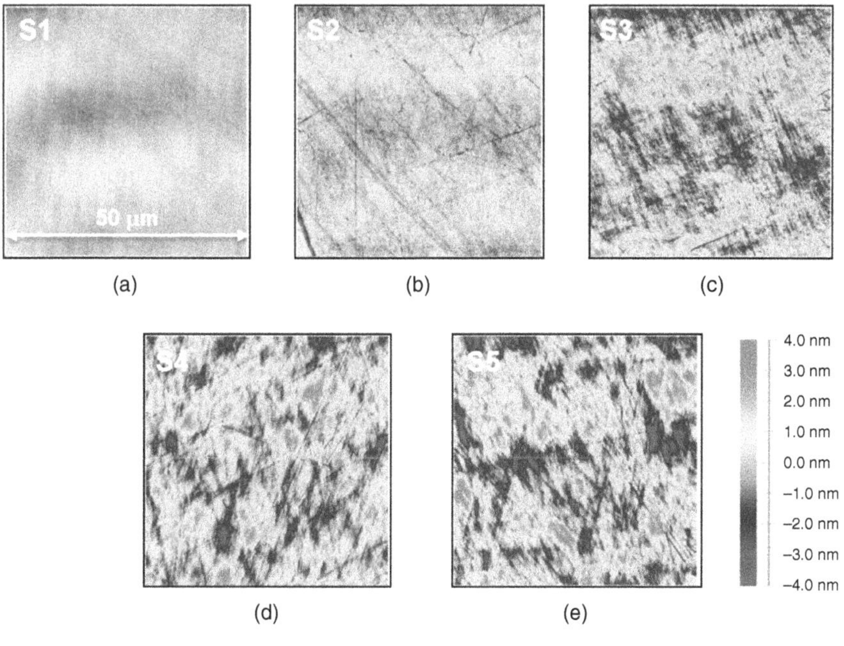

Figure 4.11 (a–e) AFM images (50 × 50 μm^2) of fused-silica surfaces after polishing with various slurries (S1–S5). All surfaces are plotted on a linear vertical scale from −4.0 to 4.0 nm. Source: Suratwala et al. 2014 [14]. Reproduced with permission from John Wiley & Sons.

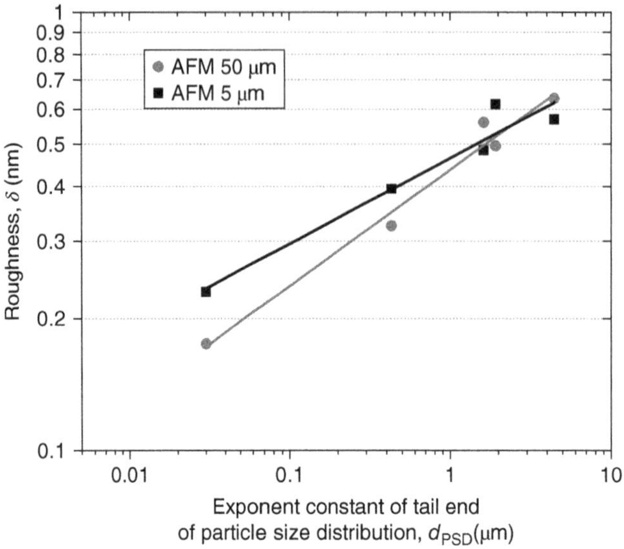

Figure 4.12 RMS roughness (δ) of fused-silica surface over various spatial-scale lengths as a function of the exponential constant (d_{PSD}) of the tail end of slurry PSD. Source: Suratwala et al. 2014 [14]. Reproduced with permission from John Wiley & Sons.

following the form,

$$d_{PSD} = d_a e^{\frac{-\delta}{\delta_o}}, \tag{4.3}$$

where d_a and δ_o are constants with values of 0.008 μm and 0.2 nm, respectively. These results imply that the largest particles (whether larger particles or agglomerates) dominate in modifying the morphology of a surface, while average sized particles (the majority of the particles in the slurry) have little or no influence on removal and resulting *surface roughness*. A similar set of concepts was proposed when investigating overall *material removal rate* during CMP wafer polishing [29, 30] and grinding processes on glass [25].

4.4 Pad Mechanical Properties and Topography

Another important parameter that affects the workpiece *surface roughness* and *material removal rate* is lap properties – the mechanical properties of the lap and its surface topography (see Figure 4.1). The basic factor relating these pad properties to *surface roughness* is largely mechanical, where an ensemble of particles at the interface under load from the workpiece will distribute the load over each contacting particle. The pad properties will influence the final load-per-particle distribution. Changes in the load-per-particle distribution will alter the amount that individual particles remove from the workpiece per the removal function described in Section 4.1, and hence alter the final workpiece *surface roughness*. A more quantitative and detailed description of this mechanism is described in Section 4.7.

At larger spatial-scale lengths, the hardness and modulus of the lap affect the conformality and the spatial-scale length of material removal, as illustrated in Figure 2.20. This affects the active contact area between the lap and workpiece. A lap with lower hardness and modulus pad yields an increase in contact area between the workpiece and lap [10, 14]. This results in the loading of more slurry particles, which reduces overall load-per-particle distribution, thus reducing *surface roughness*.

These relatively simple mechanical-interface interactions are complicated by factors such as

- The time-dependent mechanical response of the lap owing to viscoelastic and viscoplastic properties [15, 25]
- Time-dependent changes in lap properties because of chemical interactions or swelling of the lap [31]
- Pad glazing, which is the buildup of workpiece reaction products or cold flow smoothing on the pad surface
- Pad treatments, such as diamond conditioning, that affect pad topography or remove pad glazing [2, 32]
- Variations in slurry replenishment owing to pad structure, e.g., pores on the pads, fibrous surface structures, groove patterns [33–35]
- Variation in lap hardness and modulus due to pad structure or various pad layers [34]

Numerous studies specifically for CMP have evaluated these phenomena to maximize *material removal rate*, especially with respect to pad design. Some of these factors also influence *surface roughness*.

Pitch laps, which are commonly used for conventional optical polishing, exhibit viscoplastic behavior. Viscoplastic materials have a time-dependent response to deformation, and the deformation is permanent. This kind of deformation is a useful tool for adjusting the global shape of the lap and allowing opticians to create very controlled *surface figures*. At microscopic scale lengths, viscoplastic behavior also affects the load response on individual particles, as seen in Figure 4.13. After mechanical loading, the individual particles penetrate permanently into pitch, an embedding referred to as "charging." [4, 35] Larger particles penetrate more than smaller ones, resulting in a net narrowing of the load-per-particle distribution. This, in turn, narrows the removal amounts of individual particles and results in lower *surface roughness*. Historically, much lower roughness surfaces were achieved using pitch polishing, as compared with pad polishing, likely due to viscoplastic response. The rate at which particles penetrate into a viscoplastic lap depends

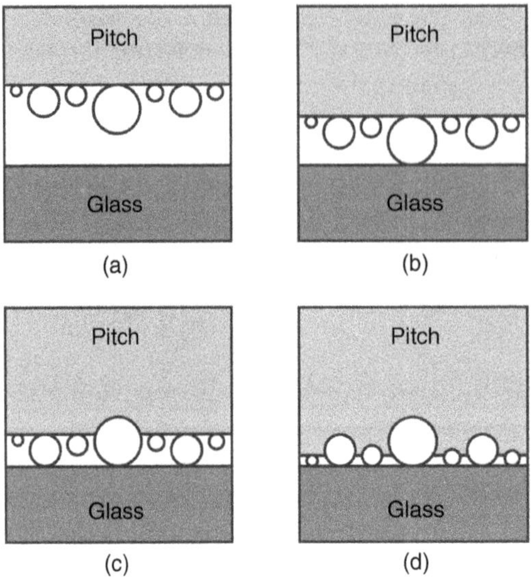

Figure 4.13 (a–d) Schematic illustrating the penetration of slurry particles into a viscoplastic-pitch lap at various stages of loading. Source: Adapted from Cook 1990 [4] and Kumanin 1967 [36].

on the loading time and pitch viscosity, given as [35]

$$\frac{h_p}{d} \propto \left(\frac{t}{\eta_2}\right)^{2/3}, \tag{4.4}$$

where h_p is amount of penetration, d is the slurry particle size, t is the time of penetration, and η_2 is the lap viscosity.

Polishing pads, by contrast, have more elastic and viscoelastic behavior. The degree of particle penetration depends on the pad's elastic modulus and the fraction that is viscoelastic. An analysis of the viscoelastic behavior of polyurethane polishing pads with respect to rogue particles is found in Section 3.1.3.

A large assortment of pads is available, many of them developed for the CMP industry. Pads can be divided into four categories, based on general properties and microstructure [34].

- Type 1 pads are felted fibers held together with a polymeric binder, leading to a continuous, channel-type microstructure. These pads tend to have medium modulus, high surface texture/topography, and a medium slurry-holding capacity. Such pads are generally suited for stock removal and rough polishing due to their high *material removal rates*.
- Type II pads are porometric pads with vertically oriented open pores. They tend to have low modulus, high surface topography, and high slurry-holding capacity and are designed to minimize scratching and achieve global planarization.

- Type III pads are microporous polymer sheets with a closed-cell-porosity microstructure. These pads tend have to high modulus, medium surface topography, and low slurry-holding capacity.
- Type IV pads are nonporous polymer sheets, generally with very high modulus, low surface texture, and very low slurry-holding capacity.

Types III and IV are generally adopted to remove only small amounts of surface irregularities, usually in the final polishing step. Because they have higher modulus and lower wear rates, they provide the greatest *surface figure* control.

Based on their mechanical mechanisms and lower modulus, Type II pads will tend to provide lower *surface roughness*. Since many other factors control roughness, however, it is possible to achieve low roughness with other types. For example, pad surface texture or topography has been shown to influence the final *surface roughness* of a workpiece. Consider a Type III polyurethane-foam pad (MHN), with a surface topography modified by various treatments and then quantitatively assessed by laser confocal microscopy (see Figure 4.14) [2]. The new MHN pad has a significant number of tall (hundreds of micrometers) asperities, which are removed with polishing time or diamond conditioning treatment. (Note the as-fabricated pad has pores breaking the surface, which are easily observable and remain after the surface treatments.) The treated pads are then used for polishing fused-silica workpieces under identical polishing conditions (i.e., same kinematics, applied load, and ceria slurry) [2]. The results show that a more uniform pad topography correlates with an increase in the observed removal rate on the polished workpiece

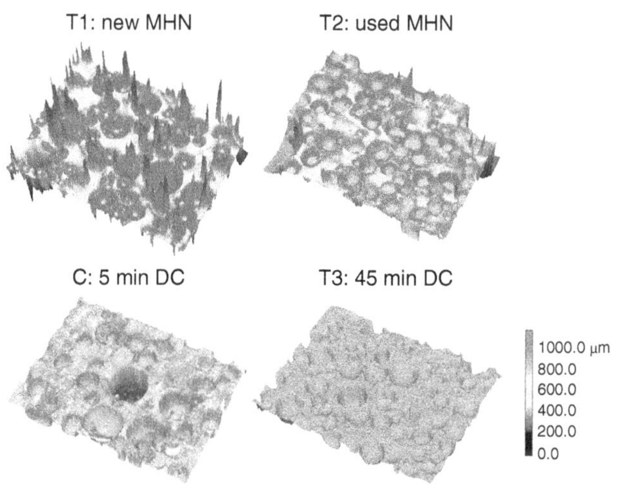

Figure 4.14 Pad-surface topographies for MHN polishing pad treated under different conditions, as measured by laser confocal microscopy over a 1400 μm × 1000 μm field of view. Source: Suratwala et al. 2016 [2]. Reproduced with permission from John Wiley & Sons.

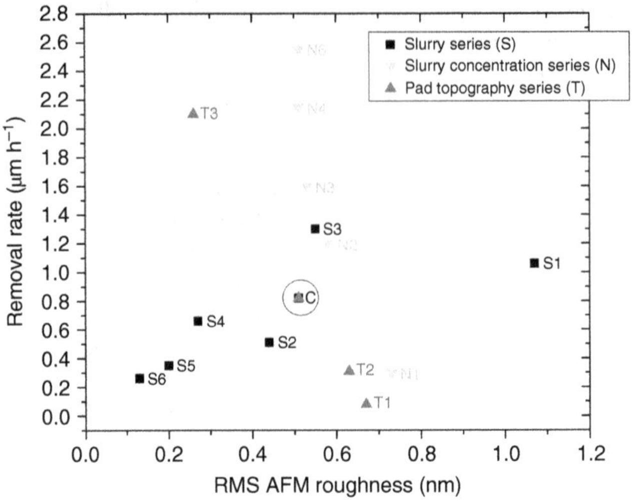

Figure 4.15 Cross-plot of measured removal rate and AFM roughness of fused-silica glass samples polished under different conditions. Source: Suratwala et al. 2016 [2]. Reproduced with permission from John Wiley & Sons.

and a reduction in *surface roughness*. This is illustrated in Figure 4.15, which shows removal rate and AFM roughness as measured on a number of polished samples, including the T series, which represents the pad treatment samples shown in Figure 4.14.

4.5 Slurry Interface Interactions

The interaction of slurry particles at the interface and how they assemble or agglomerate as slurry islands on the lap may influence resulting workpiece *surface roughness* at larger spatial-scale lengths in the μ-roughness band. This section discusses the experimental evidence for this phenomenon and the critical factors affecting agglomeration.

4.5.1 Slurry Islands and μ-roughness

Figure 4.16 compares the measured μ-roughness and AFM2 roughness for fused-silica and phosphate glass workpieces polished under a variety of conditions [14, 37]. An increase in AFM2 roughness generally corresponds to an increase in μ-roughness, though there is a large spread in the data for phosphate glass and fused silica. Earlier it was shown that specific process parameters, such as the single-particle-removal function, slurry PSD, pad mechanical properties, and pad topography influence AFM2 *surface roughness.*

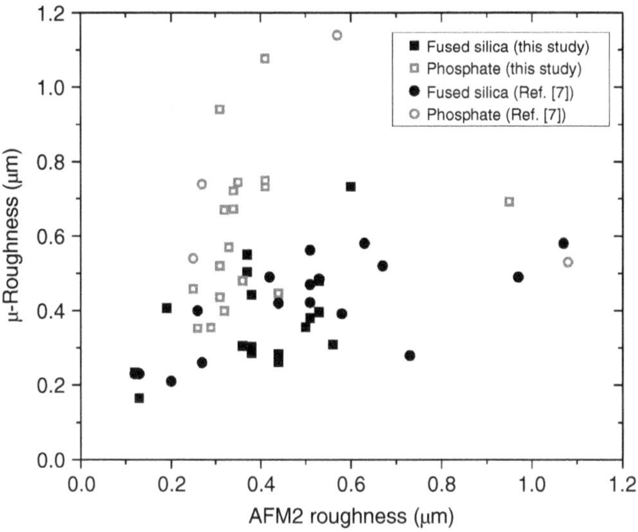

Figure 4.16 Cross-correlation between AFM2 roughness and μ-roughness for a variety of polishing conditions for fused silica and phosphate glass. These polishing runs used the same applied load and kinematics (lap rotation rate). Source: Suratwala et al. 2017 [37]. Reproduced with permission from John Wiley & Sons.

The weak relationship between the two roughness-scale lengths suggests that such parameters also influence μ-roughness. However, the large spread in this correlation suggests that another key phenomenon, operable at longer scale lengths that are not observable at AFM2-scale lengths, also affects μ-roughness.

A recent polishing study showed that the phenomena affecting the μ-roughness band may be isolated and identified [37]. A comprehensive set of polishing experiments was performed using various polishing slurries at different pHs on fused-silica and phosphate-glass workpieces. Figure 4.17 shows the resulting removal rate, μ-roughness, and AFM2 roughness. The AFM2 roughness was largely invariant with pH, suggesting the removal function of a single particle is unchanged with pH. The μ-roughness, however, changed significantly for the phosphate-glass workpieces, increasing linearly with pH for the three slurry types. This trend can be visualized from the AFM2- and μ-roughness images for phosphate glass shown on the same *z*-scale in Figure 4.18. The fused-silica glass, unlike the phosphate, did not show a systematic change in μ-roughness with pH or the various slurries. The exception was that stabilized Hastilite ceria on fused silica shows a peak in μ-roughness near pH 7–8, which is also the isoelectric point (IEP) of conventional ceria slurries; this is consistent with previous reports of ceria agglomeration near its IEP [4, 28, 38].

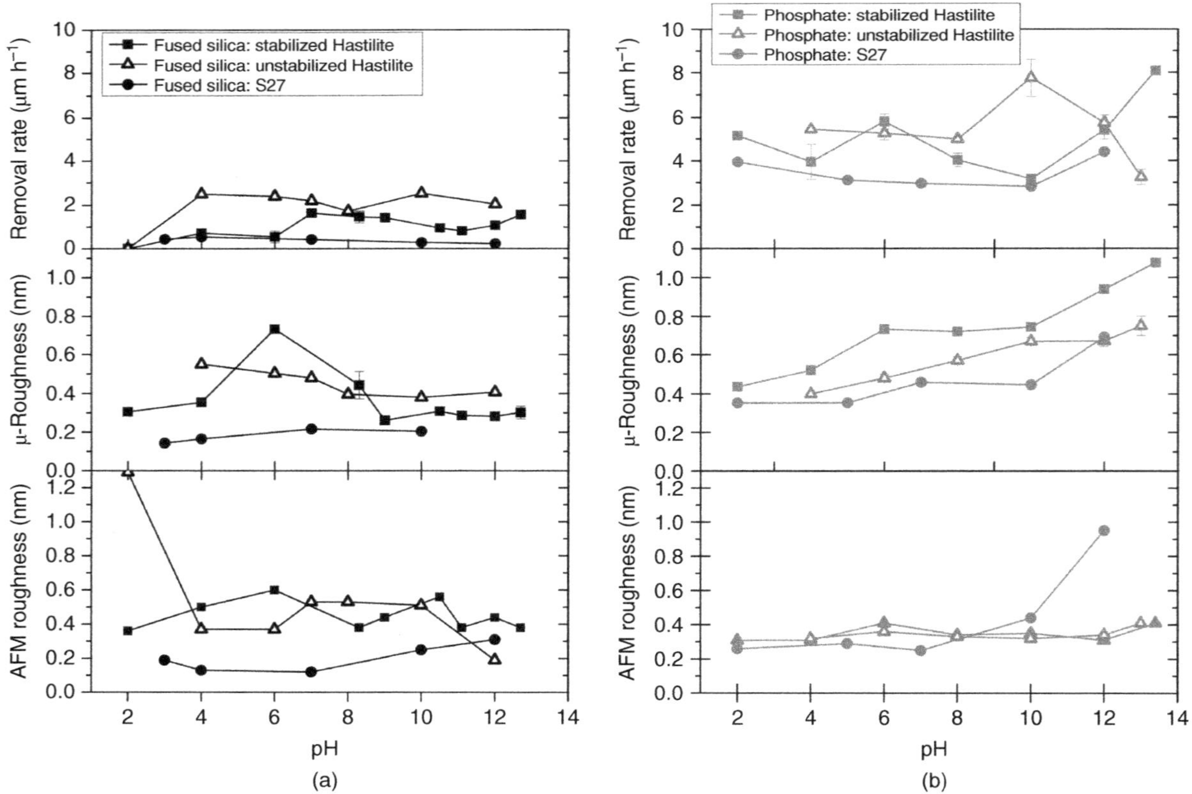

Figure 4.17 Impact of slurry pH on removal rate, μ-roughness and AFM roughness on (a) fused-silica glass and (b) phosphate glass using a variety of polishing slurries. Source: Suratwala et al. 2017 [37]. Reproduced with permission from John Wiley & Sons.

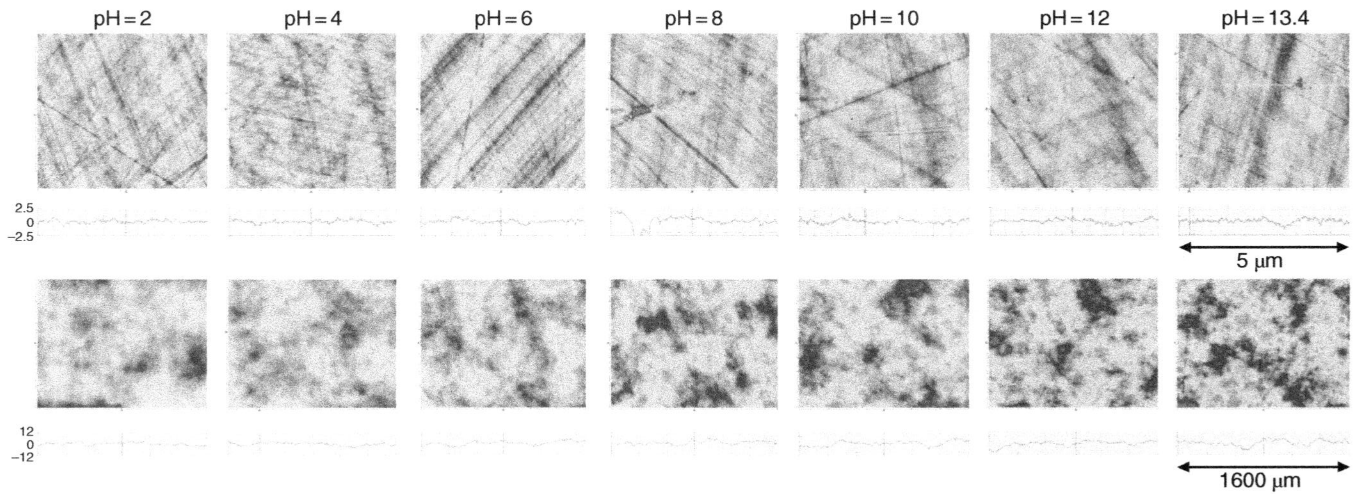

Figure 4.18 (Top) AFM2 images (5 μm × 5 μm, height scale −2.5 to 2.5 nm) of phosphate glass surface after polishing with unstabilized Hastilite ceria slurry at different pHs; (bottom) μ-roughness images (800 μm × 1600 μm, height scale −12 to 12 nm) of phosphate-glass surface after polishing with unstabilized Hastilite ceria slurry at different pHs. Source: Suratwala et al. 2017 [37]. Reproduced with permission from John Wiley & Sons.

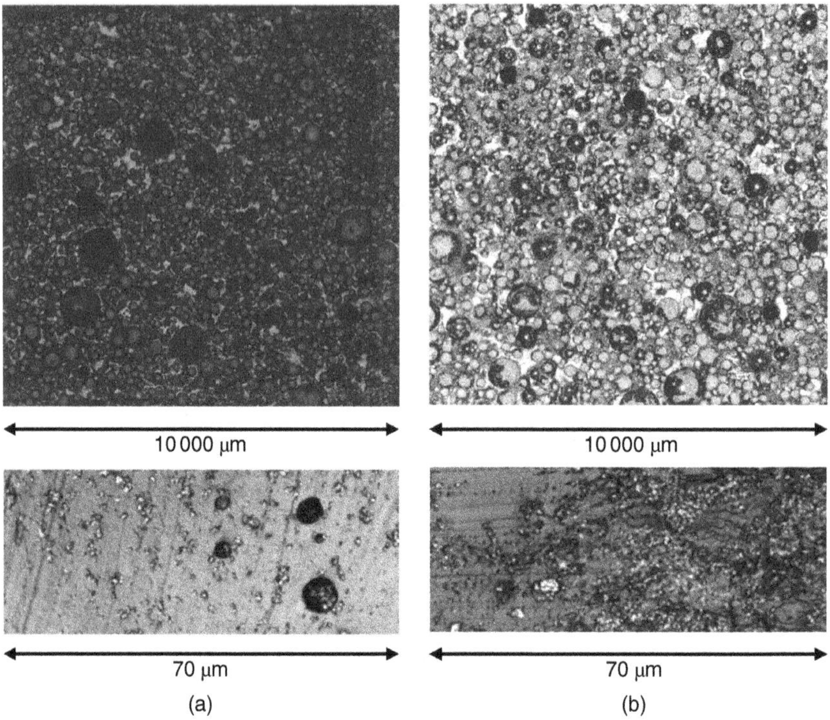

Figure 4.19 Laser confocal images of polyurethane pad after polishing phosphate glass using slurries of (a) pH = 2 and (b) pH = 13. Source: Suratwala et al. 2017 [37]. Reproduced with permission from John Wiley & Sons.

Confocal microscopy of the polishing pad used for the low- and high-pH polishing of the phosphate glass was measured and showed large differences in the quantity and distribution of the adhered slurry on the pad (see Figure 4.19). The observed differences are believed to be also present during polishing, though the characterization was performed post polishing. A large change in slurry distribution was observed with changes in slurry pH during the polishing of phosphate glass, while little change in slurry distribution was noticeable for the same variation in slurry pH during the polishing of fused-silica glass (not shown) [37]. Based on these results, the slurry attachment distribution on the pad (that is, the distribution of slurry islands) is proposed to influence the μ-roughness of the workpiece.

Using the confocal data, the pad's slurry-island size distribution and height distribution were determined (Figure 4.20a,b). In Figure 4.20a, the relative height of zero is referenced at the plateaus of the pad. Height values below zero represent the large pores on the MHN pads, and height values above zero represent the pad plateau roughness – or more important, the localized buildup

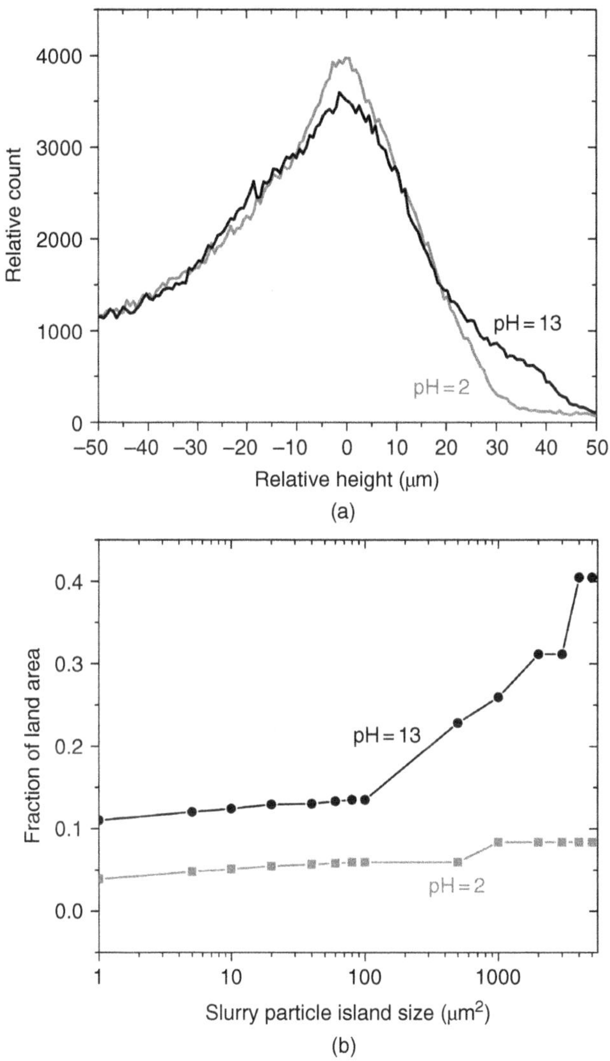

Figure 4.20 (a) Relative height histogram of pad and slurry buildup for the same pads; (b) cumulative slurry-island area distribution for the same pads. Source: Suratwala et al. 2017 [37]. Reproduced with permission from John Wiley & Sons.

of slurry particles. The pad using pH = 13 slurry had higher slurry islands (Figure 4.20a), larger slurry islands (Figure 4.20b), and a greater fractional area of pad coverage (Figure 4.20b) as compared to the pad using pH = 2 slurry.

The preferential buildup of slurry particles on the pad while polishing phosphate glass at high pH (and the lack of buildup while polishing fused silica) can be explained by electrostatic interaction between the slurry particles and pad with the addition of glass-reaction products. The remainder of this section describes this phenomenon in more detail.

4.5.2 Colloidal Stability of Particles in Slurry

It is well known in colloidal chemistry that the dispersion or agglomeration of a set of colloidal particles or surfaces in solution is determined by its surface charge, via electrostatic stabilization [39, 40]. The propensity of a particle to agglomerate or deposit onto a surface depends on a number of factors, including the surface chemistry (i.e., surface charge) of both the adsorbate and surface, the surrounding medium's pH, and the ionic strength and concentration of foreign species (e.g. cations, anions). The Derjaguin–Landau–Verwey–Overbeek (DLVO) theory can be used to describe colloidal stability under various solution conditions, accounting for electrostatic repulsion and van der Waals attraction forces [41]. This is illustrated in Figure 4.21, where two negatively charged particles are surrounded by a cloud of positive ions. The resulting forces are shown at the bottom of the figure. The interaction energy (W_t) between the two spherical colloidal particles can be described in the form,

$$W_t = \frac{rZ}{2}e^{-\kappa z} + \frac{-A_H r}{12z}, \tag{4.5}$$

where r is the particle radius, κ is the inverse of the Debye length, A_H is the Hamaker constant, z is the distance between particles, and Z is the interaction constant. The first term represents electrical double-layer interactions and the

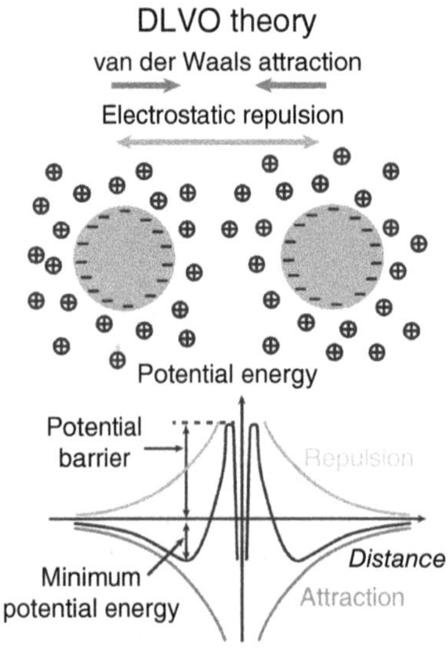

Figure 4.21 DLVO-theory schematic describing repulsive and attractive forces between two spherical colloidal particles. Source: From http://soft-matter.seas.harvard.edu/index.php/DLVO_theory.

second term, van der Waals interactions. The inverse of the Debye length is given by

$$\kappa = \left(\frac{2n_o z_c^2 e_{el}^2}{\varepsilon_4 \varepsilon_o k_b T} \right)^{1/2}, \tag{4.6}$$

where n_o is the counter-ion concentration, z_c is the valence of counter ion, e_{el} is the electron charge electron (-1.602×10^{-19} C), ε_4 is the dielectric constant of the liquid medium, ε_o is permittivity in vacuum (8.854×10^{-12} F m^{-1}), k_b is Boltzmann's constant (1.38×10^{-23} J K^{-1}), and T is temperature. The interaction constant (Z) is given by

$$Z = 64\pi \varepsilon_3 \varepsilon_o \left(\frac{k_b T}{z e_{el}} \right)^2 \tanh^2 \left(\frac{z_c e_{el} \Psi}{4 k_b T} \right), \tag{4.7}$$

where ε_3 is the dielectric constant for colloidal particle and Ψ is the Stern potential, often approximated by the measured zeta potential. The Hamaker constant is given by

$$A_H = \frac{3}{4} k_b T \left(\frac{\varepsilon_3 - \varepsilon_4}{\varepsilon_3 + \varepsilon_4} \right)^2 + \frac{3h\nu_e}{16\sqrt{2}} \frac{(n_3^2 - n_4^2)^2}{(n_3^2 + n_4^2)^{2/3}}, \tag{4.8}$$

where n_3 is the refractive index for the colloidal particle, n_4 is the refractive index of a liquid medium, ν_e is the electron-absorption frequency (3.1×10^{15} s^{-1}), and h is Plank's constant (6.626×10^{-34} J s).

From the application of the DLVO model, some general trends that can be applied the colloidal stability of a polishing slurry are apparent, including (i) an increase in the magnitude of the potential Ψ (zeta potential), which is often adjusted by pH relative to the material's IEP, result in improved colloidal stability, resisting agglomeration and (ii) an increase in counter-ion concentration (n_o) results in a greater tendency for agglomeration. Figure 4.22 illustrates the latter, showing the interaction-energy calculation for colloidal silica slurry at various salt concentrations of NaCl-supplying counter ions. Increasing the NaCl concentration decreases interaction energy (note negative values indicate attraction and positive values, repulsion).

During polishing, not only the electrostatic interaction among colloidal particles must be accounted for, but also that among the particle, workpiece surface, and lap surface. Cumbo et al. [28] illustrated how the relative difference between the IEP of the workpiece and pH of the slurry, both of which influence surface charges, can influence final workpiece roughness during polishing (see Figure 4.23).

The surface charge on a particle or planar surface can be determined by measuring its zeta potential [39]. For the polishing system of interest, the zeta potential of the slurry particles, polishing pads, and various workpieces has been measured as a function of pH, as presented in Figure 4.24 [4, 28, 37].

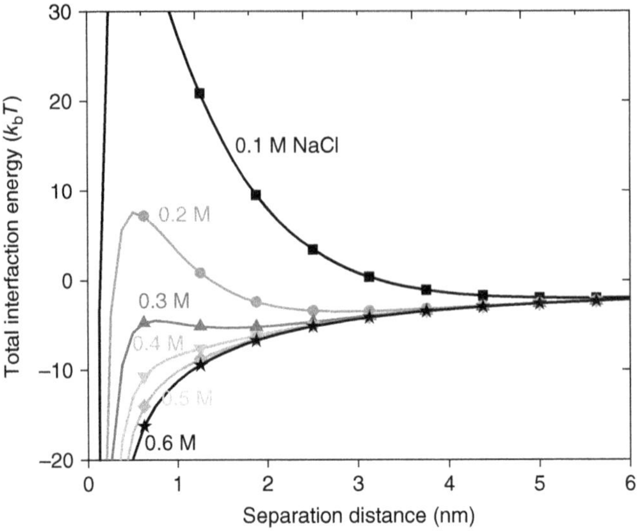

Figure 4.22 Interaction energy versus separation distance calculated by DLVO model for 100-nm silica particles at pH 8 at various NaCl concentrations. Source: Dylla-Spears et al. 2017 [42]. Reproduced with permission from Elsevier.

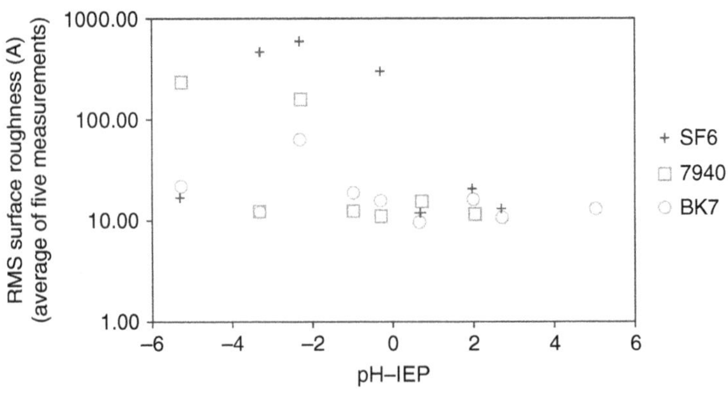

Figure 4.23 Dependence of the various polished-glass roughnesses on the difference between fluid pH and the IEP of the polishing slurry. Source: Cumbo et al. 1995 [28]. Reproduced with permission from The Optical Society.

Both the colloidal silica (S27) and the stabilized Hastilite ceria slurry particles have low IEPs (~2), while the unstabilized Hastilite PO ceria has an IEP of 10 (Figure 4.24a). The solid surfaces – fused-silica glass, phosphate glass, and the polyurethane MHN pad – have low IEPs, with various degrees of negative charge at higher pH values (Figure 4.24b).

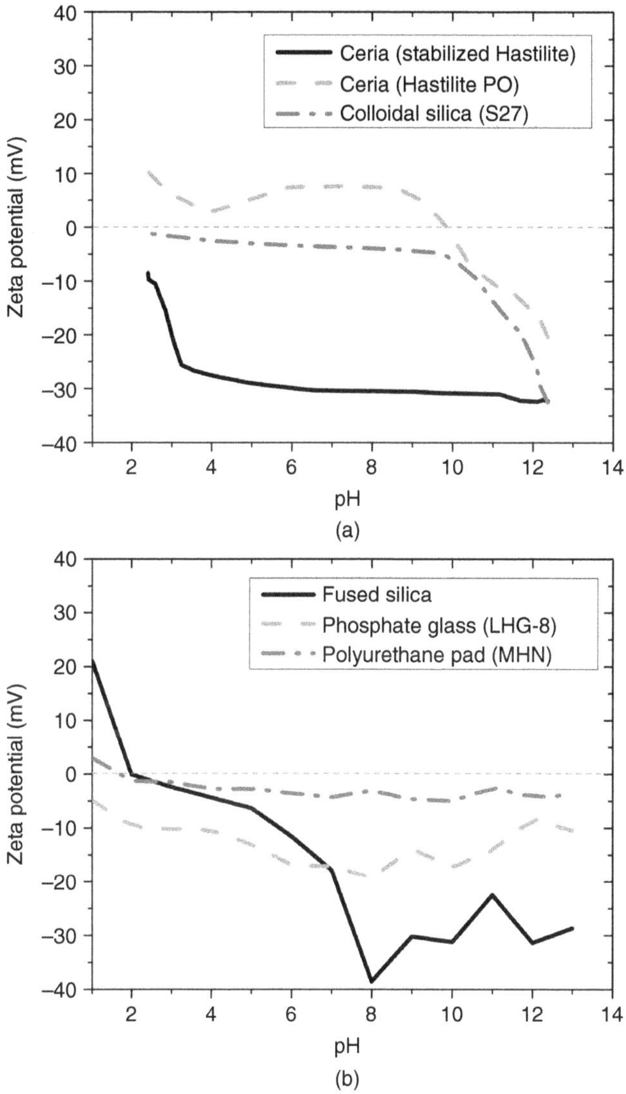

Figure 4.24 Measured zeta potential of (a) various slurries and (b) pads and workpieces as a function of pH. Source: Suratwala et al. 2017 [37]. Reproduced with permission from John Wiley & Sons.

4.5.3 Glass Reaction Product Buildup at Polishing Interface

The large difference in observed μ-roughness behavior with polishing pH for the fused silica and phosphate glasses suggests that workpiece composition influences μ-roughness (see Figure 4.17). It has been proposed that the addition of glass-reaction products at the polishing interface alters the electrostatic characteristics of slurry particles, thus influencing the spatial distribution of slurry particles on the pad [37]. It was shown earlier that glass removal during polishing can occur at the Angstrom scale (such as by the chemical-tooth mechanism proposed by Cook [4] via condensation and hydrolysis reactions) or at the nanometer scale (via nanoplastic deformation) [3, 14]. Figure 1.8 elucidates these mechanisms. For fused-silica glass, an expected polishing by-product is $Si(OH)_4$ (silicic acid), and for typical metaphosphate glasses, the major by-product is dissolved phosphate ions from K_3PO_4 [43].

In a typical polishing system, the volume of slurry is large compared to the amount of glass removed; therefore, the global concentration of glass products in the slurry is relatively low. At the polishing interface, however, where the interface gap is small and the accompanying interface slurry volume small, a much higher concentration or buildup of glass products may occur. An estimate of glass-product concentration can be obtained from a simple analysis, where the rate of glass-product introduction into a fixed-interface polishing volume is calculated during a single pass by the workpiece [37]. The molar flux of glass products (J_{gp}) removed from the workpiece is given by

$$J_{gp} = \frac{dh}{dt}\frac{\rho_1}{MW_{gp}}, \tag{4.9}$$

where dh/dt is the thickness-polishing removal rate, ρ_1 is the workpiece glass density, and MW_{gp} is the molecular weight of the glass products. The estimated flux for fused-silica glass products (using $dh/dt \sim 1\ \mu m\, h^{-1}$, $\rho_1 = 2.2\, gm\, cm^{-3}$, and $MW_{gp} = 96\, gm\, mol^{-1}$) is $6.4 \times 10^{-6}\ mol\, m^{-2}\, s^{-1}$, and for phosphate-glass products (using $dh/dt \sim 5\ \mu m\, h^{-1}$, $\rho_1 = 2.6\, gm\, cm^{-3}$, $MW_{gp} = 212\, gm\, mol^{-1}$), is $17 \times 10^{-6}\ mol\, m^{-2} s^{-1}$. Assuming that glass-product buildup occurs between grooves in the pad, with little mixing in the pores, as in the pad surface in Figure 7.16, the molar concentration of glass products (M_{gp}) at the interface is then

$$M_{gp} = J_{gp}\frac{d_g}{g_p\, v_r}, \tag{4.10}$$

where d_g is the spacing between grooves, g_p is the interface gap, and v_r is relative velocity between the workpiece and lap. Using previously calculated values for the interface gap (after pad asperity compression) of $g_p \sim 10\ \mu m$ [14, 37], $d_g = 1$ cm, and $v_r = 0.16\ m\, s^{-1}$, the estimated molar concentration at the interface is 0.04 M for $Si(OH)_4$ glass products and 0.11 M for K_3PO_4 glass products.

To mimic the chemical environment of the polishing interface, glass products consisting of K_3PO_4 surrogate for the phosphate glass, and $Si(OH)_4$ for fused silica were added/exposed at concentrations near those calculated above for a subset of slurries, pads, and workpieces [37]. Their influence on the zeta potential was then determined, as shown in Figure 4.25a–d. The addition of K_3PO_4 to stabilized Hastilite ceria particles significantly increased the zeta potential (i.e., made it more positive) with little effect on pH at all pH values measured (Figure 4.25a). However, the addition of $Si(OH)_4$ led to essentially no change in the zeta potential and a slight change in pH (Figure 4.25b). The addition of K_3PO_4 on the phosphate-glass and polyurethane-pad surfaces led to relatively small changes in the zeta potential (Figure 4.25c,d).

4.5.4 Three-Body Forces at Polishing Interface

At the polishing interface are electrostatic interactions between three major bodies (the slurry particles, workpiece surface, and pad surface). The surface charge of the three bodies and the forces resulting between them depend on the composition of the bodies and chemistry of the aqueous medium. Starting with the well-established DLVO theory [41, 44] which describes the electrical double layer of particles and flat surfaces, the interaction forces between the bodies can be calculated [37, 41]. The force (F_{ij}) between two objects, such as spheres of different radii or between a sphere and a flat surface, can be approximated using the Derjaguin approximation [45, 46] which has the following analytical form:

$$F_{ij} \approx 4\pi\varepsilon_r\varepsilon_o\kappa\left(\frac{r_i r_j}{r_i + r_j}\right)\psi_i\psi_j e^{-\kappa z}, \tag{4.11}$$

where the subscripts i and j represent the two bodies evaluated, ε_r is the dielectric constant for the slurry solution, ε_o is permittivity in vacuum (8.854×10^{-12} F m^{-1}), r is the surface radius of the body of interest, z is the average separation distance between the two surfaces, κ is the Debye–Hückel parameter, and Ψ is the surface potential of the surface of interest.

Using the measured zeta potentials for the surface potentials Ψ (see Figure 4.25), $\varepsilon_r = 78$ (water), $z = 10$ nm, $\kappa \sim 30$ nm^{-1}, estimates of the forces between the various bodies – that is, particle–particle, particle–workpiece, and particle–pad – are given in Figure 4.26 [37]. The forces are shown as a function of pH with glass products (solid lines) and without (dashed lines) and for the specific case of using stabilized Hastilite ceria slurry for polishing phosphate glass. Without glass products at high pH, all three bodies repel each other, each having large negative charges. At low pH, all three bodies are largely neutral, likely leading to weak attraction close to the IEP. However, with the addition of glass products (K_3PO_4), the change in zeta potential of the particles from largely negative to largely positive results in a strong particle–pad and particle–workpiece attractive force at high pH and weak attraction at low pH (see solid lines). With the addition of glass products, these calculations are

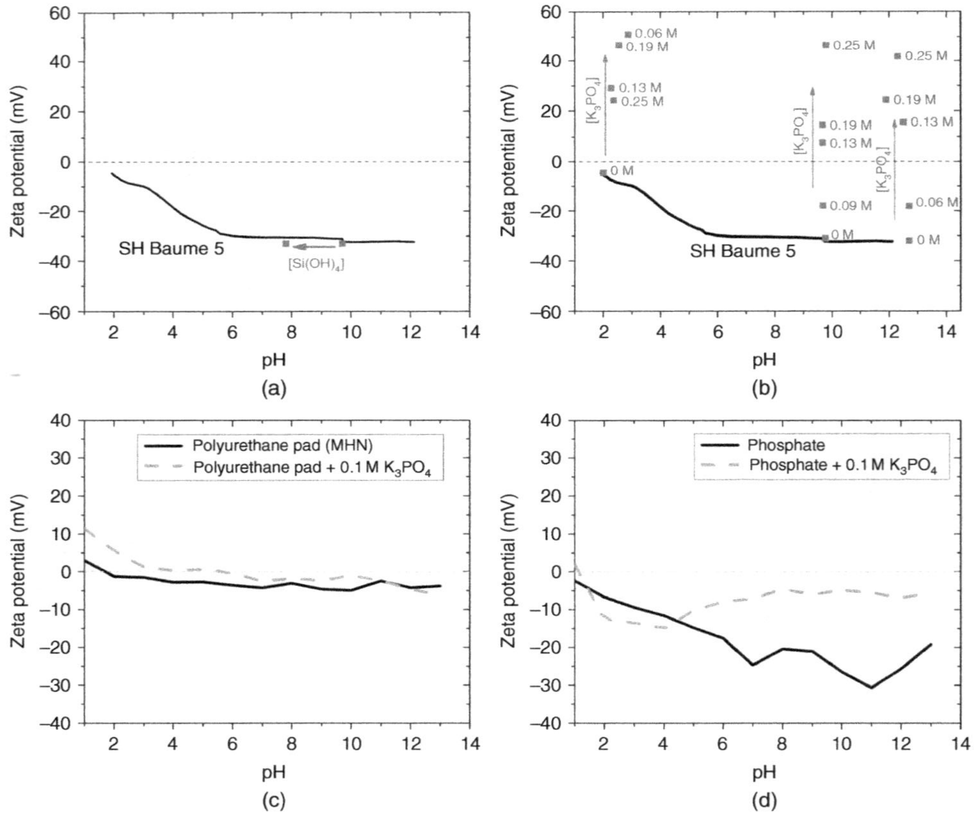

Figure 4.25 (a) Zeta potential of stabilized Hastilite slurry as a function of pH and addition of glass-product surrogate K_3PO_4 for phosphate glass; (b) zeta potential of stabilized Hastilite slurry as a function of pH and addition of glass-product surrogate $Si(OH)_4$ for fused silica; zeta potential of (c) phosphate glass and (d) polyurethane pad as a function of pH and addition of glass-product surrogate K_3PO_4 for phosphate glass. Source: Suratwala et al. 2017 [37]. Reproduced with permission from John Wiley & Sons.

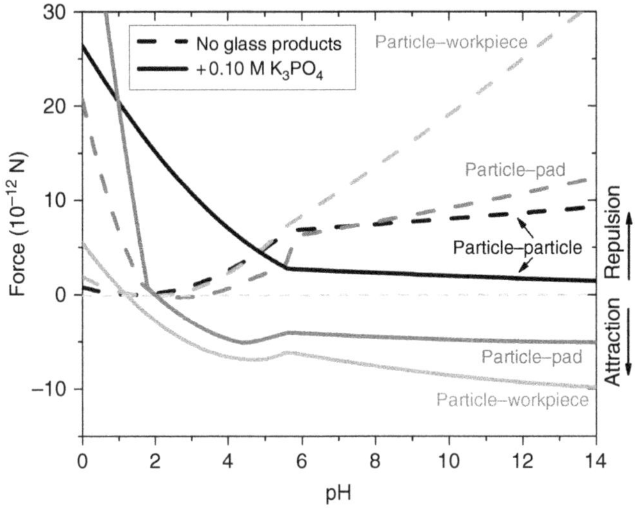

Figure 4.26 Calculated particle–particle, particle–pad, particle–workpiece forces with and without glass products for polishing phosphate glass with stabilized Hastilite, using Equation (4.12). Source: Suratwala et al. 2017 [37]. Reproduced with permission from John Wiley & Sons.

consistent with the large increase in slurry observed on the polishing pad at high pH with phosphate glass, as presented in Figure 4.19.

In sum, the slurry interface interactions at large spatial-scale lengths (microns to millimeters) have been shown to influence μ-roughness, as evidenced by large changes in μ-roughness in polishing phosphate glass, and the lack thereof with fused-silica glass, as a function of pH, despite little change in the AFM roughness. The same series also showed correspondingly large changes in the distribution of slurry attached to the pad surface with phosphate-glass polishing, but not with silica-glass polishing. Changes in electrostatic interactions between the slurry particles and pad surface with the addition of glass products can explain such changes, or the lack thereof, for both phosphate glass and fused silica. Quantitative simulation of changes in roughness due to a change in slurry spatial distribution is described by the island-distribution gap (IDG) model in Section 4.7.2.

4.6 Slurry Redeposition

In addition to slurry-interface interaction affecting islands present on the lap and influencing the *surface roughness* of the workpiece, slurry particles can directly redeposit on the workpiece surface, affecting its apparent *surface roughness*. This redeposition can be also categorized as a *surface-quality* metric, as described in Section 3.2. Governed by the similar electrostatic interactions described above, the surface charge on the workpiece and polishing particle may determine the degree of deposition. The general rule is that to obtain strong repulsion between the surfaces, pH conditions should be far from the IEP of the two species involved. A propensity for slurry particles to deposit persists even if the particle and workpiece have like charges. In a detailed study by Dylla-Spears et al. [42], the deposition of silica-slurry particles on various workpiece surfaces was examined. The adsorption was monitored in situ under various suspension conditions in the absence of surfactants or organic modifiers, using a quartz-crystal microbalance (QCM) with dissipation monitoring. Changes in surface coverage with particle concentration, particle size, pH, ionic strength, and ionic composition were quantified. An example is found in Figures 4.27 and 4.28, which show the effect of a silica-slurry particle concentration on the rate of deposition on a silica surface, in terms of frequency response on quartz crystal versus time, and the corresponding equilibrium monolayer surface coverage of particles, as measured by AFM.

A simple Monte Carlo model was developed to describe the kinetics of particle deposition [42]. In this model, particles from the suspension reaching the surface have a probability of adsorption, and particles already adsorbed on the surface desorb at a fixed rate. This model assumes that once a surface site is filled, it is no longer available for further adsorption (i.e., monolayer adsorption only). The fill fraction (f) is the ratio of the number of adsorbed particles on the surface to the total number of possible sites on the surface. The rate at which the fill fraction is changing is

$$\frac{\mathrm{d}f}{\mathrm{d}t} = p_{\mathrm{ad}} J_z A_s (1 - f) - \nu f, \tag{4.12}$$

where J_z is the flux of particles onto the surface (particles/area–time), p_{ad} is the probability of an incident particle adsorbing to the surface, ν is the rate of desorption of an adsorbed particle, and A_s is the area of a single-surface site. $J_z A_s$ is the number of particles incident on a site per unit of time. At equilibrium, the surface-fill fraction no longer changes with time and is described by the following expression:

$$f_\infty = \frac{1}{1 + \frac{\nu}{p_{\mathrm{ad}} J_z A_s}}. \tag{4.13}$$

In the simulations, the surface-fill fraction is determined as a function of time by repeatedly integrating Equation (4.12) for a time increment Δt with the number of incident particles given by the flux and number of particles removed given by the desorption rate. The initial rate of deposition, when the fill frac-

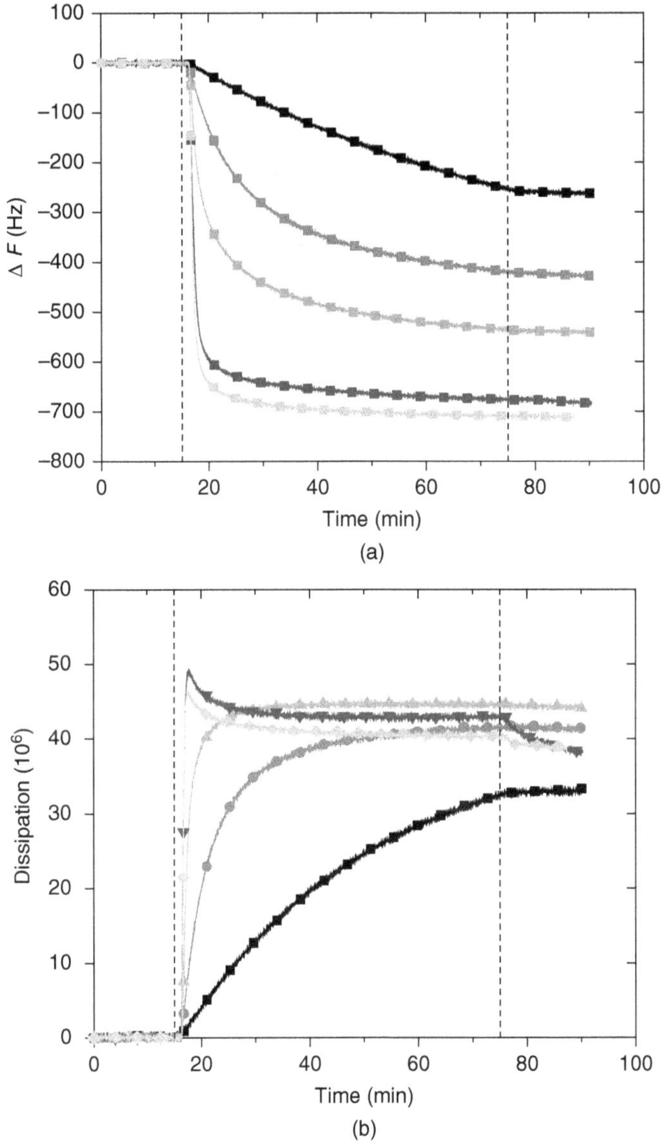

Figure 4.27 (a) Representative QCM frequency responses (third harmonic) and (b) corresponding QCM dissipation responses (third harmonic) for aqueous suspensions containing 0.5 M NaCl and various concentrations of 100-nm silica particles at pH 8, deposited onto silica-coated QCM sensors. Squares represent 0.053 nM; circles, 0.53 nM; upward triangles, 2.65 nM; downward triangles, 5.3 nM; diamonds, 13.25 nM. Dashed vertical lines in both plots denote times corresponding to the introduction of particle suspension and the reintroduction of 0.5 M NaCl background solution. Source: Dylla-Spears et al. 2017 [42]. Reproduced with permission from Elsevier.

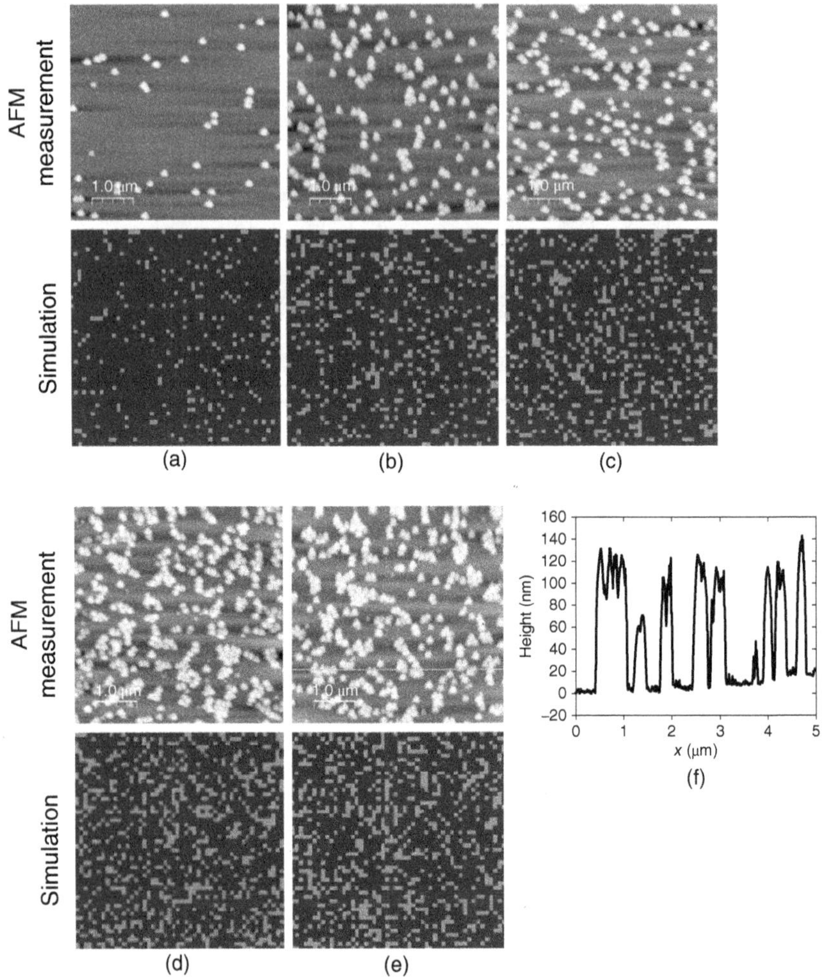

Figure 4.28 Representative 5 μm × 5 μm AFM images (top row) of silica-coated QCM sensor surfaces after adsorption of 100-nm silica particles from suspensions containing (a) 0.053 nM, (b) 0.53 nM, (c) 2.65 nM, (d) 5.3 nM, and (e) 13.25 nM silica particles. Bottom row: Monte Carlo simulation result for each experiment. (f) Height profile across line shown in AFM image in (e). Source: Dylla-Spears et al. 2017 [42]. Reproduced with permission from Elsevier.

tion is low, is governed by the deposition flux (i.e., the net adsorbed incident flux, $p_{ad}J_zA_s$). The final fill fraction at equilibrium depends on the ratio of the desorption rate to the deposition flux. Two parameters are adjusted independently for best fit to the experimental data. A comparison of this simple kinetics model compares well with the experimental data in Figure 4.28 [42].

In addition, the impact of fused-silica-workpiece *surface roughness* on the propensity for particle adsorption was studied on various spatial-scale lengths by intentionally roughening the QCM sensor surface using polishing methods [42]. It was found that a change in workpiece *surface roughness* at the AFM scale from 1.3 to 2.7 nm RMS, resulted in an increase in silica particle adsorption for 50-nm diameter particles by threefold and for 100-nm diameter particles by 1.3 times – far exceeding the adsorption observed by altering suspension conditions alone. The results for this experiment are shown in Figure 4.29. It is postulated that rougher surfaces contain higher-energy sites, leading to greater propensity for particle adsorption. Such a phenomena is known for the preferential heterogeneous nucleation of crystals or preferential condensation liquids on roughened surfaces [47, 48]. These results also imply that improving the surface roughness of the workpiece during polishing aids in reducing the propensity for deposition of particles onto the surface, leading to a lower apparent *surface roughness* and better *surface quality*.

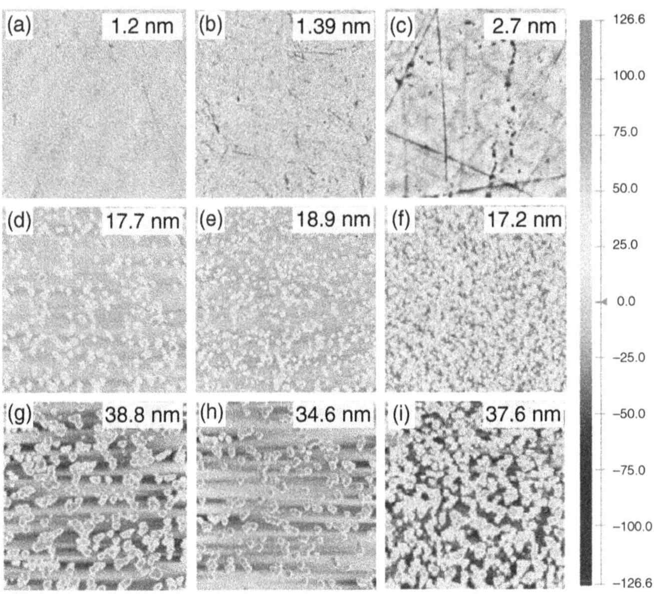

Figure 4.29 Representative 5 μm × 5 μm AFM scans showing height maps of silica-coated QCM sensor surfaces. Columns, left to right: as received, 30T-grit roughened, and ceria-slurry polished. Columns do not represent repeat scans of the same area. Sensor surfaces (a–c) before QCM adsorption experiments, (d–f) after adsorption of 21.1 nM 50-nm silica particles suspended in 0.5 M NaCl at pH 8, (g–i) after adsorption of 5.3 nM 100-nm silica particles suspended in 0.5 M NaCl at pH 8. Corresponding RMS roughness is shown on each image. Colored scale is from −125 to +125 nm. Source: Dylla-Spears et al. 2017 [42]. Reproduced with permission from Elsevier.

4.7 Predicting Roughness

At the microscopic level, the material removal amount and topographical changes that determine the roughness of the workpiece surface are generally based on Hertzian contact mechanics to describe the interactions between slurry particles and the workpiece [4, 14, 15, 29, 30, 49]. Using this formalism, the mechanical load and contact zone between each sliding polishing particle and the workpiece surface determines the removal amount and its shape. Hence, the global, macroscopic *material removal rate* and *surface roughness* observed is the summation of removals and topographical changes for each of the particles interacting with the workpiece surface. The discussion in this chapter is essentially a derivation of the global Preston *material removal rate* equation given in Equations (1.3) and (2.2), from a microscopic point of view. Because the same Hertzian mechanics influence not only *surface roughness*, but *material removal rate*, this chapter provides a foundation for the discussion of *material removal rate* in Chapter 5.

There are numerous micromechanical models that describe material removal during polishing [2, 29, 30, 37, 50–56]. Luo and Dornfeld [29, 30] introduced the concept of active and inactive particles, where only the larger, active particles load the workpiece and result in material removal. The model later extended the Hertzian contact model by accounting for adhesion forces and plastic deformation [54]. These approaches also account for pad roughness (often referred to as pad micro-asperities) and hardness, as well as the slurry's PSD. The focus of most of these approaches has been largely on understanding and predicting the average *material removal rate*, as opposed to *surface roughness* [55, 56].

In the following discussion, a pair of models based on contact mechanics is described at two spatial-scale lengths that simultaneously provide fundamental insight and predictability of both the resulting workpiece *surface roughness* and *material removal rate* [2, 14, 37]. At the particle-size scale ($<10\,\mu$m) length, the ensemble Hertzian multi-gap (EHMG) model predicts AFM roughness [2]. At the slurry-island-scale ($>10\,\mu$m) length, the IDG predicts μ-roughness [37]. The inputs for this pair of models include slurry particle size distribution, island-size distribution, the removal function of a particle or island, pad topography, and the mechanical properties of the workpiece, pad, and slurry particles.

4.7.1 EHMG – The Ensemble Hertzian Multi-gap Model

A schematic illustration of the EHMG model is shown in Figure 4.30. A pad of modulus E_2 and Poisson's ratio ν_2 has a series of asperities of height distribution $f_{pad}(h)$ compressed by a workpiece of modulus E_1 and Poisson ratio ν_1, under an applied load P or pressure σ_o, where slurry particles of modulus E_3, Poisson's ratio ν_3, and particle-size distribution $f(r)$ are present. Using the multiple

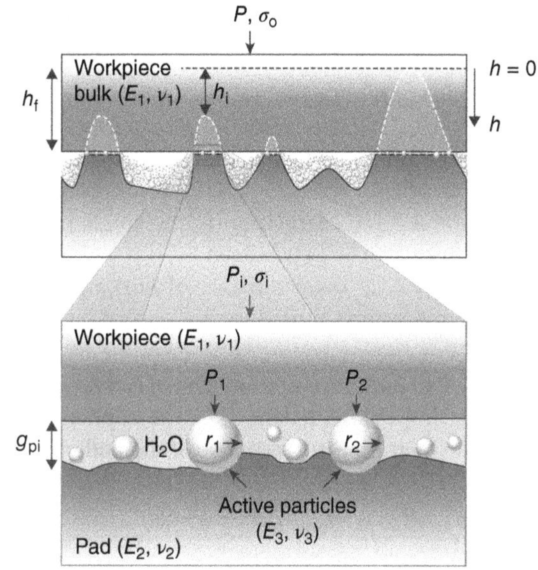

Figure 4.30 Schematic illustration of EHMG pad polishing model. Source: Suratwala et al. 2016 [2]. Reproduced with permission from John Wiley & Sons.

Hertzian particle-contact concept, the contact characteristics – load, contact zone, and elastic penetration – for each particle at the interface are determined. The workpiece penetrates a distance h_f relative to the highest asperity, as shown in Figure 4.30, top. Simultaneously, each pad asperity i is compressed a distance of h_f–h_i, resulting in a unique interface gap g_{pi} for each pad asperity (Figure 4.30, bottom). Accounting for both the slurry PSD $f(r)$ and the topography of the polishing pad $f_{pad}(h)$, two series of mechanical-load balances are used to determine overall pad compression h_f and the interface gap g_{pi} for each asperity. Evaluating h_f allows determining the load on each pad asperity, and evaluating g_{pi} allows determining the load on each particle at the interface. In some other models, the gap is referred to as the film thickness at the workpiece–lap interface [10]. The applied load is carried by only a fraction of the total number of interface particles, called the active particles, which are determined by the equilibrium gap g_{pi} at the interface. In other words, particles smaller than the gap, or nonactive particles, are not mechanically loaded and do not participate in workpiece removal.

As with any model, some key assumptions are made to simplify analysis. First, all particles are assumed spherical. Second, hydrodynamic fluid forces are considered largely negligible here, i.e., operating in contact mode (see Figure 2.5). Third, particles are distributed at the workpiece–lap interface as a single layer. In reality, the particles likely stack, which implies some smaller particles are loaded in a stack, which decreases the load on larger particles. Stacking is accounted for at larger spatial-scale lengths in the IDG model in the next section.

The EHMG model description below is divided into the following steps:

- Section 4.7.1.1: determination of pad deflection h_f and fraction of pad area making contact f_A
- Section 4.7.1.2: determination of asperity stress distribution σ_i, interface gap distribution g_{pi}, load-per-particle distribution $P(r,g)$, and fraction of active particles f_r
- Section 4.7.1.3: correlation of the single particle removal function and the load per particle distribution
- Section 4.7.1.4: determination of workpiece *surface roughness* using a Monte Carlo polishing simulation.

4.7.1.1 Pad Deflection and Fraction of Pad Area Making Contact

Using a simple load balance [2], the applied pressure σ_o on the workpiece is balanced by the elastic response of the pad given by

$$\sigma_o = E_2 \int_0^{h_f} f_{pad}(h) \frac{(h_f - h)}{t_2} \, dh, \tag{4.14}$$

where t_2 is the initial unloaded pad thickness. Equation (4.14) can be solved numerically to determine h_f, and hence the fraction of pad making contact f_A, using

$$f_A = F_{pad}(h_f) = \int_0^{h_f} f_{pad}(h) \, dh, \tag{4.15}$$

where $F_{pad}(h_f)$ is the cumulative height distribution of the pad topography at h_f.

4.7.1.2 Asperity Stress, Interface Gap, Load/Particle Distribution, and Fraction of Active Particles

The stress on each pad asperity height h_i at the equilibrium pad compression h_f is

$$\sigma_i = \frac{h_f - h_i}{t_2} E_2. \tag{4.16}$$

The stress on each pad asperity height h_i carried by slurry particles is determined as

$$\sigma_i = N_T \int_{g_{pi}/2}^{\infty} 2f(r) P(r, g_{pi}) \, dr, \tag{4.17}$$

where N_T is the areal density of particles and $P(r,g_{pi})$ is the load on a given particle on a given asperity of gap g_{pi} and particle of radius r, determined as

$$P(r, g_{pi}) = \frac{4}{3} E_{eff} \sqrt{r(2r - g_{pi})^3}, \tag{4.18}$$

where E_{eff} is the effective modulus of this three-body system, which is described as [19]

$$E_{\text{eff}} = \frac{E_{13}E_{23}}{\left(E_{13}{}^{2/3} + E_{23}{}^{2/3}\right)^{3/2}}, \tag{4.19}$$

$$E_{13} = \left(\frac{(1-\nu_1)^2}{E_1} + \frac{(1-\nu_3)^2}{E_3}\right)^{-1}, \tag{4.20}$$

$$E_{23} = \left(\frac{(1-\nu_2)^2}{E_2} + \frac{(1-\nu_3)^2}{E_3}\right)^{-1}, \tag{4.21}$$

where E_{13} is the composite modulus at the workpiece–slurry particle interface and E_{23} is the composite modulus at the pad–slurry-particle interface [2, 14].

The load on each active particle can be determined using Equation (4.18) after calculating the gap at each asperity g_{pi}, using Equations (4.16) and (4.17). Using Hertzian contact mechanics, the depth of penetration and contact zone for each slurry particle, both into the workpiece and pad, can then be determined as

$$d_{\text{t}}(r, g_{\text{pi}}) = \left(\frac{3P(r, g_{\text{pi}})}{4E_{\text{eff}}\sqrt{r}}\right)^{2/3}, \tag{4.22}$$

$$d_{13}(r, g_{\text{pi}}) = d_{\text{t}}(r, g_{\text{pi}})\left(\frac{E_{\text{eff}}}{E_{13}}\right)^{2/3}, \tag{4.23}$$

$$d_{23}(r, g_{\text{pi}}) = d_{\text{t}}(r, g_{\text{pi}})\left(\frac{E_{\text{eff}}}{E_{23}}\right)^{2/3}, \tag{4.24}$$

$$a_{13}(r, g_{\text{pi}}) = \sqrt{r\, d_{13}(r, g_{\text{pi}})}, \tag{4.25}$$

$$a_{23}(r, g_{\text{pi}}) = \sqrt{r\, d_{23}(r, g_{\text{pi}})}, \tag{4.26}$$

where d_{t} is the total penetration depth of the particle into the workpiece and lap, d_{13} and a_{13} are the penetration depth and contact radius into the workpiece, and d_{23} and a_{23} are the penetration depth and contact radius into the pad [14].

Pad asperities of different initial heights are compressed by different amounts, leading to different asperity loads, as given by Equation (4.17), illustrated in Figure 4.30. The interfacial gap is determined by the load carried by the particles, as given in Equation (4.18), the gap becoming smaller as the load increases. It is evident from Equation (4.17) that the particles can carry their maximum load when they are completely embedded into the pad (i.e., when the gap closes to zero). If the total load on the asperity is larger than this maximum, the remainder of the load is carried directly by the pad and does not contribute to material removal. This is a source of deviation from the macroscopic Preston equation (Equation (1.3)) behavior at high applied loads.

4.7.1.3 Single Particle Removal Function and Load per Particle Distribution

As discussed in Section 4.1, Figure 4.8 shows the determined single-particle removal function using ceria on a log–log plot of removal depth versus load per particle for fused silica, borosilicate, and phosphate glass. At very low loads of less than 10^{-8} N, the particles are not expected to interact with the workpiece surface and no removal occurs. At intermediate loads of 10^{-8} to 10^{-6} N, depending on the glass type, material removal occurs at via chemical reaction, with an average removal depth in the Angstrom level. Finally, at higher loads greater than 10^{-6} N, removal occurs by nanoplastic deformation in the nanometer regime, whose removal depth can change with load and glass type. If particles are removing material in the nanoplastic regime, this would result in higher roughness than if they are removing material in the chemical regime. Finally, notice the fairly significant difference in the removal depth of the various glasses in the plastic regime – 0.55 nm for fused silica, 0.85 nm for borosilicate glass, and 3.0 nm for phosphate glass [3].

In Figure 4.8, the calculated load-per-particle distributions using the EHMG model for three slurries with different PSDs are overlaid with the removal function [2]. Large differences in load-per-particle distributions result from changes in the tail end of the PSD, as presented in Figure 4.10. The narrow PSD slurry (NanoArc 6450; $d_{PSD} = 0.15\ \mu m$) leads to a lower load per particle distribution than the wider slurry (unstabilized Hastilite; $d_{PSD} = 0.92\ \mu m$). The effects of pad topography, glass type, and slurry concentration were found to have weaker effects on load-per-particle distributions [2].

4.7.1.4 Monte Carlo Workpiece Roughness Simulation

With an established formula for determining the contact characteristics and amount of material removal for each slurry particle at the polishing interface, the next step toward a simulated polished workpiece surface can be performed. Using the Monte Carlo method, slurry particles are selected from the slurry PSD $f(r)$ corresponding to a selected pad height from the pad-height distribution $f_{pad}(h)$ and randomly placed on the surface of an initially flat workpiece within a defined microscopic window [2]. The linear density of slurry particles is given by $N_t^{1/2}$, where N_t is the areal density of particles at the interface.

Each placed particle creates a trench on the workpiece surface in a randomly selected direction. The width of the trench made by the sliding particles is given by the contact zone $2a_{13}$, as seen in Equation (4.25). Removal depth is given by the removal function and the load per particle $P(r,g_{pi})$ in Equation (4.18), illustrated in Figure 4.8. The process is repeated for large number of particles until workpiece *surface roughness* is stabilized. The formation of a new polishing trench on an already-rough surface is treated by acknowledging the reality that a large particle cannot reach the bottom of a narrow trench–in other words, material removal is reduced at that location.

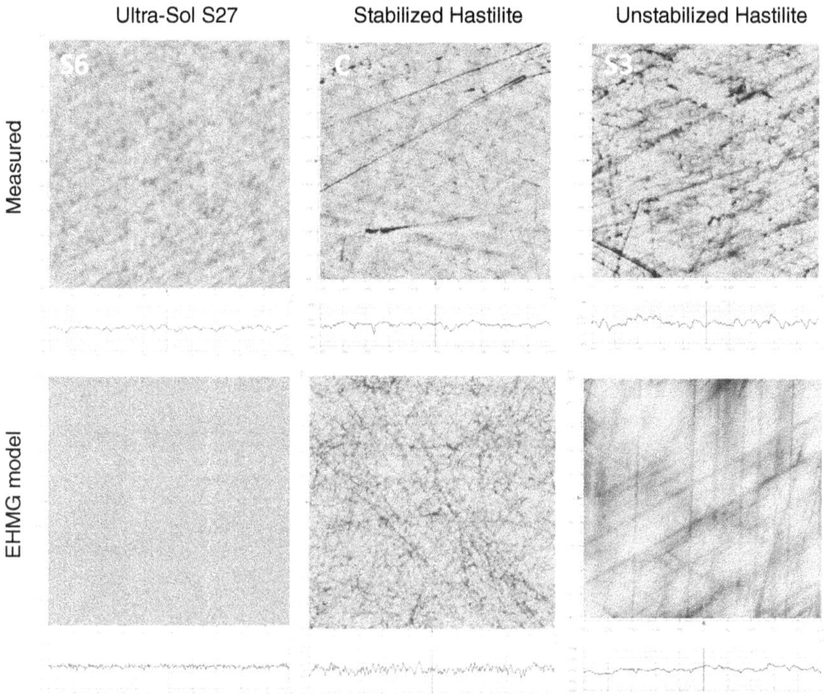

Figure 4.31 Comparison of measured AFM roughness (5 μm × 5 μm) with EHMG model simulation for selected samples in slurry PSD series. Source: Suratwala et al. 2016 [2]. Reproduced with permission from John Wiley & Sons.

Some simulated polished surfaces in the AFM roughness spatial-scale length of 5 μm × 5 μm are compared to experimental data shown in Figures 4.31–4.33 for various sets of polishing parameters. The top of the figures shows measured roughness, with corresponding simulated surfaces at the bottom [2].

Figure 4.31 compares the effects of slurry PSD where it is apparent from experimental results that widening the PSD results in an increase in workpiece *surface roughness*. The corresponding Monte Carlo simulations show a similar trend. The increase in roughness stems from the way changes in the slurry PSD influence load-per-particle distribution (see Figure 4.8). Slurries with wider PSDs result in an increase in the fraction of particles removing material by the nanoplastic mechanism, as opposed to the chemical. Since $d_p > d_m$, the polished AFM *surface roughness* increases [2].

Figure 4.32 compares the effects of pad topography. Experimentally, the workpiece *surface roughness* decreases with greater pad conditioning time, i.e., with flatter pad topography. The simulated surfaces also showed a reduction in roughness, though not as significant as in the experimental data, with

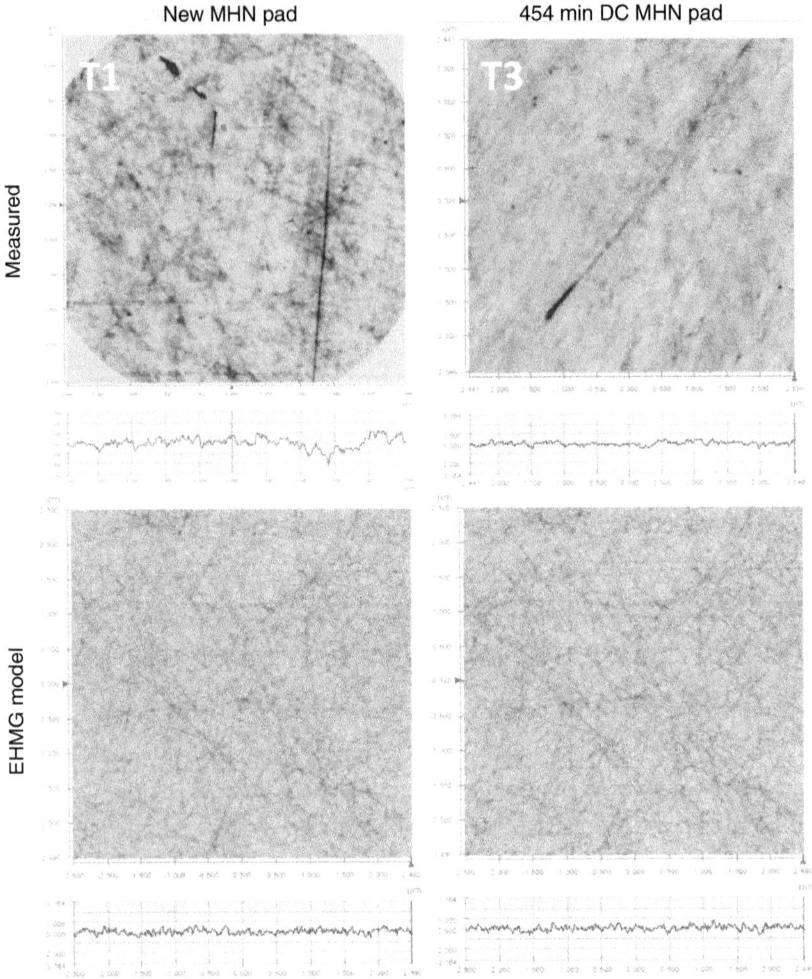

Figure 4.32 Comparison of measured AFM roughness (5 μm × 5 μm) with EHMG model simulation for selected samples in the pad-topography series. Source: Suratwala et al. 2016 [2]. Reproduced with permission from John Wiley & Sons.

increased pad flatness. A flatter pad increases the fraction of pad area making contact f_A, as describe in Equation (4.15), which results in a greater number of particles making contact and a decrease in the average load per particle. Therefore, an increase in *material removal rate* is expected, owing to the greater number of particles making contact, as well as a drop in *surface roughness* with the decrease in average load per particle. These expected trends from the EHMG model are consistent with the experimental data shown in Figure 4.15 [2].

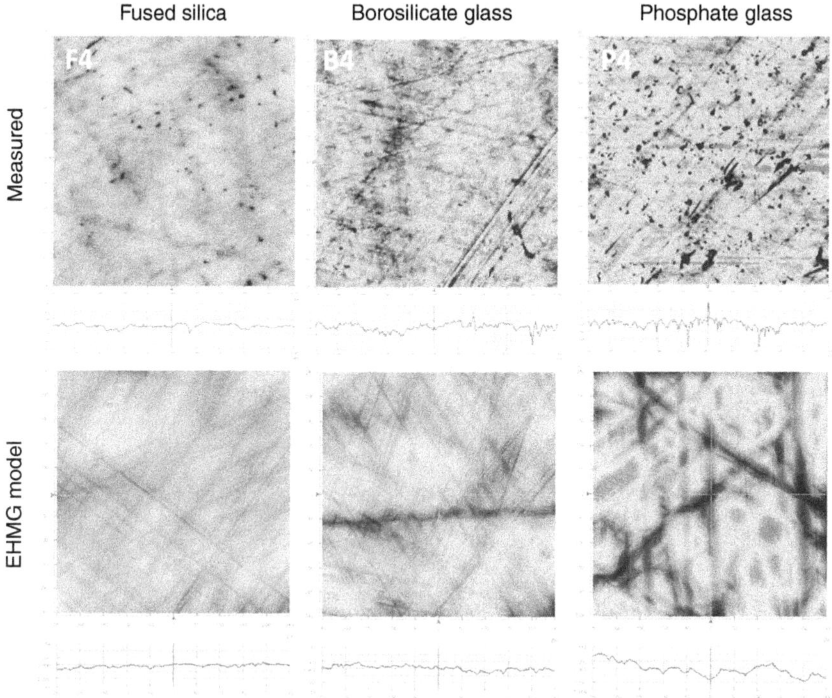

Figure 4.33 Comparison of measured AFM roughness (5 μm × 5 μm) with EHMG model simulation for selected samples in the glass-type series. Source: Suratwala et al. 2016 [2]. Reproduced with permission from John Wiley & Sons.

Finally, Figure 4.33 compares the effects of glass type. Experimentally, roughness increases with glass type such that silica < borosilicate < phosphate, matching the trend in simulation results. The slurry used in this series was unstabilized Hastilite, which has a wider PSD and hence larger load-per-particle distribution, leading to significant removal in the nanoplastic regime (see Figure 4.8). The reason for the increased roughness is an increase in the depth of nanoplastic removal d_p, with glass type (0.55 nm for fused silica, 0.85 nm for borosilicate glass, and 3.0 nm for phosphate glass).

Figure 4.34 shows the same data summarized as RMS roughness for a quantitative comparison between the simulation and experiment. The simulations capture the trends for each polishing parameter. However, the simulations systematically underestimate roughness, as compared with experiments.

4.7.2 IDG Island-distribution Gap Model

For simulating workpiece *surface roughness* at larger scale lengths for μ-roughness (10–120 μm), an IDG model has been developed, using concepts

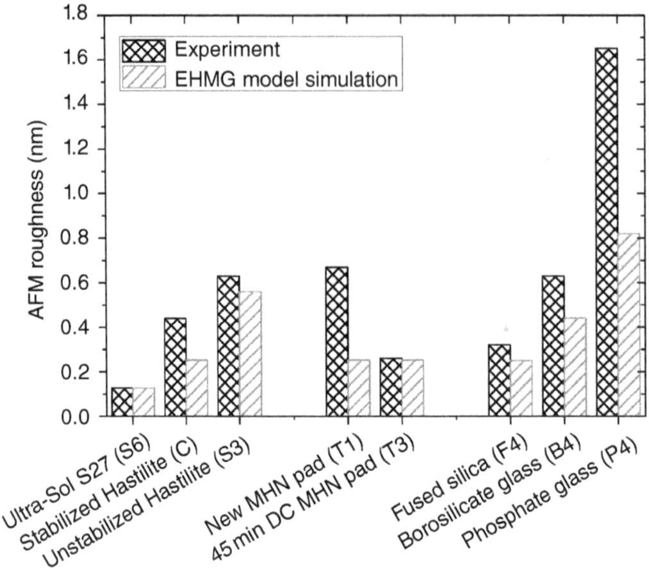

Figure 4.34 Comparison of measured RMS AFM roughness (5 μm × 5 μm) with EHMG model simulation for selected samples in the slurry PSD, pad topography, and glass type series. Source: Suratwala et al. 2016 [2]. Reproduced with permission from John Wiley & Sons.

similar to those in the EHMG model [37]. A schematic of the IDG model is shown in Figure 4.35. Here, instead of determining the load per particle directly for each particle, slurry-island height and spatial distribution are used to determine the fraction of the slurry islands that are loaded and the

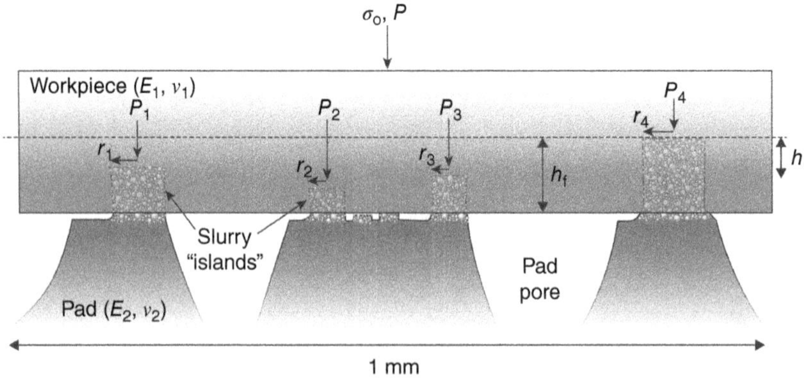

Figure 4.35 Schematic illustration of the island distribution gap (IDG) model. Source: Suratwala et al. 2017 [37]. Reproduced with permission from John Wiley & Sons.

load on each island, using a load balance similar to that above. The load on a given slurry island dictates the amount of removal and its contribution to the workpiece's overall *surface roughness*.

Using a simple load balance, the applied pressure σ_o on the workpiece is balanced by the elastic response of the pad plus the slurry-island height distribution $f_{\text{island}}(h)$ given by

$$\sigma_o = E_2 \int_0^{h_f} f_{\text{island}}(h) \frac{(h_f - h)}{t_2} \, dh, \tag{4.27}$$

where E_2 is the effective elastic modulus of the pad and slurry islands combined, h is island height, h_f is the equilibrium height, representing the combined slurry island and pad deflection, and t_2 is the initial unloaded pad thickness. For convenience, the slurry height is defined as zero at the highest island and increases toward the pad (see Figure 4.35). The fraction of contact area with the workpiece f_A is then

$$f_A = \int_0^{h_f} f_{\text{island}}(h) \, dh. \tag{4.28}$$

The load on each slurry island (i) is

$$P_i = \frac{h_f - h_i}{t_2} E_2 A_i, \tag{4.29}$$

where h_i is the initial height of the island and A_i is the island area, where the subscript i is the descriptor for each individual island. The convenience of such a formulation is that once the single value of h_f is calculated, the load on each slurry island i is easily determined using Equation (4.29). For self-consistency with the experimental data, the slurry height distribution is correlated to the slurry area distribution (Figure 4.20), whereby a higher slurry island will also have a larger island area [37].

To determine *surface roughness* with this model, the removed amount for each slurry island (d_{rem}) is needed as a function of load (i.e., its removal function). Previously, the removal function was determined for a single slurry particle, as shown by the dashed line in Figure 4.36a [3]. Note that there is an abrupt transition from molecular removal to nanoplastic removal at a load of 2×10^{-7} N. However, in the case of slurry islands, contact is made with many particles, and load per particle varies depending on the stacking of particles. Hence, it is proposed that the removal-function depth increases linearly with load for slurry islands, as shown by solid line in Figure 4.36a. In other words, the particles on a given island have a distribution of load-removing by both chemical and nanoplastic mechanisms. As the load on an island increases, the fraction of particles removing material by the nanoplastic mechanism increases. For simplicity, the IDG model assumes the removal from within each individual slurry island is considered constant. That is, for a given slurry island, only a single value of the removal depth, as opposed to multiple values, is used.

Next, the IDG model was used to calculate the workpiece *surface roughness* using a Monte Carlo simulation. First, slurry islands of radius r_i and corresponding height h_i was chosen from the measured island height distribution $f_{island}(h)$. If its calculated load is nonzero, an island removes material from the workpiece surface along a randomly selected linear track of depth d_{rem}, based on the removal function in Figure 4.36a. This process was repeated until the roughness and texture of the workpiece surface is stabilized, typically after ~150 000 islands.

The simulation results of the *surface roughness* of phosphate glass after polishing at low and high pH are compared with the measured data in Figure 4.36b. The simulation using the pH-13 slurry shows an increase in roughness compared with that using the pH 2 slurry; this is consistent with the measured data (see Figure 4.18). Close comparison of the lineouts in Figure 4.36b reveals the simulations have a similar amplitude and spatial-scale length of workpiece topography, compared to that of the measured data.

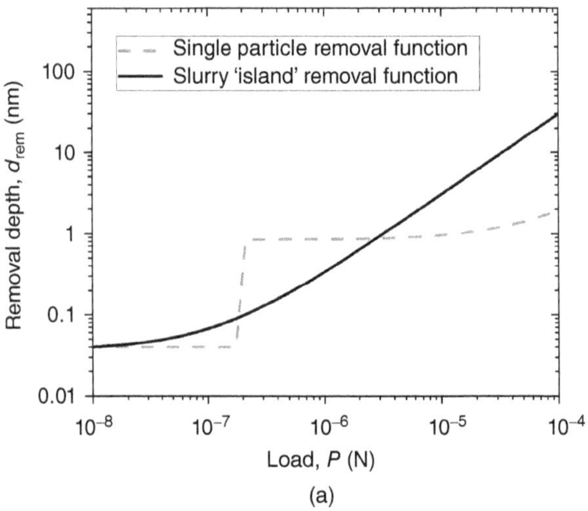

Figure 4.36 (a) Removal function determined for a single slurry particle on phosphate glass used in EHMG model compared with that postulated for the slurry islands in the IDG model. (b) Simulated workpiece-surface roughness from IDG model, using the measured-spatial-distribution adhered slurry on the pad, compared with measured roughness for polishing with pH = 2 and pH = 13 slurry on phosphate glass. All data has been bandpass filtered in the μ-roughness band (0.0083 to 0.01 μm^{-1} or 120 to 10 μm). The height scale is the same for all images, −3 to 3 nm. Source: Suratwala et al. 2017 [37]. Reproduced with permission from John Wiley & Sons.

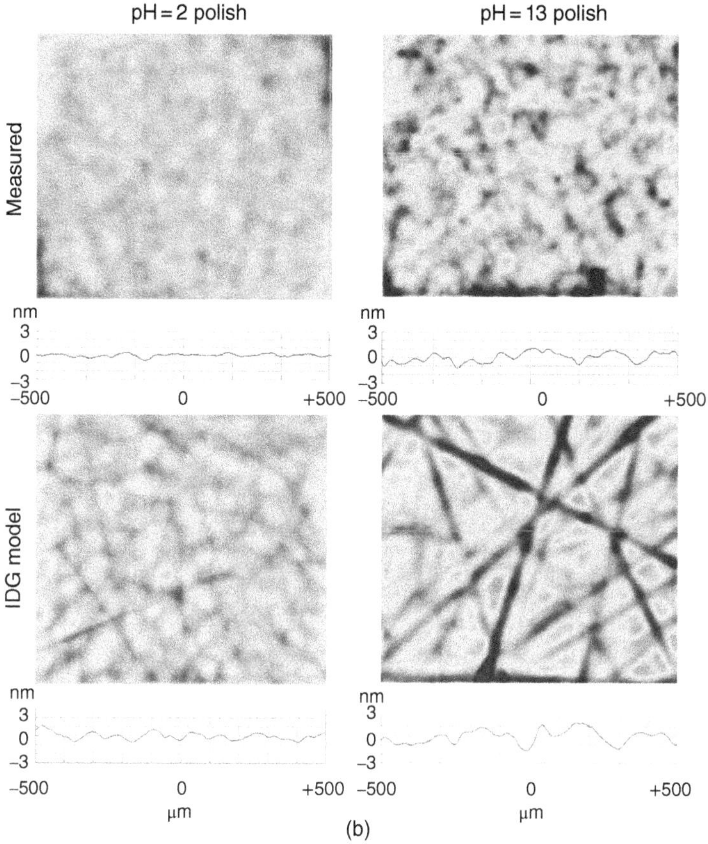

Figure 4.36 (*Continued*)

However, the overall surface texture between the measured and simulated results is slightly different. The simulations have a more linear-track texture, compared with the more random texture of the measured data. The likely reason for this difference is that the IDG model assumes a single-value removal depth for an individual island and that all islands are round (as opposed to irregular). To compensate, a more comprehensive multiscale model would be needed to reproduce the experimental surface texture after polishing, possibly combining the concepts of both the EHMG and IDG models, with additional input as to the areal shape distribution of the slurry islands. Despite the slight difference in texture, the IDG model does a decent job of simulating workpiece topography in the μ-roughness regime, which supports the basic concept that the spatial and height distribution of slurry may influence the roughness of a final polished workpiece [37].

4.8 Strategies to Reduce Roughness

The *surface roughness* of a workpiece is largely governed by the intersection of the load-per-particle distribution and the removal function of the workpiece (see Figure 4.8). Hence, two general strategies for reducing workpiece *surface roughness* are as follows:

1. Reduce or narrow the load-per-particle distribution – that is, shift the distribution to the left in Figure 4.8, which may decrease the average single-particle removal depth and decrease roughness
2. Modify the removal function of a given slurry particle, where a lower removal depth at a given load results in lower roughness.

4.8.1 Strategy 1: Reduce or Narrow the Load-per-particle Distribution

Reduce load-per-particle distribution by

- Decreasing the overall applied load
- Decreasing the average particle size of the slurry. This results in more particles per unit area and hence more particles to load.
- Increasing the fill fraction of slurry islands at the workpiece–lap interface by chemical means, as discussed in Section 4.7.2
- Increasing the fraction of pad area making contact – for example, increasing the pad flatness via diamond conditioning such that there are more active particles (see Section 4.4).
- Increasing slurry concentration (up to critical concentration)
- Transitioning from contact mode to mixed or hydrodynamic mode, by increasing velocity, or slurry viscosity or reducing the applied load. This results in increased hydrodynamic-reaction force, reducing load per particle.

Narrow load-per-particle distribution by

- Choosing a slurry with a sharp slope at the tail end of the PSD (see Figures 4.10 and 4.11)
- Preventing slurry agglomeration by chemical means (see Section 7.1.6)
- Preventing the entrance of external rogue particles
- Using a more compliant or viscoelastic/viscoplastic lap – for example, softer pads, pitch lap, or magnetorheological finishing (MRF) slurry – which allows narrowing the load distribution on all particles in the slurry.

4.8.2 Strategy 2: Modify the Removal Function of a Given Slurry

Assuming a fixed workpiece material, the removal function of a given slurry can be modified by changing the slurry composition to suit the workpiece material.

For reducing *surface roughness*, choose a slurry-particle composition that leads to a small removal depth per particle. Removal by chemical reaction tends to have a significantly smaller removal depth per particle, as compared to removal by a nanoplastic mechanism (see Figure 1.9 and Section 4.1), which can lead to lower roughness. For many glasses, cerium oxide is chosen because removal depth is small and dominated by chemical reaction. Details of cerium-oxide behavior and selection strategies for slurry-particle composition are discussed in Section 5.2.3.2.

The objective of polishing, however, is not only to reduce *surface roughness,* but also to achieve a desired *surface figure* and *surface quality* with a high *material removal rate* (see Figure 1.6). Unfortunately, many of the strategies described above lead to lower *surface roughness* at the expense of *material removal rate.* For example, decreasing particle size and reducing the applied pressure can reduce *surface roughness,* but will also significantly reduce the *material removal rate.* Listed below are some specific strategies which allow for both low *surface roughness* and high *material removal rate.*

1. *Increase applied pressure* σ_0 *until the load-per-particle distribution just reaches the chemical–nanoplastic removal-regime transition.* For many slurry-particle–workpiece materials, there is a sharp transition in the removal function depth from the chemical regime to the nanoplastic (see Figure 4.8). Also, the single-particle removal functions tend to be largely load independent in the chemical regime. Hence, *material removal rate* can be increased by increasing applied pressure, largely owing to more particles being loaded with a smaller interface gap, as described by the EHMG model (see Section 4.7.1), up to the point where the largest load per particle in distribution does not transition into the nanoplastic regime. Thus, under such conditions, *surface roughness* does not increase.
2. *Increase the relative velocity* V_r *using kinematics, while staying within the contact mode.* Relative velocity can be increased, for example, by increasing lap and optic rotation rate. This leads to higher material removal without much decrement to *surface roughness,* per the concepts in the EHMG model. However, relative velocity can be increased only to the point where the system is still in contact mode (see Figure 2.5). Higher relative velocities will result in a transition to the mixed and hydrodynamic modes, resulting in more applied load being carried by the slurry fluid, as opposed to the slurry particles – leading to a significant drop in *material removal rate.* An example of exploiting both the higher-applied-pressure and relative-velocity strategies is shown in Figure 4.37, using the same data originally shown in Figure 4.15. The figure presents a cross-plot of the *surface roughness* and *material removal rate* for a variety of fused-silica workpieces polished under various conditions. Two of the polishing conditions that led to low roughness, samples T3 and S6, where repeated,

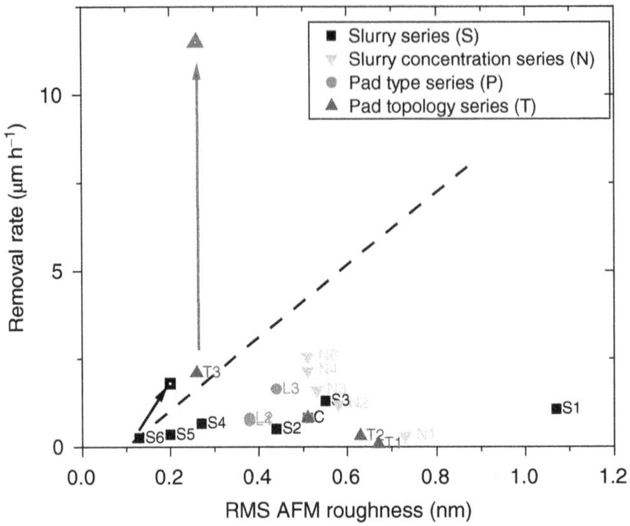

Figure 4.37 Cross-plot of measured removal rate and AFM roughness of fused-silica glass workpieces polished under different conditions. Major changes to processing conditions include slurry-particle size distribution, slurry concentration, pad type, and pad topography. The results of two additional samples polished at three-times-higher applied pressure and twice the relative velocity are indicated by arrows.

but with an applied pressure three times higher and double the relative velocity. The EHMG model calculations confirm that such conditions lead to a load-per-particle distribution within the chemical material-removal regime. The Hershey number figure of merit (FOM) was also still within contact mode, as described in Section 2.3. The resulting *surface roughness* stayed low and the *material removal rate* increased (see Figure 4.37). In one case, it was demonstrated that fused-silica glass can be polished at $>10\,\mu\text{m}\,\text{h}^{-1}$ and still have an AFM roughness of only 2–3 Å.

3. *Obtain or modify the slurry PSD so that it has a larger slope at the tail of the distribution without changing the average particle size.* As discussed in Sections 4.3 and 4.7.1, the slope of the tail end of the slurry PSD significantly influences the load per particle distribution and resulting *surface roughness*. The larger the slope, the lower the *surface roughness* (see Figures 4.10 and 4.11). A technique to change the slope of the tail end using chemical stabilization is discussed in Section 7.1.6.
4. *Alter slurry chemistry such that slurry islands have a larger fill fraction at the workpiece–lap interface.* Section 4.5 discusses how slurry–chemical-interface interactions can influence island distribution of slurry particles at the interface. If the slurry fill-area fraction can be increased without an increase in island height distribution, a significant

increase in the material removal rate may be realized without degrading the load-per-particle distribution and surface roughness.

5. *Increase the fraction of pad area making contact f_A*. Another method to decrease *surface roughness* and increase *material removal rate* is to increase the fraction of pad area making interface contact. For pad polishing using a MHN, it was determined that this fraction is very small: 10^{-4} [14]. Increasing the fraction of pad area making contact results in more slurry particles becoming load bearing and contributing to material removal. A nice example is shown in Figure 4.15. A series of fused-silica workpieces were polished on a polyurethane pad with different pad topographies (T-series samples). Increasing the pad flatness increased the fraction of pad area making contact. The *material removal rate* increased and the *surface roughness* decreased with increasing pad flatness.

References

1 Bennett, J.M. and Mattsson, L. (1999). *Introduction to Surface Roughness and Scattering*, 2e, viii, 130. Washington, DC: Optical Society of America.
2 Suratwala, T., Steele, W., Feit, M. et al. (2016). Mechanism and simulation of removal rate and surface roughness during optical polishing of glasses. *J. Am. Ceram. Soc.* 99 (6): 1974–1984.
3 Shen, N., Suratwala, T., Steele, W. et al. (2016). Nanoscratching of optical glass surfaces near the elastic-plastic load boundary to mimic the mechanics of polishing particles. *J. Am. Ceram. Soc.* 99 (5): 1477–1484.
4 Cook, L. (1990). Chemical processes in glass polishing. *J. Non-Cryst. Solids* 120 (1–3): 152–171.
5 Partridge, P., Fookes, A., Nicholson, E. et al. (1996). Nanoscale ductile grinding of glass by diamond fibres. *J. Mater. Sci.* 31 (19): 5051–5057.
6 Ong, N. and Venkatesh, V. (1998). Semi-ductile grinding and polishing of Pyrex glass. *J. Mater. Process. Technol.* 83 (1): 261–266.
7 Zhong, Z. and Venkatesh, V. (1995). Semi-ductile grinding and polishing of ophthalmic aspherics and spherics. *CIRP Ann. Manuf. Technol.* 44 (1): 339–342.
8 Fielden, J.H. and Rubenstein, C. (1969). Grinding of glass by a fixed abrasive. *Glass Technol.* 10 (3): 73–83.
9 Mairlot, H. (1972). Texture of ground glass surfaces. *Ind. Diamond Rev.* FEB: 61.
10 Moon, Y. (1999). Mechanical aspects of the material removal mechanism in chemical mechanical polishing (CMP).
11 Puttick, K., Rudman, M., Smith, K. et al. ed. (1989). Single-point diamond machining of glasses. *Proc. R. Soc. London, Ser. A* 426: 19–30.

12 Fu, G., Chandra, A., Guha, S., and Subhash, G. (2001). A plasticity-based model of material removal in chemical–mechanical polishing (CMP). *IEEE Trans. Semicond Manuf.* 14 (4): 406–417.

13 Komanduri, R. (1996). On material removal mechanisms in finishing of advanced ceramics and glasses. *CIRP Ann. Manuf. Technol.* 45 (1): 509–514.

14 Suratwala, T., Feit, M., Steele, W. et al. (2014). Microscopic removal function and the relationship between slurry particle size distribution and Workpiece roughness during pad polishing. *J. Am. Ceram. Soc.* 97 (1): 81–91.

15 Brown, N. (1981). *A Short Course on Optical Fabrication Technology.* Lawrence Livermore National Laboratory.

16 Shorey, A.B., Kwong, K.M., Johnson, K.M., and Jacobs, S.D. (2000). Nanoindentation hardness of particles used in magnetorheological finishing (MRF). *Appl. Opt.* 39 (28): 5194–5204.

17 Yu, J., Yuan, W., Hu, H. et al. (2015). Nanoscale friction and wear of phosphate laser glass and BK7 glass against single CeO_2 particle by AFM. *J. Am. Ceram. Soc.* 98 (4): 1111–1120.

18 Pergande, S.R., Polycarpou, A.A., and Conry, T. (2001). Use of Nano-Indentation and Nano-Scratch Techniques to Investigate Near Surface Material Properties Associated with Scuffing of Engineering Surfaces. Air Conditioning and Refrigeration Center. College of Engineering. University of Illinois at Urbana-Champaign.

19 Johnson, K.L. and Johnson, K.L. (1987). *Contact Mechanics.* Cambridge University Press.

20 Campbell, J.H., Suratwala, T.I., Thorsness, C.B. et al. (2000). Continuous melting of Nd-doped phosphate laser glasses. *J. Non-Cryst. Solids* 263&264: 342–357.

21 Yoshida, S., Sanglebœuf, J.-C., and Rouxel, T. (2005). Quantitative evaluation of indentation-induced densification in glass. *J. Mater. Res.* 20 (12): 3404–3412.

22 Gadelrab, K., Bonilla, F., and Chiesa, M. (2012). Densification modeling of fused silica under nanoindentation. *J. Non-Cryst. Solids* 358 (2): 392–398.

23 Yang, J., Yi, K., Yang, M. ed. (2014). Surface structure of fused silica revealed by thermal annealing. *Pacific-Rim Laser Damage.* International Society for Optics and Photonics.

24 Mittal, K.L. (1994). *Particles on Surfaces: Detection: Adhesion, and Removal.* CRC Press.

25 Suratwala, T., Steele, R., Feit, M.D. et al. (2008). Effect of rogue particles on the sub-surface damage of fused silica during grinding/polishing. *J. Non-Cryst. Solids* 354 (18): 2023–2037.

26 Basim, G., Adler, J., Mahajan, U. et al. (2000). Effect of particle size of chemical mechanical polishing slurries for enhanced polishing with minimal defects. *J. Electrochem. Soc.* 147 (9): 3523–3528.

27 Kim, H., Yang, J.C., Kim, M. et al. (2011). Effects of ceria abrasive particle size distribution below wafer surface on in-wafer uniformity during chemical mechanical polishing processing. *J. Electrochem. Soc.* 158 (6): H635–H640.

28 Cumbo, M., Fairhurst, D., Jacobs, S., and Puchebner, B. (1995). Slurry particle size evolution during the polishing of optical glass. *Appl. Opt.* 34 (19): 3743–3755.

29 Luo, J. and Dornfeld, D.A. (2001). Material removal mechanism in chemical mechanical polishing: theory and modeling. *IEEE Trans. Semicond. Manuf.* 14 (2): 112–133.

30 Luo, J. and Dornfeld, D.A. (2003). Effects of abrasive size distribution in chemical mechanical planarization: modeling and verification. *IEEE Trans. Semicond. Manuf.* 16 (3): 469–476.

31 Li, W., Shin, D.W., Tomozawa, M., and Murarka, S.P. (1995). The effect of the polishing pad treatments on the chemical–mechanical polishing of SiO_2 films. *Thin Solid Films* 270 (1–2): 601–606.

32 Elmufdi, C.L. and Muldowney, G.P. ed. (2007). The impact of diamond conditioning on surface contact in CMP pads. In: *MRS Proceedings*. Cambridge University Press.

33 Berman, M.J. and Kalpathy-Cramer, J. (1999). Chemical mechanical polishing pad slurry distribution grooves. Google Patents.

34 Wang, C., Paul, E., Kobayashi, T., and Li, Y. (2007). Pads for IC CMP. In: *Microelectronic Applications of Chemical Mechanical Planarization* (ed. Y. Li), 123. Wiley Interscience.

35 Brown, N. (1977). *Optical Polishing Pitch*. Livermore, CA: Lawrence Livermore National Laboratory, University of California.

36 Kumanin, K. (1967). *Generation of Optical Surfaces*. New York: Focal Library.

37 Suratwala, T., Steele, R., Feit, M. et al. (2017). Relationship between surface μ-roughness & interface slurry particle spatial distribution during glass polishing. *J. Am. Ceram. Soc.* 100: 2790–2802.

38 Abiade, J.T., Choi, W., and Singh, R.K. (2005). Effect of pH on ceria–silica interactions during chemical mechanical polishing. *J. Mater. Res.* 20 (05): 1139–1145.

39 Hunter, R.J. (2013). *Zeta Potential in Colloid Science: Principles and Applications*. Academic Press.

40 Hiemenz, P.C. and Hiemenz, P.C. (1986). *Principles of Colloid and Surface Chemistry*. New York: Marcel Dekker.

41 Israelachvili, J.N. (2011). *Intermolecular and Surface Forces: Revised Third Edition*. Academic Press.

42 Dylla-Spears, R., Wong, L., Shen, N. et al. (2017). Adsorption of silica colloids onto like-charged silica surfaces of different roughness. *Colloids Surf., A* 520: 85–96.

43 Tischendorf, B.C. (2005). *Interactions Between Water and Phosphate Glasses*. Rolla, MO: University of Missouri.

44 Werway, E. and Overbeek, J.T.G. (1948). *Theory of Stability of Lyophobic Colloids*, 34. Amsterdam, New York: Elsevier.

45 Carnie, S.L., Chan, D.Y., and Gunning, J.S. (1994). Electrical double layer interaction between dissimilar spherical colloidal particles and between a sphere and a plate: the linearized Poisson–Boltzmann theory. *Langmuir* 10 (9): 2993–3009.

46 Hogg, R., Healy, T.W., and Fuerstenau, D. (1966). Mutual coagulation of colloidal dispersions. *Trans. Faraday Soc.* 62: 1638–1651.

47 Koutsky, J., Walton, A., and Baer, E. (1965). Heterogeneous nucleation of water vapor on high and low energy surfaces. *Surf. Sci.* 3 (2): 165–174.

48 Liu, Y.-X., Wang, X.-J., Lu, J., and Ching, C.-B. (2007). Influence of the roughness, topography, and physicochemical properties of chemically modified surfaces on the heterogeneous nucleation of protein crystals. *J. Phys. Chem. B* 111 (50): 13971–13978.

49 Preston, F.W. (1922). The structure of abraded glass surfaces. *Trans. Opt. Soc.* 23 (3): 141.

50 Zhao, Y. and Chang, L. (2002). A micro-contact and wear model for chemical–mechanical polishing of silicon wafers. *Wear* 252 (3): 220–226.

51 Nanz, G. and Camilletti, L.E. (1995). Modeling of chemical–mechanical polishing: a review. *IEEE Trans. Semicond. Manuf.* 8 (4): 382–389.

52 McGrath, J. and Davis, C. (2004). Polishing pad surface characterisation in chemical mechanical planarisation. *J. Mater. Process. Technol.* 153: 666–673.

53 Oh, S. and Seok, J. (2008). Modeling of chemical–mechanical polishing considering thermal coupling effects. *Microelectron. Eng.* 85 (11): 2191–2201.

54 Bastaninejad, M. and Ahmadi, G. (2005). Modeling the effects of abrasive size distribution, adhesion, and surface plastic deformation on chemical–mechanical polishing. *J. Electrochem. Soc.* 152 (9): G720–G730.

55 Sampurno, Y.A., Borucki, L., Zhuang, Y. et al. (2005). Method for direct measurement of substrate temperature during copper CMP. *J. Electrochem. Soc.* 152 (7): G537–G541.

56 Wang, C., Sherman, P., Chandra, A., and Dornfeld, D. (2005). Pad surface roughness and slurry particle size distribution effects on material removal rate in chemical mechanical planarization. *CIRP Ann. Manuf. Technol.* 54 (1): 309–312.

5

Material Removal Rate

This chapter covers the last of the four major characteristics of optical fabrication, *material removal rate* (see Figure 1.6). As discussed in Chapter 1, the macroscopic *material removal rate* is governed by the Preston equation (Equation (1.3)), where removal rate largely scales linearly with applied pressure and relative velocity, and all the process and material parameters are lumped into the Preston coefficient k_p. The Preston equation can be applied to both grinding (which is discussed in Section 5.1) and polishing (Section 5.2). The parameters that govern *material removal rate* and resulting *surface roughness* are intimately connected. Hence, the principles of the ensemble Hertzian multi-gap (EHMG) and island distribution gap (IDG) models, as discussed in Chapter 4, can be largely applied when discussing polishing *material removal rate.*

5.1 Grinding Material Removal Rate

The grinding of brittle materials can be described microscopically as the removal of workpiece particles created from an ensemble of intersecting brittle fractures, which are caused by an ensemble of normally loaded, hard indenters or abrasives sliding or rolling across the surface of the workpiece. For the grinding *material removal rate,* governed by the Preston equation (Equation (1.3)), the Preston coefficient incorporates many process parameters related to the grinding particle (material properties, particle size distribution (PSD), and shape), the lap or tool (material properties), the kinematics (depth of cut, feed rate, tool rotation), the contact conditions (lubricant, temperature, friction), and the workpiece (material properties).

Various statistical grinding models, requiring the determination of empirical parameters experimentally, have been proposed to predict *material removal rates*. For example, one model using a breaking function from Poisson statistics has set the basis for predicting *material removal rate* for various grinding conditions, but requires the measurements for determining the breakage distribution parameters from experimental data [1, 2]. In another approach, the

Materials Science and Technology of Optical Fabrication, First Edition. Tayyab I. Suratwala.

spatial and time dependence of crack distribution was used to determine the material removal rate using a modified Preston equation approach [3].

Owing to the complex nature of all these interactions, a single comprehensive model starting from fundamental fracture mechanics is not well established to date to predict the *material removal rate* as a function of the combined process parameters. However, useful correlations of some of the important process variables have been established.

Historically, the influence of workpiece mechanical properties on the grinding rate has been characterized by several glass companies that established material-removal figure of merits (FOMs) by comparing the grinding rates of many glasses processed under a standardized, fixed grinding process. Later, Aleinikov [4], Izumitani [5, 6], Lambropoulos [7], and Buijs and Korpel-van Houten [8] correlated these grinding FOMs to basic indentation crack-growth behavior, ultimately giving rise to the term lapping hardness (LH). Lapping hardness is derived from the principle that the volume of material removed is dominated by chips removed from the workpiece, created by lateral cracks from mechanically loaded abrasives sliding across the workpiece (see Figure 5.1) [7]. In contrast, subsurface mechanical damage (SSD) is determined largely by trailing indent cracks (see Section 3.1). Hence, the contribution of the workpiece-material properties to the grinding Preston coefficient should scale with the LH as follows:

$$\mathrm{LH} = \frac{{E_1}^{7/6}}{K_{1c}{H_k}^{23/12}} \sim \frac{E_1}{K_{1c}{H_k}^{2}}, \tag{5.1}$$

where E_1 is the workpiece elastic modulus, K_{1c} is the workpiece fracture toughness, and H_k is the workpiece Knoop hardness. Intuitively, the lapping hardness FOM can be thought of as the combination of three mechanical properties of the workpiece: the elastic modulus, which determines the stress generated in the material; hardness, which determines the degree of plastic deformation to absorb the stress; and fracture toughness, which describes the ability for fractures to propagate within the material. Figure 5.2 shows that lapping hardness scales well with the grinding rate FOMs for a wide variety of glasses [7]. A related metric that scales with grinding rate is the

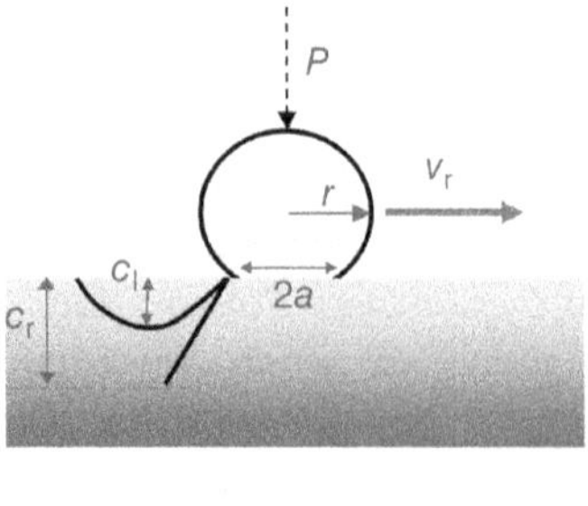

Figure 5.1 A schematic illustration of sliding abrasive particle creating trailing indent and lateral fractures.

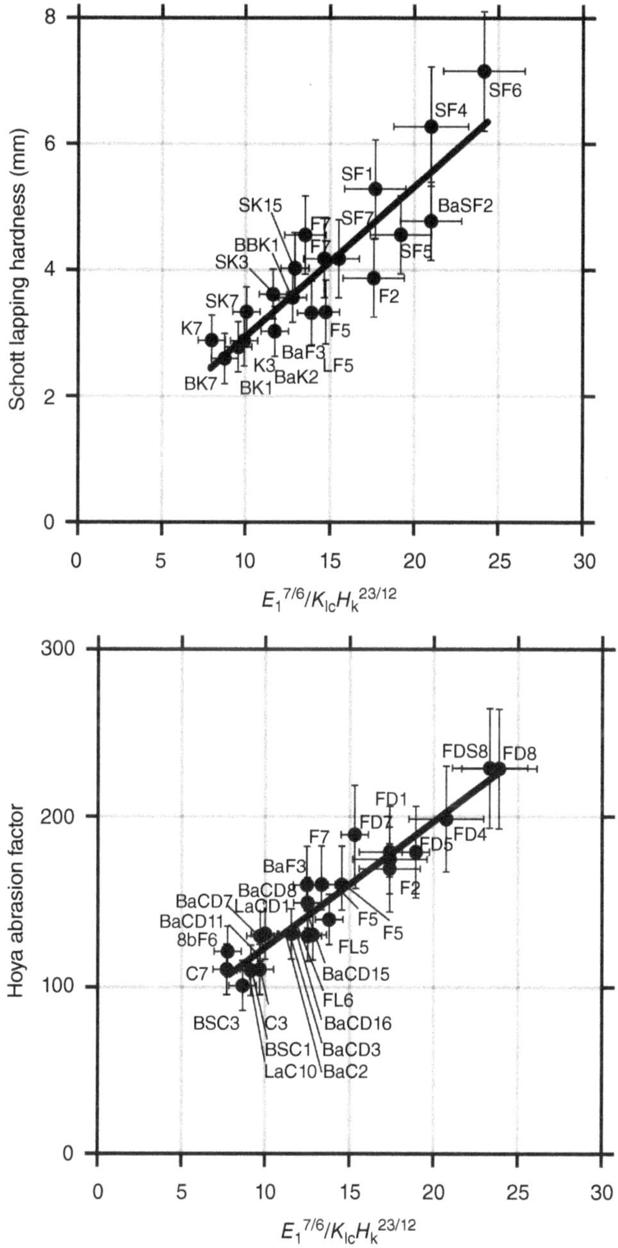

Figure 5.2 Correlation of two FOMs of grinding rate with that of lapping hardness, performed by Schott and Hoya. Source: Lambropoulos et al. 1997 [7]. Reproduced with permission from The Optical Society.

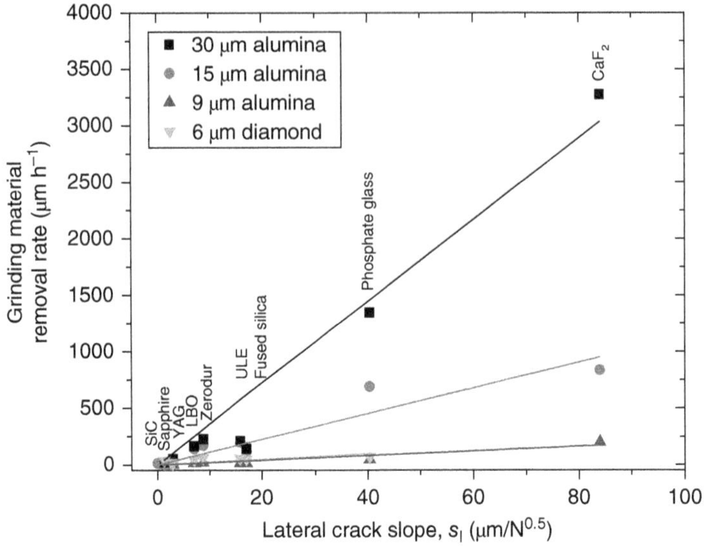

Figure 5.3 Measured material-removal rates of various optical materials as a function of the lateral-crack slope (s_ℓ). Process conducted as loose-abrasive grinding (1 psi applied pressure, 50 mm sample, 20 rpm workpiece and lap rotation, and granite lap) using various abrasives (Microgrit 30T (alumina), Microgrit 15T (alumina), and 6 μm diamond). Source: From After Suratwala et al. 2018 [9].

lateral-crack growth slope, $s_\ell = \chi_\ell (E_1/H_1)^{2/5}/H_1{}^{1/2}$, shown in Equation (3.6) and Figure 3.3c. For a variety of optical materials, including glasses, single crystals, and glass–ceramic, the grinding rate for several processes has been shown to scale linearly with the lateral-crack slope (see Figure 5.3) [9]. Lapping hardness and lateral-crack growth slope serve as useful metrics, using known material properties of workpieces to predict the relative grinding removal rate of a given workpiece material.

From the data in Figure 5.3, the material-removal rate increases roughly linearly with abrasive size over this large range of workpiece materials. In a more specific study performed on polycrystalline alumina workpieces, the material removal rate scaled as the square root of the mean particle size of the abrasive, $d^{1/2}$ (see Figure 5.4) [10].

In some materials and under the right conditions, grinding removal may also occur in the ductile regime. Using the balance between the energy needed to propagate a crack and that needed for plastic yielding, Bifano et al. [11] determined the transition from ductile to brittle removal for different materials, defined in terms of critical depth of cut d_c:

$$d_c = 0.15 \left(\frac{E_1}{H_1}\right) \left(\frac{K_{1c}}{H_1}\right)^2, \tag{5.2}$$

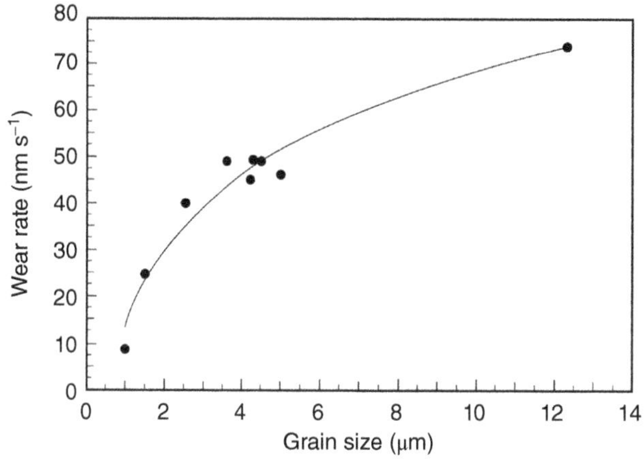

Figure 5.4 Measured grinding material-removal rate for polycrystalline alumina as a function of the mean abrasive-particle size. Source: Davidge and Riley 1995 [10]. Reproduced with permission from Elsevier.

where E_1, H_1, and K_{1c} are the elastic modulus, hardness, and fracture toughness of the workpiece. The critical depth of cut for a variety of materials was determined by measuring the number of surface microfractures (versus ductile tracks) as a function of the infeed rate on a fixed abrasive grinder – i.e. the *material removal rate.* For example, the transition from ductile to brittle fracture is shown for fused-silica glass in Figure 5.5 as the infeed rate was increased from 2 to 37.5 nm/revolution. The transition from ductile to brittle fracture can also be described as an indentation load, as shown in Figure 4.2, for fused-silica glass. Figure 5.6 shows that Equation (5.2) works well for a variety of materials, comparing the measured cut depth to that calculated. When the critical depth is small, as with most glasses, the material is more prone to removal by fracture under equivalent grinding conditions. Intuitively, the critical depth of cut can be largely thought of as the ratio of fracture toughness to hardness, where the fracture toughness describes the energy needed to propagate a crack and the hardness describes the energy needed to result in plastic deformation. The competition between fracture and plastic deformation is also described by the index of brittleness B_i of a material, given by [12]:

$$B_i = \frac{H_1 E_1}{K_{1c}}. \tag{5.3}$$

One important challenge in linking fundamental lateral crack growth to the global grinding-material removal rate is understanding the interactions among neighboring cracks. Demirci et al. [13] ran three simultaneous particle-scratch tests at various separation distances to understand the

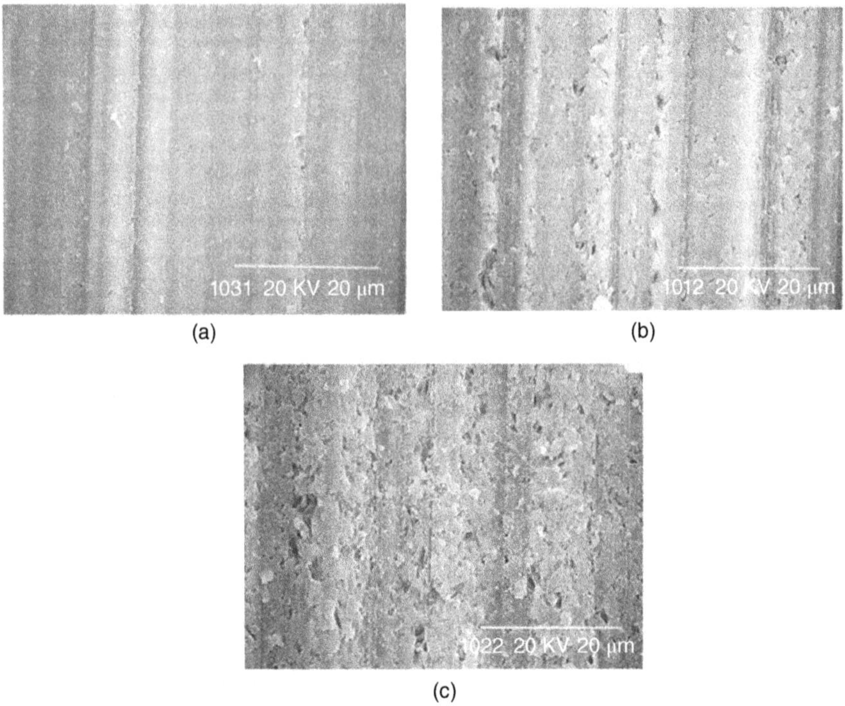

Figure 5.5 Scanning electron microscope (SEM) images of fused-silica glass after fixed-abrasive grinding at different infeed rates, illustrating the transition from ductile to fracture removal. (a) 2 nm/revolution, (b) 7.5 nm/revolution, and (c) 27.5 nm/revolution. Source: Bifano et al. 1991 [11]. Reproduced with permission from The American Society of Mechanical Engineers.

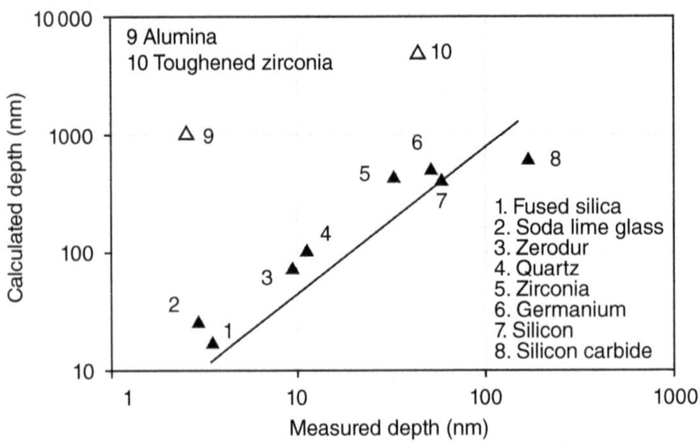

Figure 5.6 A comparison of calculated depth of cut (Equation (5.2)) and measured depth of cut for a variety of workpiece materials. Source: Bifano et al. 1991 [11]. Reproduced with permission from The American Society of Mechanical Engineers.

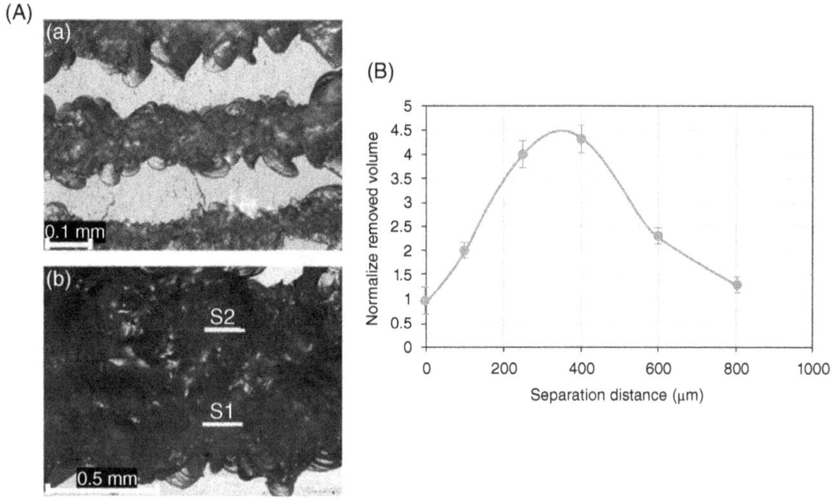

Figure 5.7 (A) Optical micrographs of glass scratches from left to right at a velocity of 25 m min^{-1} for nonintersecting cracks (a) and intersecting cracks (b); (B) relative removal volume of three simultaneous scratch tests as a function of the separation distance between abrasive particles. Source: Demirci et al. 2014 [13]. Reproduced with permission from Elsevier.

influence of neighboring cracks. Figure 5.7A shows images of the scratches with and without crack intersection. An approximately fourfold increase in volumetric material removal was found at an optimal spacing between abrasive particles. Particles too far apart do not have intersecting cracks, and particles too close do not get the full extent of the lateral crack growth from individual particles (see Figure 5.7B).

5.2 Polishing Material Removal Rate

5.2.1 Deviations from Macroscopic Preston Equation

There is much literature experimentally describing the dependence of polishing *material removal rate* with abrasive size, PSD, load, velocity, slurry concentration, and pad topography [14–19]. In many of these studies, a deviation is shown from the expected linear pressure and relative velocity dependence in the macroscopic Preston equation (Equation (1.3)). Pal et al. [20] reviewed a number of these variations and summarized them with various proposed modified Preston equations. An example by Tseng et al. [17] showed the following dependence on material removal (see Figure 5.8):

$$\frac{dh}{dt} = k_p \sigma_o^{5/6} V_r^{1/2}. \tag{5.4}$$

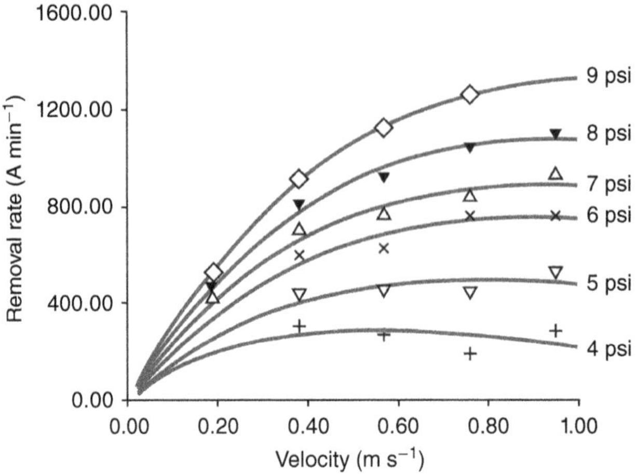

Figure 5.8 Effect of velocity and pressure on the material-removal rate during CMP polishing. Solid lines are predictions using a modified Preston equation (Equation (5.4)). Source: Tseng et al. 1999 [17]. Reproduced with permission from The Electrochemical Society.

The belief is that such nonlinear dependencies of applied pressure and relative velocity stem from different microscopic interactions at the interface. Changes in the polishing regime pertaining to pressure and velocity range and in the polishing parameters determine specific pressure and velocity dependence. To understand such variations, a microscopic approach to global material removal is required, as explored in Section 5.2.2.

Another example of a deviation from the Preston equation is associated with the time dependence. The original Preston equation provides a time-averaged or time-independent *material removal rate*. However, *material removal rate* has been shown to change with time. For example, Cumbo et al. [21] showed that during ceria polishing of BK7 glass, the workpiece *material removal rate* decreased with polishing time due to a change in the slurry PSD. Also, as discussed in Chapter 2, slurry-particle reactivity, without replenishment, decreases with distance, and hence time, across the workpiece–lap interface (see Figure 2.2).

Temperature rise due to frictional heating also changes the reactivity of the slurry, increasing the *material removal rate* (see Figure 2.31). Finally, the local deposit of workpiece reaction products on the lap with polishing time also leads to temporal and spatial changes in the *material removal rate* (see Figure 2.39). Just as deviations in the Preston equation occur with pressure and relative velocity, deviations in time dependence stem from changes in the microscopic interactions at the interface.

5.2.2 Macroscopic Material Removal Trends from Microscopic/Molecular Phenomena

Using Equation (2.1), a modified version of the Preston equation, may be sufficient to describe spatial and even temporal variations in the macroscopic *material removal rate,* as discussed in Chapter 2. To understand the influence of many other process parameters on the macroscopic *material removal rate,* however, it is advantageous to examine microscopic interactions, that is, the sum of removals from each active interface particles. Using concepts from the EHMG model discussed in Section 4.7.1 [22], the macroscopic *material removal rate* can be estimated using average microscopic parameters.

In this formulation, the thickness *material removal rate* is a summation of the removal by all particles in the interface, which can be described as follows:

$$\frac{\mathrm{d}h}{\mathrm{d}t} = \frac{v_\mathrm{r}}{A_\mathrm{pad}} \sum_{j=1}^{N_\mathrm{t}A_\mathrm{pad}} d_j 2a_j, \tag{5.5}$$

where v_r is the relative velocity of the particle, A_pad is the contact area between workpiece and pad, d_j is the depth of removal for a given particle j, and a_j is the contact radius for a given particle j [22]. Depending on the load per particle, the removal d_j can be 0 for no contact, d_m for chemical or molecular removal, or d_p for nanoplastic removal (see Figure 4.8). The contact radius for each particle size and interface gap is given by

$$a(r, g_\mathrm{p}) = \sqrt{r}\left(\frac{3}{4}\frac{P(r, g_\mathrm{p})}{E_{13}\sqrt{r}}\right)^{1/3}, \tag{5.6}$$

where g_p is the interface gap, r is the particle radius, E_{13} is the composite particle–workpiece modulus in Equation (4.20), P is load per particle [22]. The interface gap is determined from a series of load balances described using the EHMG model in Section 4.7.1.

Rewriting Equation (5.5) mentioned above, the *material removal rate* can be estimated using average microscopic material removal parameters in the form [23]:

$$\frac{\mathrm{d}h}{\mathrm{d}t} \approx N_\mathrm{t} f_\mathrm{r} f_\mathrm{A} f_\mathrm{L} V_\mathrm{r} (f_\mathrm{p} \bar{d}_\mathrm{p} 2\bar{a}_\mathrm{p} + f_\mathrm{m} \bar{d}_\mathrm{m} 2\bar{a}_\mathrm{m}), \tag{5.7}$$

where N_t is the areal number density of particles at the workpiece–pad interface, f_r is the fraction of active particles contributing to removal in the PSD, f_A is the fraction of pad area making contact with the workpiece, f_L is the fraction of applied load carried by particles (rather than by the pad directly), f_p is the fraction of active particles that contribute to plastic removal, $\bar{d}_\mathrm{p}$ is the average depth of particles that contribute to plastic removal, $\overline{a}_p$ is the average contact zone radius of particles that contribute to plastic removal, f_m is the fraction of active particles that contribute to chemical removal, $\overline{d}_m$ is the average depth of particles that contribute to chemical removal, and $\overline{a}_m$ is the average contact-zone radius of particles that contribute to chemical removal. The terms in the parentheses in Equation (5.7) represent the cross-sectional area of material removed from the workpiece surface; the first four variables before the parentheses represent the number of loaded particles per unit area that contribute to removal.

Unlike the macroscopic Preston equation, the microscopic *material removal rate* equation, Equation (5.7), does not have the applied pressure σ_o specifically identified. Instead, it is embedded in the microscopic parameters. As a check, the EHMG model was used to calculate *material removal rate* as a function of applied pressure and results compared with the measured *material removal rate* for a given set of polishing conditions (see Figure 5.9). Both the measured and calculated *material removal rate* show a largely linear dependence consistent with the original Preston equation (Equation (1.3)) [23].

Equation (5.7), the microscopic *material removal rate* equation, also provides significant insights as to the type of microscopic interactions that may influence the *material removal rate* during polishing. Hence, this expression can be used as a general tool for optimizing polishing conditions, complementing

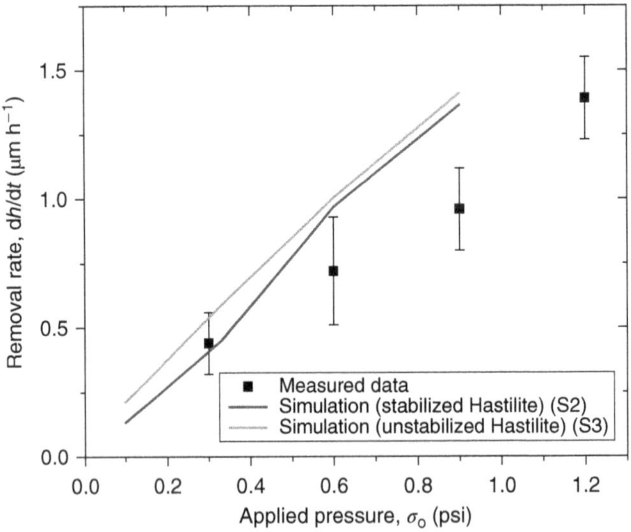

Figure 5.9 Measured and calculated macroscopic material-removal rate using EHMG model for two Hastilite slurries, unstabilized (S2) and stabilized (S3), as a function of applied pressure. Source: Suratwala et al. 2014 [23]. Reproduced with permission from John Wiley & Sons.

the standard practice of simply adjusting load and velocity as suggested by the Preston equation (Equation (1.3)). To test this, the EHMG model was used to calculate the critical parameters in Equation (5.7) for a representative set of polishing experiments [24]. The results are summarized in Figure 5.10. The top of Figure 5.10 shows the measured *material removal rate* for four experimental series in which the removal rate increases with a different manipulated polishing variables, namely, widening PSD, increasing slurry concentration, flattening pad topography, and changing the glass type. For each series, the bottom row of Figure 5.10 shows which microscopic parameters in Equation (5.7), as calculated by the EHMG model, change the most in response to changes in the manipulated variable. (The other calculated parameters in Equation (5.7) varied comparatively slowly and, for simplicity, are not plotted.) For each series, the model predicts that a different microscopic parameter dominates the observed trends. Hence, the reason for an increased removal rate is significantly different for each of the series at the microscopic scale.

Starting with the slurry PSD series (Figure 5.10, column 1), the change in PSD has a large effect on load-per-particle distribution, as shown in Figure 4.8 for the same slurries. Changes in the tail end of the slurry PSD influence the number of particles loaded, i.e. active particles, and the fraction of particles that remove material in the chemical f_{m} versus plastic f_{p} regime [23]. The 30 nm CeO_2 slurry (NanoArc 6450) resulted in the loading of essentially all particles in the chemical removal regime ($f_{\mathrm{m}} \sim 1$ and $f_{\mathrm{p}} \sim 0$). The 200 nm CeO_2 slurry (stabilized Hastilite) resulted in particle loading in both the chemical and nanoplastic removal regimes. Finally, the agglomerated 200 nm CeO_2 slurry (unstabilized Hastilite) yielded a larger fraction of particles in the nanoplastic regime. The increase in the *material removal rate* is dominated by the greater number of particles removing material in the plastic regime, which has a larger removal depth per particle [24].

For the slurry-concentration series (Figure 5.10, column 2), the *material removal rate* initially increases and then plateaus at high slurry concentrations. The EHMG model that predicts this is dominated by two factors: the number of active particles $N_{\mathrm{t}} \times f_{\mathrm{r}}$, and the fraction of load carried by the pad f_{L}. At low concentrations, both $N_{\mathrm{t}} \times f_{\mathrm{r}}$ and f_{L} increase, thus increasing the *material removal rate.* At higher concentrations, however, the number of active particles starts to decrease, owing to a slight increase in load per particle from sampling more, larger particles in the distribution. In addition, the fraction of load carried by the particles approaches and plateaus at one (i.e. there are so many particles at the interface that all the load is carried by particles, and none by the pad). Both experimental and calculated data show a similar plateau effect [24].

For the pad-topography series (Figure 5.10, column 3), the increase in *material removal rate* with the diamond conditioning treatment, which causes pad flattening, is dominated by an increase in the fraction of pad area making contact with the workpiece f_{A} (see pad images in Figure 4.14). Using the measured

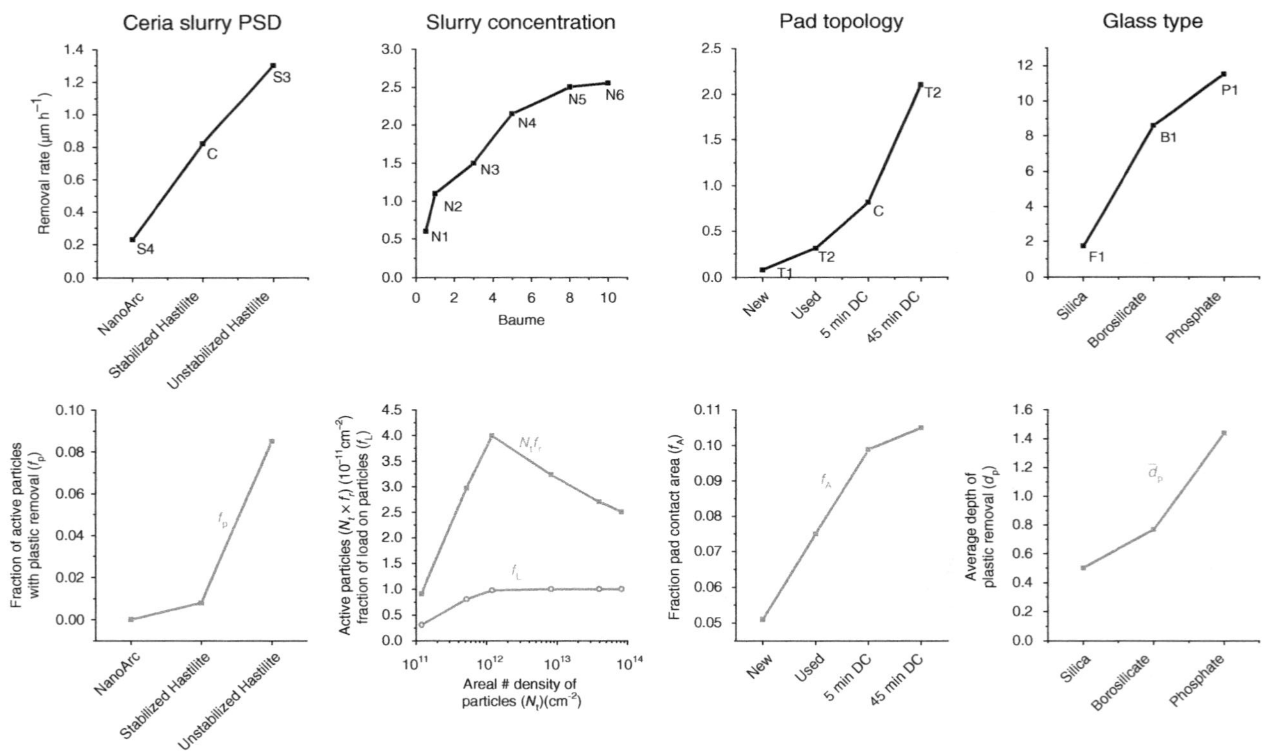

Figure 5.10 A comparison of measured removal rate (top) for key polishing variables for various series (PSD, slurry concentration, pad topography, and glass type) with key EHMG model parameter calculated or defined, and (bottom) for the same corresponding polishing conditions. Source: Suratwala et al. 2016 [24]. Reproduced with permission from John Wiley & Sons.

height histograms in Figure 4.20a, the calculated load balance, using Equation (4.15), determines f_A. The pad with tall asperities (a new MHN pad) results in less overall contact between the pad and workpiece, i.e. smaller f_A, and hence, fewer particles at the interface contributing to removal. As a result, flatter pad topographies lead to a larger f_A and higher calculated *material removal rate*, correlating well with experimental data [24].

Experimentally, the fraction of pad area making contact has been determined to be quite low. Figure 5.11 shows f_A as a function of applied pressure for various pads from different studies. In the pressure range for optical polishing, f_A is on the order of $<10^{-3}$ for fibrous pads and <0.1 for polyurethane foam pads [23, 25, 26]. Pad topography and effective pad elastic modulus strongly influence f_A. For example, Moon [27] showed an increase in *material removal rate* with decrease in mass density and increase in compressibility; both trends are consistent with the influence of f_A with *material removal rate.* Warnock [28] proposed a polishing model that incorporates pad asperities, especially for fibrous pads and discussed how their bending upon loading may contribute to increased f_A and *material removal rate.* If further strategies were developed to make the small f_A value larger, potentially much higher *material removal rates* might be achieved.

Finally, for the glass-type series (Figure 5.10, column 4), the *material removal rate* increase is dominated by a change in the single-particle removal function, namely the depth of nanoplastic removal (d_p). Since the agglomerated 200 nm CeO_2 slurry (unstabilized Hastilite) was used, a significant number of particles were removing material in the nanoplastic removal regime, as opposed to the chemical removal. The measured removal rate and d_p increased by approximately the same amount going from silica to borosilicate to phosphate glass. Note that changes in the mechanical properties of the glass yield only slight changes in the load-per-particle distribution [24].

In a microscopic-model approach by Luo and Dornfeld [16] based on Hertzian mechanics (like the EHMG model), the relative *material removal rate* was correlated to the characteristics of the slurry PSD as follows:

$$\frac{dh}{dt} \propto \frac{(\bar{r} + 3\sigma_{part})^2}{\bar{r}^3} \tag{5.8}$$

where $\bar{r}$ is the average slurry-particle size, and σ_{part} is the standard deviation in the PSD. A plot of the relative *material removal rate* as a function of the $\bar{r}$ and σ_{part} is shown in Figure 5.12a. At a fixed average particle size, increasing σ_{part} can result in a minimum in the *material removal rate.* Increase in σ_{part} results in (i) a reduction in the number of active particles (reducing *material removal rate*), and (ii) a larger load per particle leads to more removal per particle (increasing the *material removal rate*).

Generally, *material removal rate* **increases** with particle size, owing to increases in load per particle and removal-function depth (see Figure 1.3).

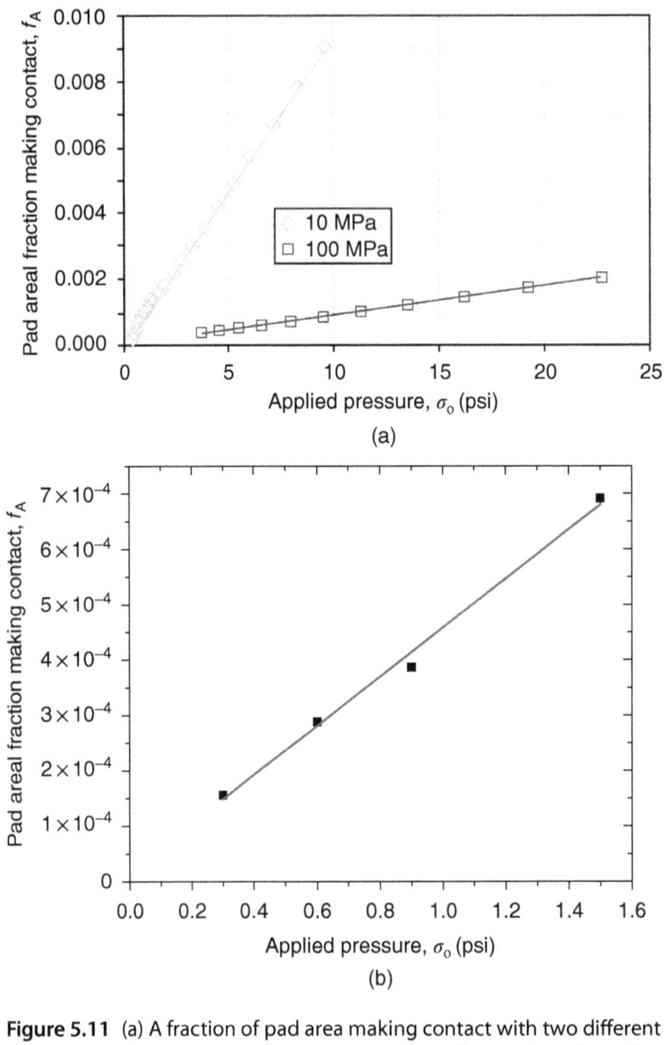

Figure 5.11 (a) A fraction of pad area making contact with two different pads on a Si wafer. Source: Li 2007 [25]. Reproduced with permission from John Wiley & Sons. (b) A fraction of pad area making contact on a polyurethane pad on a fused-silica workpiece. Source: Suratwala et al. 2014 [23]. Reproduced with permission from John Wiley & Sons.

Figure 5.12b shows a unique case where an increase in average particle size (with little change in σ_{part}) results in a **decrease** in the *material removal rate,* because fewer particles are loaded and contributing to material removal. This effect has been commonly observed with colloidal-silica slurries in which a change in $\bar{r}$ results in little change in the σ_{part}. With most other polishing and grinding slurries, an increase in the average particle size is accompanied by an

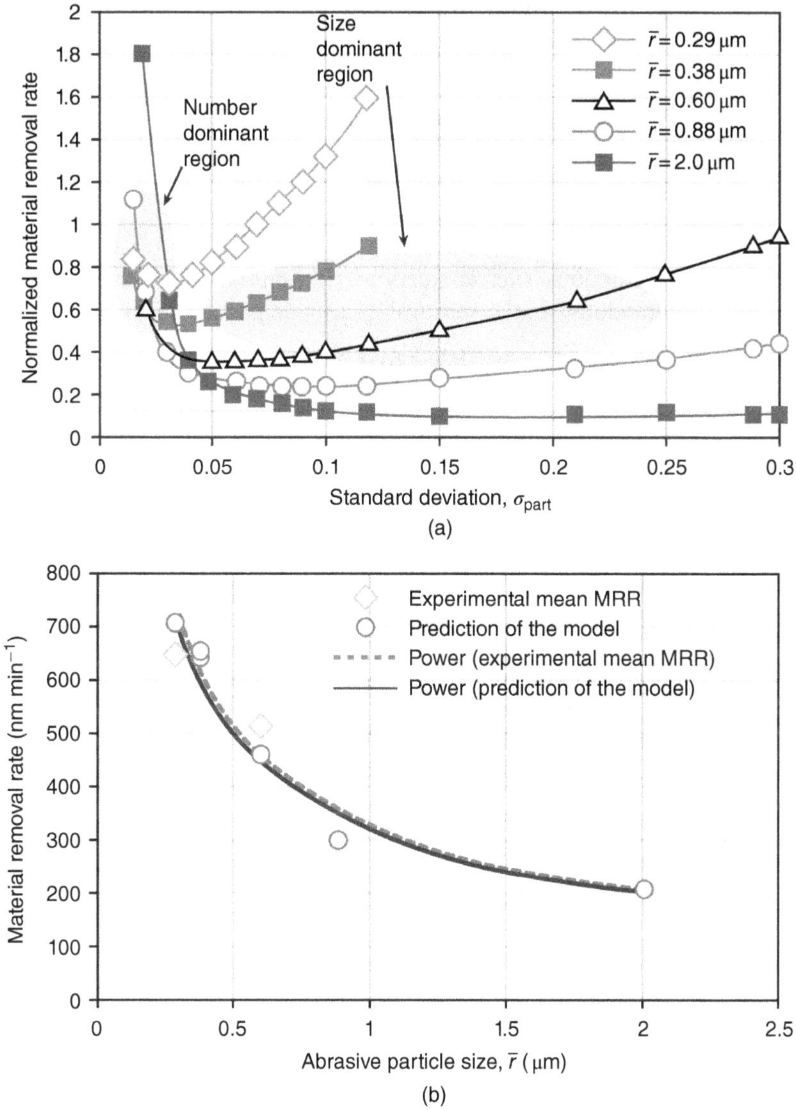

Figure 5.12 (a) The relative removal rate as a function of average particle size and standard deviation using Equation (5.8). (b) A comparison of model, represented by Equation (5.8), and measured data for material removal rate as a function of abrasive-particle size in the number-dominant region. Source: Luo and Dornfeld 2003 [16]. Reproduced with permission from IEEE.

equivalent change in PSD width, usually resulting in an increase in *material removal rate* (see Figure 1.3).

5.2.3 Factors Affecting Single-particle Removal Function

To review, the *material removal rate* during polishing is governed by the following:

- Large-scale interactions, such as pressure distribution (Chapter 2)
- Microscopic interactions, such as slurry particle interactions (Section 5.2.2)
- Molecular interactions, that is, removal by a single particle (Section 4.1).

The amount of material a particle removes from the workpiece surface depends on the mechanisms outlined in Figure 1.9. For polishing, the removal mechanisms are

- Chemical dissolution
- Nanoplastic removal
- Chemical reactions.

Since chemical dissolution does not permit local removal control, it is generally not used for polishing, but recognizing its potential occurrence is important. This section first discusses nanoplastic trends, which are dominated by the mechanical properties of the particle and workpiece (Section 5.2.3.1), and then chemical reaction, which is dominated by the factors that control the condensation rate between the particle and workpiece (Section 5.2.3.2).

5.2.3.1 Nanoplastic Effects: Workpiece Hardness

As evidenced by the nanoscratching discussed in detail in Section 4.1, nanoplastic material removal during polishing may occur at the appropriate load-per-particle regime during polishing (see Figure 4.8) or if little or no chemical reaction takes place between the particle and the workpiece. For a fixed polishing process where nanoplastic material removal is dominant, the workpiece material microhardness is expected to dominate. Izumitani observed this dependence in a series of optical glasses where the global polishing rate scaled inversely with Vickers microhardness when nonaqueous slurries were used, which inhibit the chemical interactions from occurring (Figure 5.13) [5].

In a more recent study, the nanoscratching method for measuring nanoplastic removal function was measured on a variety of optical materials (see Figure 5.14) [29]. The effective material-removal depth d_p was found to scale linearly with polishing rates when the optical materials were polished with colloidal-silica slurry (Figure 5.15). These results suggest that colloidal-silica polishing at room temperature is dominated by nanoplastic material removal and chemical reaction is minimal. Also illustrating the mechanical nature of

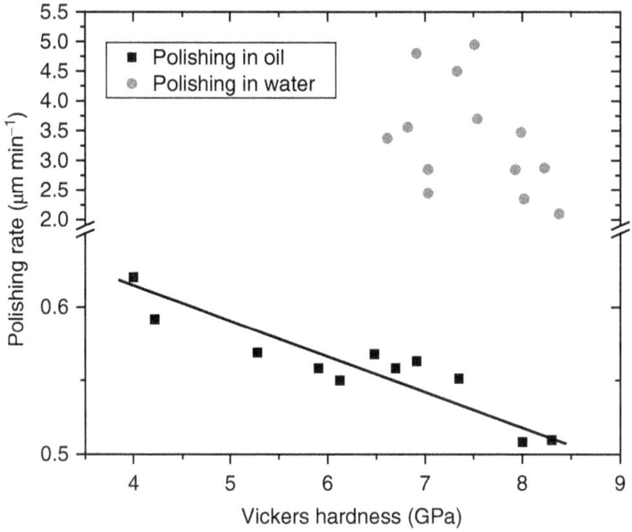

Figure 5.13 Polishing rate of various glasses using an oil-based slurry and an aqueous slurry as a function of Vickers hardness. Source: Data from Izumitani 1986 [5].

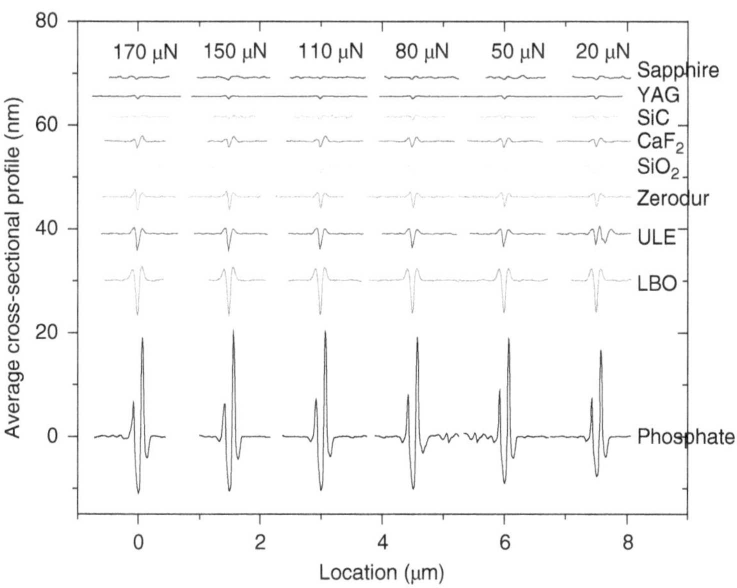

Figure 5.14 Cross-section lineout of nanoscratches of various optical materials performed at various applied loads in an aqueous environment. Source: From Shen et al. 2018 [29].

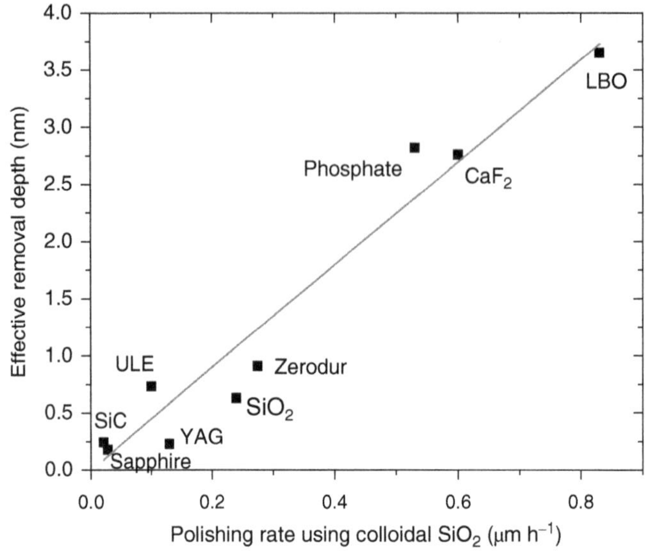

Figure 5.15 From data in Figure 5.14, correlation between effective nanoscratch depth d_p of various optical materials and their measured polishing material-removal rate using colloidal silica slurry. Source: From Shen et al. 2018 [29].

removal during colloidal-silica polishing, Figure 5.16a shows the correlation between the polishing and grinding rates for various optical materials, which are both determined by the mechanical properties of the workpiece [30]. The corresponding mechanical properties that scale with polishing and grinding rates are also correlated, as seen in Figure 5.16b.

For the specific case of colloidal-silica polishing of sapphire (Al_2O_3), several studies have reported different material-removal mechanisms. Namba and Tsuwa [31] propose that removal was mechanical, consistent with the results mentioned above. Other studies, however, have reported a chemical reaction leading to the formation of aluminosilicate ($A1_2Si_2O_7$-H_2O) on the workpiece surface [32, 33]. These studies showed chemical-reaction evidence at elevated temperatures during polishing. A possible explanation consistent with both data sets is that mechanical removal dominates at room temperature and chemical reaction at elevated temperatures.

5.2.3.2 Chemical Effects: Condensation Rate and Partial-charge Model

As discussed in Sections 2.2, 3.3.2, and 4.1, a chemical reaction may occur between the workpiece and polishing particle as a mechanism for material removal during polishing. The nature of the chemical reaction, specifically for oxide glasses, has been proposed as governed by condensation and hydrolysis

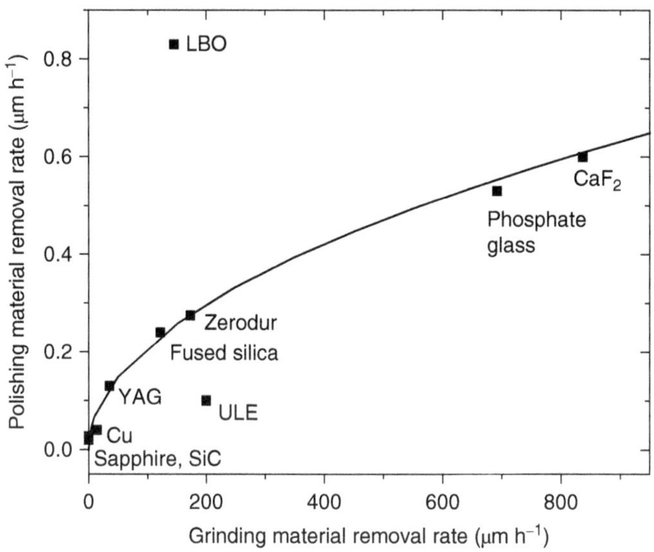

(a)

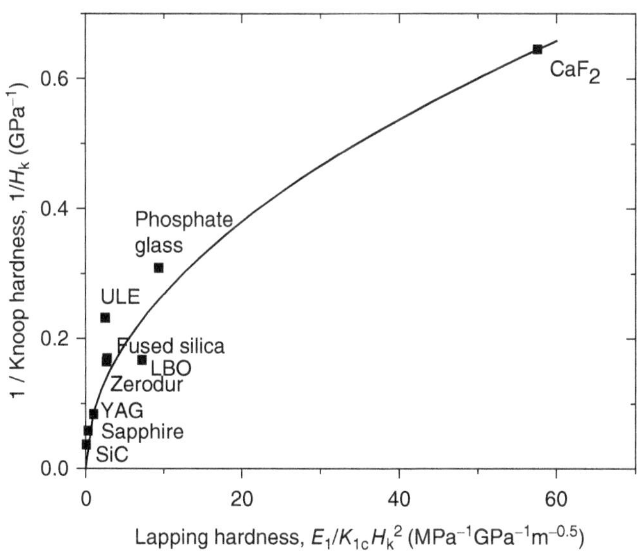

(b)

Figure 5.16 (a) The correlation between the measured polishing material-removal rate of various optical materials using colloidal-silica slurry with the grinding material-removal rate using 15 μm alumina particles. (b) Replot of data in (a), but replacing polishing removal rate with the expected hardness scaling and replacing the grinding removal rate with expected lapping hardness scaling. Source: After Suratwala et al. 2018 [9] and Suratwala et al. 2018 [30].

reactions, as illustrated for silica glass using cerium oxide in Equations (3.61)–(3.63) [34]. The chemical nature of the removal was first demonstrated by a need for the presence of H_2O. Izumitani showed a significant increase in the material removal rate on a series of glasses polished in an aqueous slurry, as compared to oil-based slurries. This difference is shown in Figure 5.13 – note the change in y-axis scale [5, 35]. In other studies, after polishing in various solvent environments such as alcohol and glycol–water mixtures, the material removal rate scaled strongly with the OH or H_2O content in the slurry [36, 37].

For the specific case of silicate glasses, Cook showed that the relative material removal rate using different polishing-slurry compounds was correlated to the slurry particle's isoelectric point (IEP), as shown in Figure 5.17 [34]. The maximum polishing material removal rate occurred with cerium oxide (IEP = 7), with rates decreasing significantly at lower or higher values of the slurry particle's IEP. The lowest material removal rate occurred with silicon dioxide,

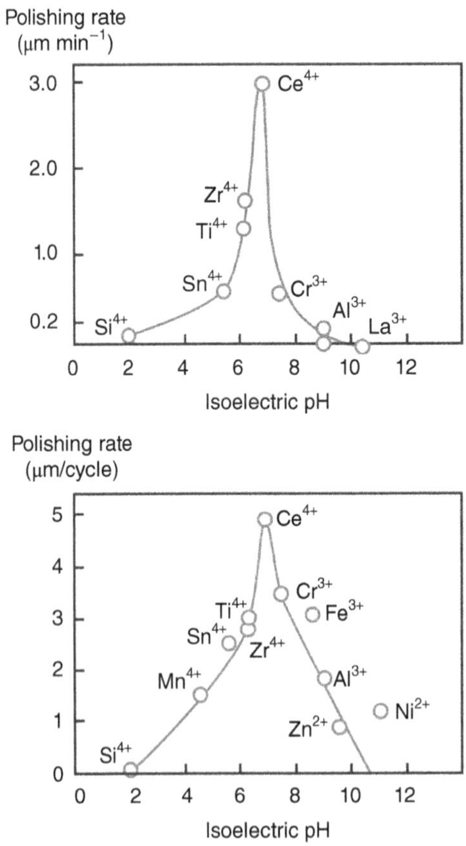

Figure 5.17 Relative polishing material-removal rate as a function of IEP of various slurry compositions on two different silica-based glass workpieces. Source: Cook 1990 [34]. Reproduced with permission from Elsevier.

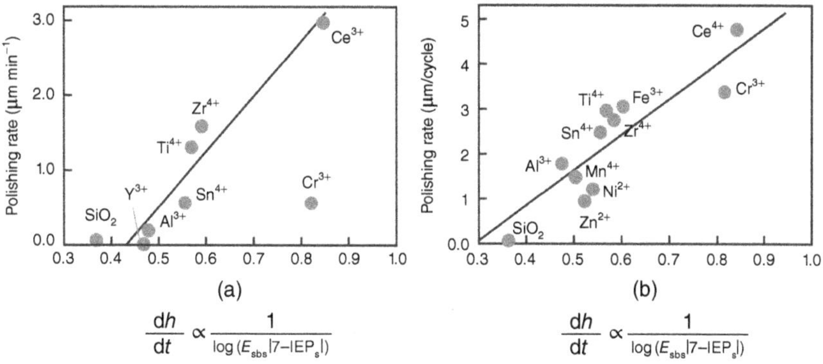

Figure 5.18 Polishing material-removal rate as a function of reaction-rate factor Equation (5.9) for the same data in Figure 5.17. Source: Cook 1990 [34]. Reproduced with permission from Elsevier.

consistent with the mechanical nature of colloidal silica proposed in Section 5.2.3.1. Using the same data, Cook demonstrated that the polishing material removal rate is proportional to an empirically derived reaction-rate factor given by

$$\frac{\mathrm{d}h}{\mathrm{d}t} \propto \frac{1}{\log(E_{\mathrm{sbs}}|7 - \mathrm{IEP_s}|)}, \tag{5.9}$$

where E_{sbs} is the single bond strength and $\mathrm{IEP_s}$ is the isoelectric point of the slurry particle (Figure 5.18).

Cook's proposed reason for the increase in *material removal rate* under the conditions of CeO_2 (IEP = 7) slurry on a SiO_2 workpiece (IEP = 2) is that a condensation reaction is favored when the workpiece has a negative surface, and the particle surface is largely neutral [34, 38]. Hence, the condensation reaction given by Equation (3.62) is better written as follows:

$$\equiv \mathrm{Si{-}O^-} + \mathrm{HO{-}M_p} \equiv \rightarrow \equiv \mathrm{Si{-}O{-}M_p} \equiv + \mathrm{OH^-} \tag{5.10}$$

From the perspective of colloidal stability, however, keeping a slurry near the particle's IEP may lead to agglomeration, as discussed in Section 4.5.2. Agglomeration can be prevented using various slurry additives or operating at a pH slightly off the IEP of the slurry particle.

When slurry pH is above the IEP of the polishing particle or workpiece, the surface charge is negative; when the slurry pH is below the IEP, the surface charge is positive. The charge of surface hydroxyls on the polishing particle or workpiece can be described as follows:

$$-\mathrm{M_s{-}OH_2^+} = -\mathrm{M_s{-}OH} + \mathrm{H^+} \quad K_{\mathrm{s1}} \tag{5.11}$$

$$-\mathrm{M_s{-}OH} = -\mathrm{M_s{-}O^-} + \mathrm{H^+} \quad K_{\mathrm{s2}} \tag{5.12}$$

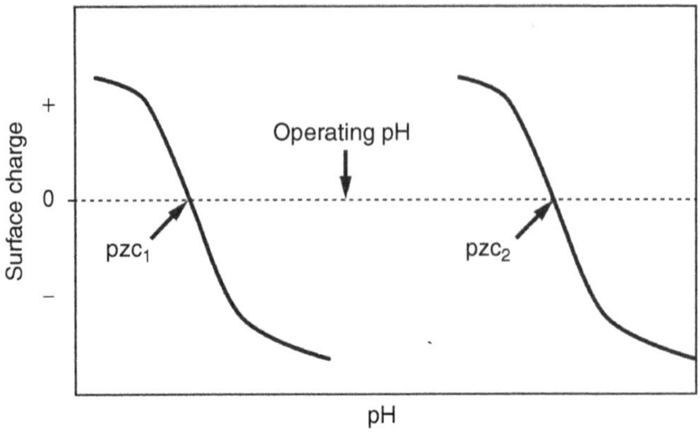

Figure 5.19 A comparison of the surface charge on a particle with PZC above or below the pH of the slurry. Source: Osseo-Asare 2002 [38]. Reproduced with permission from The Electrochemical Society.

where K_{s1} and K_{s2} are the equilibrium constants, and M_s is the atomic species of the polishing particle or workpiece surface. The point of zero charge (PZC), i.e. the pH at which the surface charge is neutral, is approximately the IEP [38]. These are quantitatively related to the equilibrium constants as [39]

$$\text{IEP} \sim \text{PZC} = -\frac{1}{2}\log(K_{s1}K_{s2}) = \frac{1}{2}(pK_{s1} + pK_{s2}), \tag{5.13}$$

where pK_{s1} is simply $-\log(K_{s1})$ and pK_{s2} is $-\log(K_{s2})$. Hence, the IEP of the material scales linearly with pK_{s1} or pK_{s2}. A schematic surface charge as a function of pH, and IEP is illustrated in Figure 5.19.

The studies above are largely limited to silicate-glass workpieces. To develop a broader understanding of the solid-state chemical-removal mechanism, a comprehensive model is needed that accounts not only for slurry-particle composition and slurry pH but also the composition of the workpiece. Izumitani showed evidence that polishing reactivity scales with the chemical durability of the workpiece material [5]. In a recent study, the polishing *material removal rate* was correlated for polishing slurry materials, slurry pH, and a matrix of optical-workpiece materials while largely keeping the other microscopic interactions, as discussed in Section 5.2.2, constant [30]. Using this data, an expanded chemical-reaction-rate model, using partial charge and the IEP and accounting for both slurry-particle and workpiece composition, was developed, as described below.

Using the chemical-tooth model as a foundation [34], the general reaction mechanism can be described by the following reaction steps [30]:

(1) Formation of reactable surface moieties on the workpiece and polishing particle surface, such as the formation of hydroxides.
(2) Approach of particle to the workpiece surface .

(3) Formation of an interface bond, such as between the surface hydroxides of the workpiece and the polishing particle via a condensation reaction.
(4) Strain-induced bond breakage of the workpiece or particle, such as by a hydrolysis reaction

The specific reaction pathway outlined is likely for most oxide glasses and possibly for other workpiece materials. Other pathways are possible, following the basic steps outlined here, such as by stress corrosion, some other type of direct bonding, or a more complex reaction scheme, such as the formation of an interfacial reaction layer that then chemically reacts or dissolves (see for example [32, 33]).

The condensation and hydrolysis reactions, described by Equations (3.61)–(3.63) for cerium-oxide particles on fused-silica glass, can generally be written without accounting for pH and surface charge as

$$\equiv \mathrm{M_{wp}{-}OH + HO{-}M_p} \equiv \ \rightarrow\ \equiv \mathrm{M_{wp}{-}O{-}M_p} \equiv + \mathrm{H_2O}, \tag{5.14}$$

$$\begin{aligned} &\equiv \mathrm{M_{wp}{-}O{-}M_{wp}{-}O{-}M_p{-}O{-}M_p} \\ &\equiv + \mathrm{H_2O} \rightarrow\ \equiv\ \mathrm{M_p{-}OH + HO{-}M_{wp}{-}O{-}M_p{-}O{-}M_p} \equiv, \end{aligned} \tag{5.15}$$

where M_{wp} is a metal atom of the workpiece (e.g. Si for SiO_2 and Al for Al_2O_3) and M_p is the metal atom for the polishing particle (e.g. Ce for CeO_2). The slowest of these reactions will determine the overall reaction rate and polishing *material removal rate*. For most oxide glasses, hydroxide formation is fast. For other nonoxide optical materials (e.g. CaF_2 and SiC), however, the rate at which surface hydroxides are formed may be rate limiting. As proposed by Cook [34], the condensation rate generally limits the rate.

When rate limited by step 4, the partial charge and the partial-charge difference for a given polishing particle and workpiece material has been shown to be related to the condensation rate, and hence the polishing *material removal rate* [30]. It is important to distinguish between the net charge on a species – the total surface charge of the compound – and the partial charge, which describes the charge distribution around the species of interest. Net charge is determined by the electron balance of all the ions in a compound, the IEP of the material, and the pH of the slurry environment. By contrast, the partial charge of a species is largely determined by the electronegativity of the metal atom and the surrounding moieties. In nonpolishing-related studies, Livage et al. used the partial-charge model to describe the relative condensation rates of various metal alkoxides in sol–gel chemistry [40]. Later, Reynolds quantitatively correlated the condensation rates of various silicate species to the partial-charge difference between two reacting species [41]. Both showed that the larger the partial charge and partial-charge difference between the reactants, the greater the rate of condensation reaction.

When charge transfer occurs between two bonding atoms, each atom will acquire a partial charge δ_i, largely determined by the electronegativity χ_i° of the neutral atom. The electron transfer stops when the electronegativities of all the atoms equal the mean electronegativity of the compound $\overline{\chi}$, which gives

$$\overline{\chi} = \frac{\sum_i p_{si}\sqrt{\chi_i^o} + 1.36z}{\sum_i \frac{p_{si}}{\sqrt{\chi_i^o}}}, \tag{5.16}$$

where p_{si} is the stoichiometry of atom i and z is the valence of the compound [40]. The partial charge of a specific atom δ_i in the compound is then defined as follows:

$$\delta_i = \frac{(\overline{\chi} - \chi_i^o)}{1.36\sqrt{\chi_i^o}}. \tag{5.17}$$

Figure 5.20 shows the calculated partial charge for various hydroxide surfaces for metal cations of different oxidation states. $Ce(OH)_4$ or Ce^{4+} has the greatest partial charge and would be expected to have the highest reaction rate. Note that Ce^{4+} has a higher partial charge than Ce^{3+}. The partial-charge value is dominated by the electronegativity of the metal atom, where a smaller electronegativity leads to a higher partial charge. Ce has an electronegativity of 1.12. There are elements on the periodic table with lower electronegativity, but most are Group I alkali metals (e.g. Li, Na, K, and Rb) or Group II alkaline earth metals (e.g. Ca, Sr, and Ba), which are mostly water soluble. La has lower electronegativity, but Ce in the 4^+ state has the edge over La, which can only be in a 3^+ oxidation state. This partial-charge analysis appears to answer

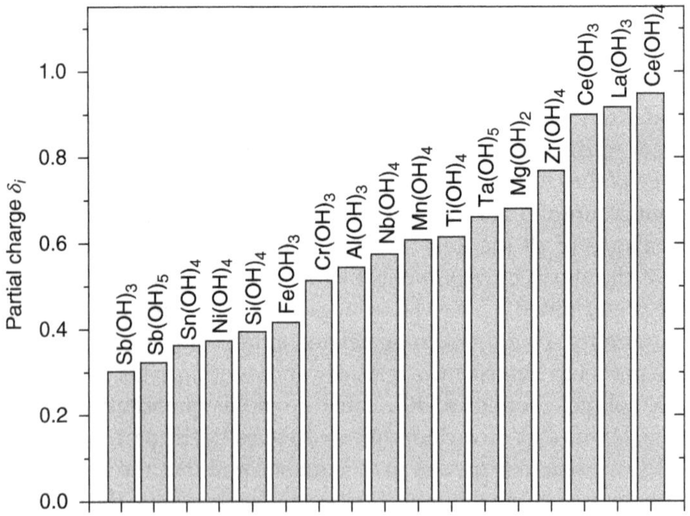

Figure 5.20 Calculated partial charge using Equations (5.16) and (5.17) for various hydroxides, as a representation for polishing particles. Source: From Suratwala et al. 2018 [30].

why, historically, cerium oxide has been the polishing slurry of choice for oxide glasses [30].

The partial charges of the other metal hydroxides in Figure 5.20 also generally follow the trend shown by Cook as to the polishing rate of various polishing particles (see Figure 5.17) [34]. A small change in partial charge has been shown to have a large influence on condensation rate, as shown for various alkoxides. For example, the hydrolysis-condensation rate of the silicon alkoxide $Si(OC_2H_5)_4$ with a partial charge of 0.32 is 10^{-4} mol^{-1} s^{-1}, whereas the rate for a titanium alkoxide is with a partial charge of 0.63 is 30 mol^{-1} s^{-1}, varying by five orders of magnitude [40, 42]. Since Si has a low partial charge, it is not expected to have a strong chemical component in the *material removal rate*. In fact, the room-temperature-polishing removal-rate data of colloidal-silica polishing slurry scales with the mechanical properties of the workpiece, i.e. linked to nanoplastic removal, as shown in Figures 5.15 and 5.16.

Expanding the partial-charge concept to include both the slurry particle and the workpiece material, the condensation rate – and hence polishing *material removal rate* – should be proportional to the partial-charge difference $\delta_{\text{wp-s}}$ between the workpiece and slurry as

$$\delta_{\text{wp-s}} = \delta_{\text{wp}} - \delta_{\text{s}}, \tag{5.18}$$

where δ_{wp} is the partial charge of the workpiece surface and δ_{s} is the partial charge of the slurry particle [30]. Figure 5.21 shows the measured polishing *material removal rate* as a function of the partial-charge difference for polishing eight workpiece materials: fused silica (SiO_2), laser-phosphate glass -LHG8 ([56–60]P_2O_5-[8–12]Al_2O_3-[13–17]K_2O-[10–15]BaO-[0–2]Nd_2O_3), sapphire (Al_2O_3), silicon carbide (SiC), ULE (SiO_2:TiO_2 glass), calcium fluoride (CaF_2), YAG ($Y_3Al_4O_{12}$), Zerodur (SiO_2 based glass ceramic), and LBO (LiB_3O_5). Four polishing slurries are used with approximately the same PSD and particle size of 30–50 nm: (Sb_2O_5, MgO, ZrO_2, and CeO_2) [30]. The slurry materials were selected for a wide range of partial charge (see Figure 5.20), narrow PSDs, and a small average-particle size, such that the load-per-particle distribution would overlap with the removal function of the workpiece in the solid-state chemical regime only (i.e. nanoplastic material removal is negligible). The polishing rate for most of the data generally follows the partial charge difference, with the following empirical exponential dependence [30]:

$$\frac{dh}{dt} \approx 0.35\,\mu\text{m}\,\text{h}^{-1} \exp\left(\frac{\delta_{\text{wp-s}}}{0.18}\right), \tag{5.19}$$

suggesting that condensation rate is the controlling reaction mechanism for these specific slurry – workpiece combinations. From Figure 5.21, it is also clear, however, that there are two sets of outliers where the *material removal rate* did not correlate with the partial-charge difference. The first is with the CaF_2 and LBO workpieces polished with Sb_2O_5 slurry (see dashed circles) where the

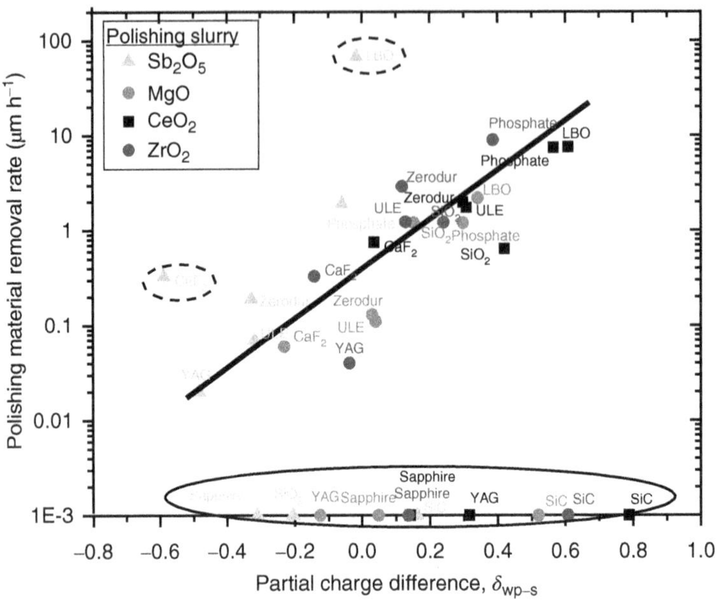

Figure 5.21 The material-removal rate of various optical material workpieces polished using various colloidal slurries as a function of partial-charge difference, given by Equations (5.16)–(5.18). Samples shown with a 0.001 μm h^{-1} material-removal rate had an immeasurable rate and were assigned this value to identify them on the log scale. Source: From Suratwala et al. 2018 [30].

measured *material removal rate* was significantly higher than expected from the partial-charge difference relationship. The probable explanation is that a different material-removal mechanism is dominating, namely dissolution, as evidenced by both workpiece materials' dissolving by just soaking in the slurry.

The second set of outliers is largely with the sapphire, YAG, and SiC workpieces, identified by the solid circles in Figure 5.21. Regardless of the slurry used, the material-removal rate was essentially zero. These three materials have IEPs between 4.5 and 6.5, as illustrated in Figure 5.22. The workpiece material-removal rate increases significantly when the IEP is either greater than 6.5 or less than 4.5. Only colloidal silica slurries polished these materials, albeit at a slow rate, because they are effecting nanoplastic removal (refer to Figure 5.15) [30].

The interpretation is that workpiece materials with IEPs in the 4.5–6.5 range are inherently resistant to condensation reactions. As proposed by sol–gel researchers [42], the condensation reaction rate of metal alkoxides may also be influenced by surface charge (not to be confused with partial charge) as controlled by the pH. Figure 5.23 illustrates this for silicon alkoxides, where the condensation rate, measured as the inverse of gel time, increases dramatically above and below the IEP = 2 of SiO_2. Because the IEP of sapphire, YAG, and

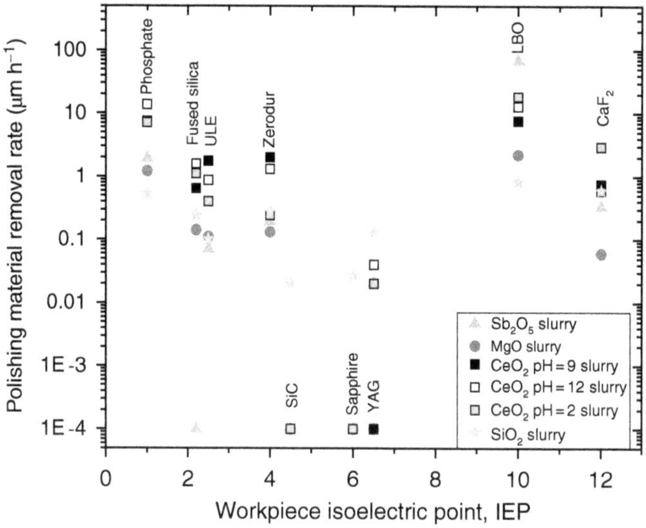

Figure 5.22 The material-removal rate of the data in Figure 5.21 as a function of workpiece IEP. Samples shown with a 0.0001 μm h^{-1} material-removal rate had an immeasurable rate and were assigned this value to identify them on the log scale. Source: From Suratwala et al. 2018 [30].

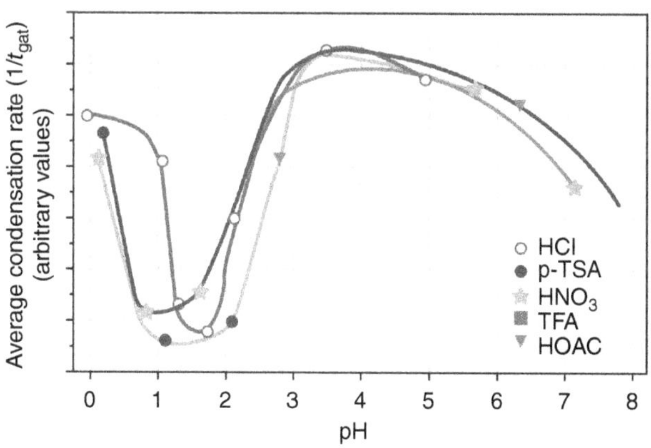

Figure 5.23 Condensation rate versus pH for silicon alkoxides. Source: Brinker and Scherer 1990 [42]. Reproduced with permission from Academic Press.

SiC is neutral, they are prone to a more neutral surface charge at neutral pH environments. Hence, it is proposed that these workpiece materials are inherently less prone to condensation reactions, and thus produce little or no solid-state chemical material removal. Another possibility, especially with SiC, is the formation of the surface hydroxide may be limiting the overall set of reactions (Step 1).

Various combinations of surface charge on the particle and workpiece may exist, depending on the IEPs of the particle and workpiece and the slurry pH. The change in surface charge not only influences the condensation rate but also affects the interface slurry-particle distribution, as discussed in Section 4.5. Both these results may influence the overall polishing material-removal rate, possibly as competing effects. The influence of slurry pH changes on the material-removal rate of the matrix of polishing conditions is also shown in Figure 5.22.

In sum, the polishing *material removal rate*, when limited by the condensation rate, can be correlated to the partial-charge difference between the workpiece and polishing particle, as well as the IEP of the workpiece. These concepts expand the original chemical-tooth model proposed by Cook [34] and offer utility in developing new polishing processes for various workpiece materials.

References

1 Austin, L.G. (1971). Introduction to the mathematical description of grinding as a rate process. *Powder Technol.* 5 (1): 1–17.

2 Klimpel, R. and Austin, L. (1977). The back-calculation of specific rates of breakage and non-normalized breakage distribution parameters from batch grinding data. *Int. J. Miner. Process.* 4 (1): 7–32.

3 Wiese, G. and Wagner, R. (1974). Physical model for predicting grinding rates. *Appl. Opt.* 13 (11): 2719–2722.

4 Aleinikov, F. (1957). The effect of certain physical and mechanical properties on the grinding of brittle materials. *Sov. Phys. Tech. Phys.* 2 (12): 2529–2538.

5 Izumitani, T. (1986). *Optical Glass*, x, 197. New York: American Institute of Physics.

6 Izumitani, T. and Suzuki, I. (1973). Indentation hardness and lapping hardness of optical glass. *Glass Technol.* 14 (2): 35–41.

7 Lambropoulos, J.C., Xu, S., and Fang, T. (1997). Loose abrasive lapping hardness of optical glasses and its interpretation. *Appl. Opt.* 36 (7): 1501–1516.

8 Buijs, M. and Korpel-van Houten, K. (1993). A model for lapping of glass. *J. Mater. Sci.* 28 (11): 3014–3020.

9 Suratwala, T., Steele, R., Wong, L. et al. (2018). Grinding and SSD correlations in various optical materials. *J. Am. Ceram. Soc. (in preparation for)*

10 Davidge, R. and Riley, F. (1995). Grain-size dependence of the wear of alumina. *Wear* 186: 45–49.

11 Bifano, T.G., Dow, T., and Scattergood, R. (1991). Ductile-regime grinding: a new technology for machining brittle materials. *J. Eng. Ind.* 113 (2): 184–189.

12 Lawn, B. and Marshall, D. (1979). Hardness, toughness, and brittleness: an indentation analysis. *J. Am. Ceram. Soc.* 62 (7–8): 347–350.

13 Demirci, I., Mkaddem, A., and El Khoukhi, D. (2014). A multigrains' approach to model the micromechanical contact in glass finishing. *Wear* 321: 46–52.

14 Forsberg, M. (2005). Effect of process parameters on material removal rate in chemical mechanical polishing of Si(100). *Microelectron. Eng.* 77 (3): 319–326.

15 Park, K., Kim, H., Chang, O., and Jeong, H. (2007). Effects of pad properties on material removal in chemical mechanical polishing. *J. Mater. Process. Technol.* 187: 73–736.

16 Luo, J. and Dornfeld, D.A. (2003). Effects of abrasive size distribution in chemical mechanical planarization: modeling and verification. *IEEE Trans. Semicond. Manuf.* 16 (3): 469–476.

17 Tseng, W.T., Chin, J.H., and Kang, L.C. (1999). A comparative study on the roles of velocity in the material removal rate during chemical mechanical polishing. *J. Electrochem. Soc.* 146 (5): 1952–1959.

18 Tseng, W.T. and Wang, Y.L. (1997). Re-examination of pressure and speed dependences of removal rate during chemical-mechanical polishing processes. *J. Electrochem. Soc.* 144 (2): L15–L17.

19 Zhang, Z., Yan, W., Zhang, L. et al. (2011). Effect of mechanical process parameters on friction behavior and material removal during sapphire chemical mechanical polishing. *Microelectron. Eng.* 88 (9): 3020–3023.

20 Pal, R.K., Garg, H., and Karar, V. (2016). Full aperture optical polishing process: overview and challenges. In: *CAD/CAM, Robotics and Factories of the Future*, 461–470. Springer.

21 Cumbo, M., Fairhurst, D., Jacobs, S., and Puchebner, B. (1995). Slurry particle size evolution during the polishing of optical glass. *Appl. Opt.* 34 (19): 3743–3755.

22 Suratwala, T., Steele, R., Feit, M. et al. (2017). Relationship between surface μ-roughness and interface slurry particle spatial distribution during glass polishing. *J. Am. Ceram. Soc.* 100: 2790–2802.

23 Suratwala, T., Feit, M., Steele, W. et al. (2014). Microscopic removal function and the relationship between slurry particle size distribution and workpiece roughness during pad polishing. *J. Am. Ceram. Soc.* 97 (1): 81–91.

24 Suratwala, T., Steele, W., Feit, M. et al. (2016). Mechanism and simulation of removal rate and surface roughness during optical polishing of glasses. *J. Am. Ceram. Soc.* 99 (6): 1974–1984.

25 Li, Y. (2007). *Microelectronic Applications of Chemical Mechanical Planarization.* Wiley.

26 Yu, T-K., Yu, C., and Orlowski, M. ed. (1993). A statistical polishing pad model for chemical-mechanical polishing. Electron Devices Meeting, 1993 IEDM'93 Technical Digest, International. IEEE.

27 Moon, Y. (1999). Mechanical aspects of the material removal mechanism in chemical mechanical polishing (CMP). PhD thesis, University of California Berkeley, 193 pages.

28 Warnock, J. (1991). A two-dimensional process model for chemimechanical polish planarization. *J. Electrochem. Soc.* 138 (8): 2398–2402.

29 Shen, N., Feigenbaum, E., Suratwala, T. et al. (2018). *Single Particle Nanoplastic Removal Function of Optical Materials.* J. Am. Cer. Soc.; Submitted.

30 Suratwala, T., Steele, R., Wong, L. et al. (2018). Influence of partial charge on the material removal rate during chemical polishing. *J. Am. Ceram. Soc.*; Submitted.

31 Namba, Y. and Tsuwa, H. (1977). Ultra-fine finishing of sapphire single crystal. *Ann. CIRP* 26 (1): 325.

32 Gutsche, H.W. and Moody, J.W. (1978). Polishing of sapphire with colloidal silica. *J. Electrochem. Soc.* 125 (1): 136–138.

33 Vovk, E., Budnikov, A., Dobrotvorskaya, M. et al. (2012). Mechanism of the interaction between Al_2O_3 and SiO_2 during the chemical–mechanical polishing of sapphire with silicon dioxide. *J. Surf. Invest.* 6 (1): 115–121.

34 Cook, L. (1990). Chemical processes in glass polishing. *J. Non-Cryst. Solids* 120 (1–3): 152–171.

35 Steigerwald, J.M., Murarka, S.P., and Gutmann, R.J. (2008). *Chemical Mechanical Planarization of Microelectronic Materials.* Wiley.

36 Cornish, D. (1961). The Mechanism of Glass Polishing. *Res. Rep. 267.* Chislehurst, Kent: British Scientific Instrument Research.

37 Silvernail, W. and Goetzinger, N. (1971). The mechanics of glass polishing: part one. *The Glass Ind.* 52: 130–133.

38 Osseo-Asare, K. (2002). Surface chemical processes in chemical mechanical polishing relationship between silica material removal rate and the point of zero charge of the abrasive material. *J. Electrochem. Soc.* 149 (12): G651–G655.

39 Stumm, W. (1992). *Chemistry of the Solid-Water Interface: Processes at the Mineral-Water and Particle-Water Interface in Natural Systems.* Wiley.

40 Livage, J., Henry, M., and Sanchez, C. (1988). Sol-gel chemistry of transition metal oxides. *Prog. Solid State Chem.* 18 (4): 259–341.

41 Reynolds, J.G. (2007). Thermodynamic evaluation of a partial charge model assumption for the dissolved silica system. *Silicon Chem.* 3 (5): 267–269.

42 Brinker, C.J. and Scherer, G.W. (1990). *Sol–Gel Science: The Physics and Chemistry of Sol-Gel Processing*, 638–672. San Diego, CA: Academic Press.

Part II

Applications – Materials Technology

6

Increasing Yield: Scratch Forensics and Fractography

6.1 Fractography 101

If a workpiece falls and breaks catastrophically, the cause of failure is obvious. But a workpiece may fracture catastrophically (or noncatastrophically) for many reasons far more difficult to diagnose and hence, prevent or mitigate. For such purposes, the field of fractography – the science of means and methods of characterizing a fractured component by examining its fracture pattern and surface – can be very useful. The theory and practice of fractography are well described by Frechette [1], Varner and Frechette [2], Quinn [3], Freiman and Mecholsky [4], and Bradt and Tressler [5]. This chapter first provides a fractography introduction, describing the major features observed on the fracture surfaces and what information they provide about failure. Next, the fractography is extended to diagnosing yield loss due to scratching a workpiece, using a technique called scratch forensics and to insights on the time-dependent fracture failure caused by slow crack growth. Finally, some relevant case studies of the fractured workpieces during optical fabrication analyzed by fractography and scratch forensics are presented.

The strength of brittle workpieces such as glasses, single crystals, and polycrystalline ceramics is almost always limited by the flaw distribution on the workpiece, and not the inherent, intrinsic bond-strength of the material [6, 7]. Hence, how the workpiece material is processed, especially surface processing, such as that used in optical fabrication, determines its flaw distribution and ultimate strength. The critical stress to failure σ_f is given by Griffith's law,

$$\sigma_f = \sqrt{\frac{2\gamma_f E_1}{\pi a_c}} = \frac{K_{Ic}}{\sqrt{\pi a_c}}, \qquad (6.1)$$

where E_1 is the workpiece elastic modulus, K_{Ic} is the workpiece fracture toughness, γ_f is the fracture surface energy, and a_c is the flaw size (the radius of a

Materials Science and Technology of Optical Fabrication, First Edition. Tayyab I. Suratwala.

half-penny flaw) [8]. The flaws are typically microscopic fractures, which are the same as the microfractures referred to as sub-surface mechanical damage (SSD) (see Chapter 3). Flaws are usually present on the edges of the workpiece and are the initiation point for failure. To make a stronger workpiece, the size of the flaws present during fabrication and end use must be minimized.

The stress applied on a given workpiece material results in a stress distribution in the workpiece that can be spatially, directionally, and temporally dependent. In other words, the stress is different in different directions at various positions in the workpiece and can change with time (especially for a propagating crack). A fracture in a brittle material will propagate perpendicular to the maximum principal mode-I tension; hence, the crack pathway can be determined if the stress distribution is known. An example is shown for trailing-indent cracks in Figure 3.9.

Table 6.1 defines some major markings observable on a fracture surface, generally seen under a stereo microscope with a proper lighting or perhaps with a scanning electron microscope for smaller features. Table 6.2 outlines the information that can be gathered using fractography.

Table 6.1 The various types of fracture surface markings observed on glass and their definitions.

Marking	Definition
(1) Mirror	Smooth region with a mirrorlike surface, typically near fracture origin
(2) Mist	Hazy or misty surface caused by crack approaching terminal velocity
(3) Wake hackle	Ridge extending from a singularity Caused by crack front propagating through a pre-exiting inclusion/defect
(4) Twist hackle	Daggerlike marking running parallel to the direction of crack propagation Caused by twist in local axis of principal tension
(5) Arrest lines	Sharp curved line on fracture surface indicating fracture stopped and restarted
(6) Wallner lines	Wavelike lines on fracture surface caused by intersection of an elastic pulse with propagating crack
(7) Cavitation scarp	A line marking the interface between a mirror region and a very fine mist-hackle region, caused by liquid-filled crack accelerating and becoming dry
(8) Sierra scarps	Arc-shaped line resulting from a decelerating crack that was overtaken by liquid water
(9) Stress wave	Sinusoidal lines on surface caused by external vibrations within workpiece

Table 6.2 Major data fractography may reveal on a fractured workpiece.

Information obtained	How?
(1) Fracture origin	• Locating mirror/mist/hackle • Tracing crack propagation
(2) Direction of fracture, sequence of events	• Reading specific fracture markings, especially Wallner lines and twist hackle
(3) Stress distribution	• Observing symmetry or asymmetry of mirror/mist/hackle and shape of Wallner lines
(4) Failure stress, stored energy	• Measuring mirror radius, flaw size, branching distance • Finding fracture area
(5) Crack velocity	• Measuring the intersection of Wallner lines, insight from other fracture markings
(6) Stress type (thermal, indent/impact, mechanical, chemical)	• Observing overall fracture pattern, evidence of indent damage
(7) Environment during fracture (wet or dry)	• Finding scarp markings on fracture surface

Figure 6.1 shows an example of the fracture surface near the origin point on a broken borosilicate glass workpiece, illustrating the mirror, mist, and twist-hackle markings. The circled region indicates the origin and the dashed line that outlines the direction of crack propagation. The origin is determined by identifying the markings such as mirror and mist regions on the fracture surface and/or by tracing back the direction of crack propagation (based on the characteristics on the fracture surface or crack patterns). At the origin, the mirror radius r_m is particularly useful, since it can be used to determine the stress at failure σ_f as follows:

$$\sigma_f = \frac{A_m}{\sqrt{r_m}}, \tag{6.2}$$

where A_m is the mirror constant for a given material and r_m is the mirror radius. For the example in Figure 6.1, $r_m = 1820\ \mu m$ and $A_m = 2.0\ MPa\ m^{1/2}$ for borosilicate glass [3], leading to a failure stress of 47 MPa.

The other markings, including Wallner lines, which aid in determining the propagation direction and sometimes velocity (Figure 6.2); cantilever curl, which indicates a large change in stress distribution and propagation direction (Figure 6.2); Sierra scarps and cavitation hackle, which identify

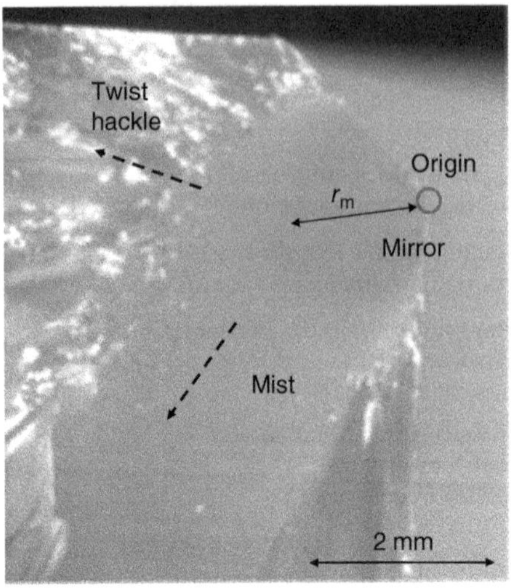

Figure 6.1 The fracture surface of borosilicate glass, illustrates fracture origin and mirror, mist, and twist-hackle markings. Dashed arrow indicates crack propagation direction.

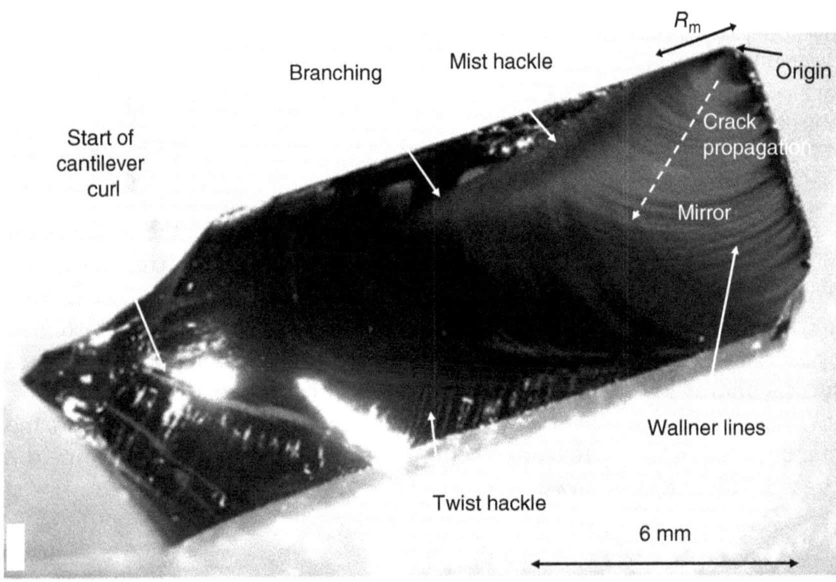

Figure 6.2 The fracture surface of an absorbing glass with various fracture markings labeled.

crack propagation in the presence of liquid water (Figure 6.3); and stress-wave markings, which indicate fracture in the presence of some periodic cyclic loading (Figure 6.4). All provide important characteristics of the fracture event.

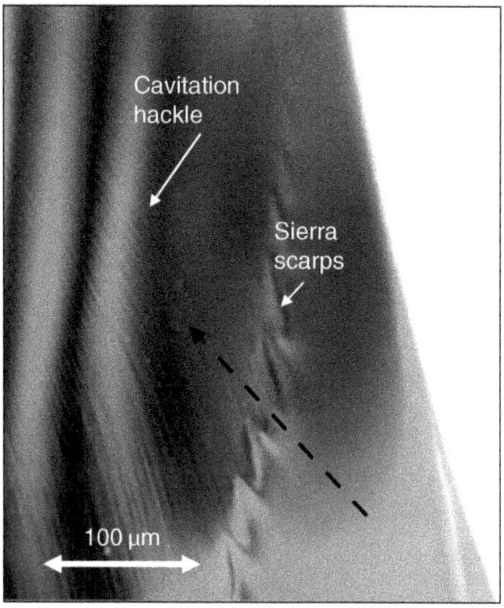

Figure 6.3 The fracture surface of a fractured laser phosphate glass, illustrating Sierra scarps and cavitation hackle that occurred in the presence of liquid water. Crack propagation is from bottom right to top left.

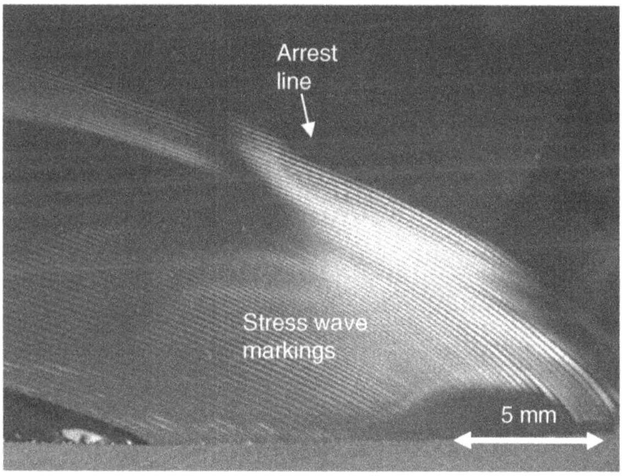

Figure 6.4 The fracture surface of a fractured laser phosphate glass, illustrates arrest-line and stress-wave markings indicative of cyclical loading of the workpiece during fracturing. Crack propagation is from bottom left to top right.

Crack patterns are also useful forensics for determining the characteristics of a failure. These may include the following:

- The angle of crack branching from which to estimate the ratios of the principal stresses (Figure 6.5a).
- The total fracture area, which is proportional to the stored energy, allows estimation of the total stress the workpiece was exposed to (Figure 6.5b).
- Waviness in the fracture propagation, used to identify thermally induced stress (as opposed to mechanical) on the workpiece (Figure 6.5c) [1, 9].

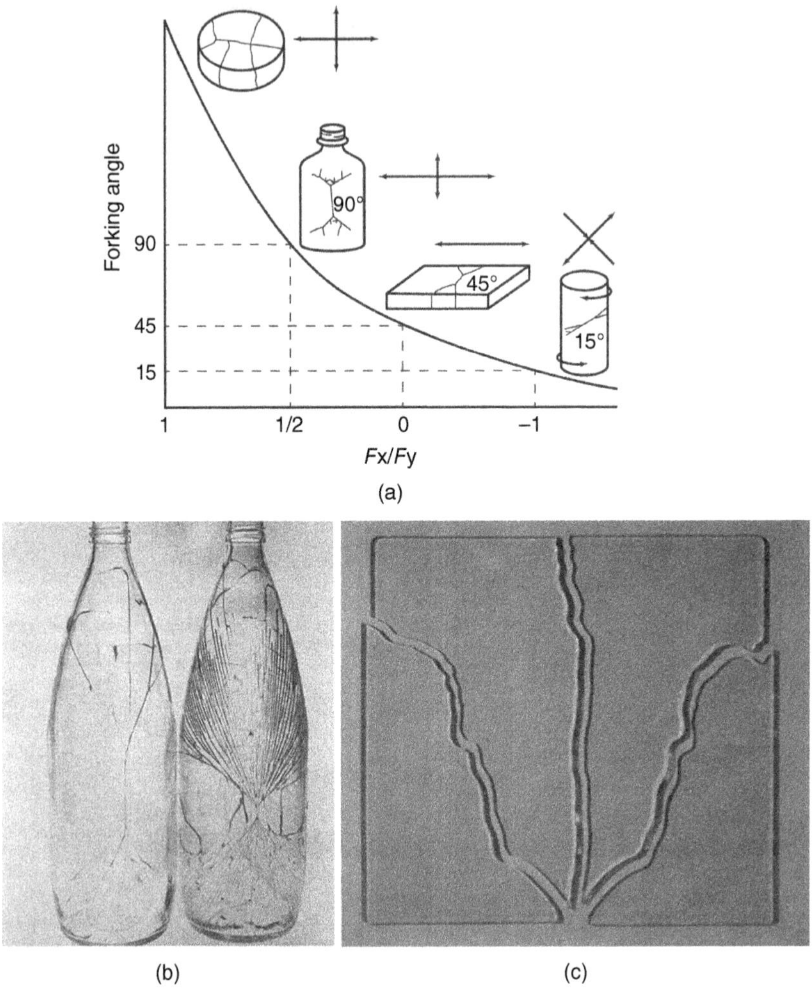

Figure 6.5 Examples of the types of fracture patterns. Source: From Fréchette 1990 [29].

6.2 Scratch Forensics

As discussed in Chapter 3, scratches are a form of SSD caused by rogue particles which bear loads high enough to lead to permanent deformation or microfractures on the surface. Scratches can occur at any step in the fabrication process (grinding, polishing, cleaning, handling, transport, inspection, etching, etc.) as well as during its end use. Scratches incurred in and between polishing steps can be particularly detrimental because *material removal rates* during polishing are typically very small relative to the depth of a typical scratch. Thus, to remove the scratch, the workpiece is often returned to grinding. Understanding the cause of the scratch – for example, the responsible rogue particles and the fabrication step involved – aids tremendously in identifying the mitigation(s) required to prevent similar failures from happening again, leading to improved manufacturing yields and significant increases in throughput and cost savings.

The evaluation and interpretation of scratch characteristics, including dimensions, type, location, shape, and orientation, can be powerful techniques for identifying the cause of the scratch. With the development of fracture mechanics and tribology, new rules of thumb have emerged, allowing the estimation of

- The size and shape of the asperity or rogue particle causing the scratch
- The load on the rogue particle
- The depth of the trailing indent fractures of a scratch
- The mechanical properties of the material housing and applying load to the rogue particle.

The employment of these new techniques is referred to as "scratch forensics" [10]. In the discussion below, the various characteristics of scratches and what they can reveal about the scratching event are described. The correlations presented here were largely derived for fused-silica glass workpieces, but in many cases can be extended to other brittle materials as well.

6.2.1 Scratch Width

The width of a scratch provides insight about the size of the rogue particle (see Figure 3.10b). For a spherical rogue particle at a given applied load, Hertzian fracture mechanics can be used to predict the size of the contact zone [11]. In practice, rogue particles are usually not spherical and the applied load is unknown; thus predictions of the particle size become less certain. However, in controlled polishing and grinding experiments using rogue particles of known size that were intentionally added to the slurries, the characteristic crack length $\overline{L}_t$ (i.e. the width of scratch) was found to correlate with the particle average diameter $\overline{d}$ as: [12]

$$0.30\,\overline{d} \leq \overline{L}_{\mathrm{t}} \leq 0.50\,\overline{d} \quad \text{for polishing, and} \tag{6.3}$$

$$0.15\,\overline{d} \leq \overline{L}_{\mathrm{t}} \leq 0.30\,\overline{d}, \quad \text{for grinding.} \tag{6.4}$$

In other words, for polishing, the width of a scratch is roughly 30–50% of the apparent size of the rogue particle and, for grinding, 15–30% of the apparent size of the rogue particle. An example set of determined-characteristic crack lengths after different grinding processes on fused-silica workpieces is shown

Table 6.3 Measured relationship between various grinding processes and resulting surface-crack length.

Sample	$\bar{L}_t$ (μm)
A: Sandblast	27.1
B: 120 grit generator grind	28.3
C: 320 grit generator grind	14.9
D: 15 μm loose abrasive	4.6
E: 15 μm fixed abrasive	4.5
F: 9 μm loose abrasive	1.9
G: 7 μm fixed abrasive	8.4

Source: Suratwala et al. 2006 [13]. Reproduced with permission from Elsevier.

in Table 6.3, which uses the same data as in Figure 3.19. This is a valuable lookup table to identify which process steps have contributed to observed scratches or surface fractures [10, 13].

The width of the scratch has been related to its depth. Crack depth is ultimately determined by the load on the rogue particle, as in Equations (3.4)–(3.6). However, knowing that the load range is limited for a rogue particle, the following rules of thumbs to probabilistically predict depth have been established, based on experimentally measured crack-depth distributions (see Section 3.1.2.4): [10]

$$c_{90} \cong 0.9L_t, \tag{6.5}$$

$$c_{max} \cong 2.8L_t. \tag{6.6}$$

To illustrate the utility of these expressions, take for example, a case where a 30 μm wide scratch is observed on polished workpiece. According to Equation (6.5), also known as the 90 rule, there is 90% probability that the scratch will be removed from the workpiece if 0.9 × 30 or 27 μm from the surface has been ground or polished through. Using Equation (6.6), there is 100% probability that the scratch will be removed if 280 μm is removed from the surface. Knowing how much material to remove allows economical decisions, such as to whether to continue polishing or return the workpiece to an earlier step and reduces the number of scratch-removal iterations needed to verify that the scratch has been removed. Note the proportionality constants in these equations were determined for fused silica, based on material properties. The proportionality constant for different materials can be estimated using the fracture mechanics-based-grinding model. See Equations (3.24)–(3.25).

6.2.2 Scratch Length

As discussed in Section 3.1.3, the length of a scratch is related to lap properties, rogue-particle size, and applied load. Restating Equation (3.35), the average length of scratch $\overline{L}_s$ is approximated by

$$\overline{L}_s \cong 2.2\frac{V_r \eta_2 d^2}{P}, \tag{6.7}$$

where V_r is the average relative velocity of the rogue particle relative to the workpiece surface, η_2 is the effective viscosity of the viscoelastic lap, d is the rogue-particle size diameter, and P is applied load on the rogue particle. This relationship stems from the fact that the length of the scratch during polishing is governed by the time it takes for the rogue particle to penetrate into the lap (Figure 3.35). When the rogue particle has penetrated completely, its load is reduced to the point where it cannot initiate fracture (i.e. $P < P_{ct}$), and the scratch ends.

Equation (6.7) can be used in many ways forensically, depending on the information available. For example, if the viscosity of a lap is known and typical loads on rogue particles are used during polishing (1–5 N), the size of the particle causing the scratch can be estimated using the observed scratch length. If this value is consistent with the particle size estimated from Equation (6.3), one may conclude with confidence that the scratch occurred in polishing, and not in another step or handling. This relationship is also useful where very different kinematics (i.e. relative velocities) are used between and within polishing steps. The length of the scratch can be used to determine the polishing step or sequence by identifying the relative velocity causing the scratch.

6.2.3 Scratch Type

The various types of scratches are shown in Figure 3.11 [10]. Scratches may be caused by rogue particles or asperities that are either blunt or sharp. In general, blunt particles create scratches that are a line of trailing indent fractures, and sharp particles may create a combination of plastically deformed material and trailing indent fractures.

The types of fractures present on the scratch, and their extent, largely depend on the load impressed on the workpiece by the rogue particle at

- *Low loads ($P < 0.05$ N).* Particularly in the presence of a sharp particle, a plastic trench is formed without fractures.
- *Intermediate loads (0.1 N $< P < 5$ N).* Well-defined radial (or trailing indent) fractures along with lateral cracks are observed.
- *High loads ($P > 5$ N).* The plastically deformed track fractures into a rubble-like appearance and lateral and trailing indent cracks are less pronounced [14, 15].

These scratch characteristics in the forensics toolbox can determine the shape and approximate load during scratch formation.

If the fracture observed is an isolated fracture, it could be (i) a scratch that has been partially polished through and only a deep trailing indent remains; (ii) a deep SSD fracture that has not been polished through (see Section 3.1.2), or (iii) a static indent (also, commonly referred to as a dig). If the latter, static-indentation-fracture mechanics can be used to determine the shape of the indenter (or its contacting radius of curvature) and the load of indentation based on observed crack depth. See the discussion in Section 3.1.1.1.

6.2.4 Scratch Number Density

The scratch number density may indicate the size and number of rogue particles present. As shown in Figure 3.32, where the rogue diamond particles were added to a cerium-oxide slurry during fused-silica polishing, the scratch number density increased exponentially with rogue-particle size at fixed concentration. By contrast, an increase in rogue-particle concentration generally did not increase the scratch number density. This means that larger rogue particles result in more damaging scratches and in creating more scratches.

6.2.5 Scratch Orientation and Trailing-indent Curvature

The direction in which a scratch propagates can be discerned using the orientation of the trailing-indent fractures. The rogue particle travels (or the scratch propagates) in the concave direction of the trailing-indent fractures. For example, in Figure 3.10b, the scratch started from the top of the image and propagated downward.

Determination of the scratch-propagation direction can be useful for determining the source of the rogue particles. For example, if most of the scratches appear to be propagating in from the edge of the workpiece, it is telltale that the source of particles is from the edge of the workpiece, such as tiny chips on the workpiece edge or dried slurry dislodging from the ground edge. Scratch direction may also line up with standard relative motions during polishing, cleaning,, or wiping. Consider the case in which scratches are diagnosed to travel in the same direction, parallel to an edge of the workpiece. This orientation may line up with the wiping direction used by an operator during final cleaning or inspection of the workpiece, suggesting the wipes were contaminated with rogue particles.

6.2.6 Scratch Pattern and Curvature

The forensic information from scratch pattern and curvature is well known in the optical-fabrication field and very valuable. Scratches with large, sharp

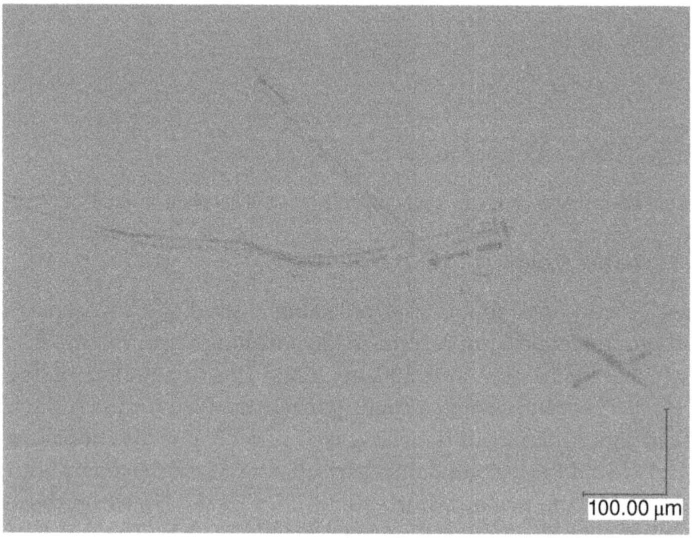

Figure 6.6 Example of a handling scratch on fused-silica workpiece.

changes in direction are typically handling scratches, such as from placing a workpiece on a flat surface with rogue particles or asperities (see scratch example in Figure 6.6). The directional changes are driven by the random relative microscopic motions between the workpiece and the rogue particle as the two surfaces meet. Another example is when a scratch propagates along a curve with a distinct radius. If the curvature matches the kinematic path of the relative motion of the lap or tool with respect to the workpiece, it is a telltale indicator of the processing step at which the scratch was inflicted.

6.2.7 Location on Workpiece

Sometimes the location of scratches on the workpiece can identify important characteristics of the scratching event. For example, for scratches or fractures that are randomly oriented all along the edges of a workpiece or randomly orientated in the center, it is likely that SSD from grinding has not been polished through. This is because a workpiece going into polishing is either convex or concave after grinding, relative to the polishing-lap shape. During polishing, the SSD is nonuniformly removed from the surface, leading either to the center or edge with a "gray" surface that is removed last. See workpiece–lap mismatch due to workpiece shape in Section 2.5.6.1. In such cases, the characteristic length of the fractures, as noted in Table 6.3, should match that of the previous grinding step.

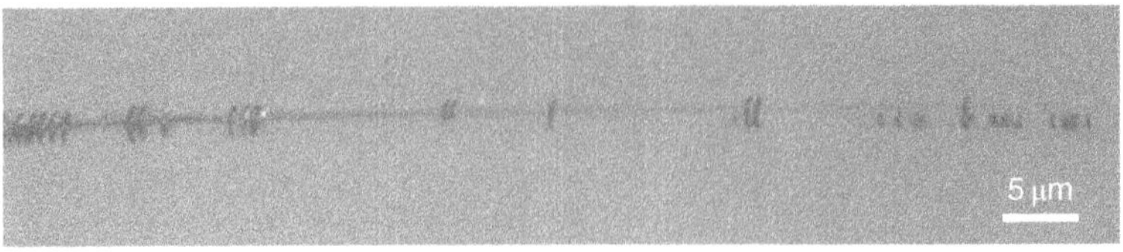

Figure 6.7 An optical microscope image of a scratch on a fused-silica glass workpiece.

6.2.8 Scratch Forensics Example

Consider a scratch on an optic found after polishing a fused-silica workpiece using a polyurethane lap and 0.5 μm cerium-oxide polishing slurry (Figure 6.7) [10]. The scratch is 1.9 μm wide and 130 μm long. The orientation of the trailing-indent fractures indicates that a rogue particle traveled (or the scratch propagated) left to right, relative to the workpiece surface. It is also apparent that the rogue particle has both a blunt and sharp character, as inferred by both trailing indent fractures and plastic deformation (i.e. a sleek). Based on these features, the load causing the scratch is approximately 0.1–1 N. The width of the trailing-indent fracture is different along the scratch, suggesting the particle has a complex shape and is possibly rotating or twisting, making contact with various asperities on the particle as it propagates along the workpiece. The largest width of the scratch is 1.9 μm; hence the size of the rogue particle is likely 3.8–5.7 μm, per Equation (6.3). Finally, using Equation (6.7), the rogue-particle size can be calculated as 4.5 μm, using the measured scratch length (130 μm), known relative velocity (15.7 cm s^{-1}), and known lap viscosity (9.7×10^7 poise) [12]. This calculated size matches that calculated using Equation (6.3), confirming that the scratch most likely occurred in the polishing step.

6.3 Slow Crack Growth and Lifetime Predictions

In Section 3.1.1, crack propagation is discussed as resulting from stress distribution and inherent mechanical properties of the workpiece. For oxide glasses and some other materials, there is also a chemical component, via reaction of the workpiece with water, by which a crack may propagate in the presence of stress at a stress-intensity level K_I below the fracture toughness K_{Ic} of the workpiece. This slow-crack growth, or stress-corrosion cracking [16], may have important consequences, for example, when SSD grows deeper than expected or a workpiece exhibits delayed failure when stored in humid environments.

As first demonstrated by Wiederhorn [16], in the presence of a humid environment, H_2O can migrate to the crack tip and react with an oxide workpiece via a stress-enhanced hydrolysis reaction given by

$$H_2O + M_{wp} - O - M_{wp} \rightarrow 2(M_{wp} - OH), \tag{6.8}$$

where M_{wp} is a network-forming metal atom of the glass. For silica glasses, M_{wp} is Si, and for phosphate, P. The rate of crack propagation or velocity can depend on many factors, including the concentration of H_2O in the environment, the rate of transport of H_2O to the crack tip, the inherent rate of reaction for Equation (6.8), residual stress at the crack tip (i.e. the stress intensity), and crack-tip geometry.

A number of studies have measured and modeled slow-crack behavior in a variety of glasses and oxides [17–21]. Typically, measured data is presented as the log of the crack velocity v_c as a function of stress intensity K_I, which describes the stress enhancement at the crack tip, accounting for applied and residual stress and the geometry of the crack [6]. Figure 6.8 shows an example of such data for phosphate glass [17]. Crack velocity may vary by many orders of magnitude and increase significantly with increases in stress intensity, water-vapor pressure,, and temperature. Three different regimes are often observed in plots where the crack-growth rate is limited by different processes.

- In Region I, crack growth is reaction-rate limited and displays a linear relation between $\log v_c$ and K_I. In other words, transport of H_2O to the crack tip is rapid such that the rate of reaction between H_2O and the workpiece is controlling the crack growth rate (Reaction (6.8)).
- In Region II, crack velocity becomes mass-transport limited, that is, by the rate of H_2O transport to the crack tip. Consequently, Region-II behavior is usually observed at smaller H_2O vapor pressures and lower temperatures, where mass transport rates are slower. The measured crack velocities in Region II remain nearly constant with increasing K_I.
- In Region III, crack velocity becomes independent of the chemical environment and is limited by the intrinsic fracture toughness of the glass [22]. In Figure 6.8, the fracture toughness of the phosphate glass is 0.43 MPa $m^{1/2}$.

Slow-crack growth behavior is generally modeled using either an empirical power law or a chemical-reaction rate model. The empirical power law has the form

$$v_c = A_o K_I^{N_c}, \tag{6.9}$$

where v_c is crack velocity, A_o is a pre-exponential constant, and N_c is the power constant. The stressed-enhanced chemical mass transport or the reaction-rate model has the form [16, 17]:

$$v_c = \frac{v_I v_{II}}{v_I + v_{II}} \tag{6.10}$$

$$v_I = A_o \exp\left[\left(-Q_I + \frac{2V_a K_I}{3\sqrt{\pi \rho_c}}\right)/RT\right] \tag{6.11}$$

$$v_{II} = C_o \frac{p}{p_o} \exp\left(-\frac{Q_{II}}{RT}\right), \tag{6.12}$$

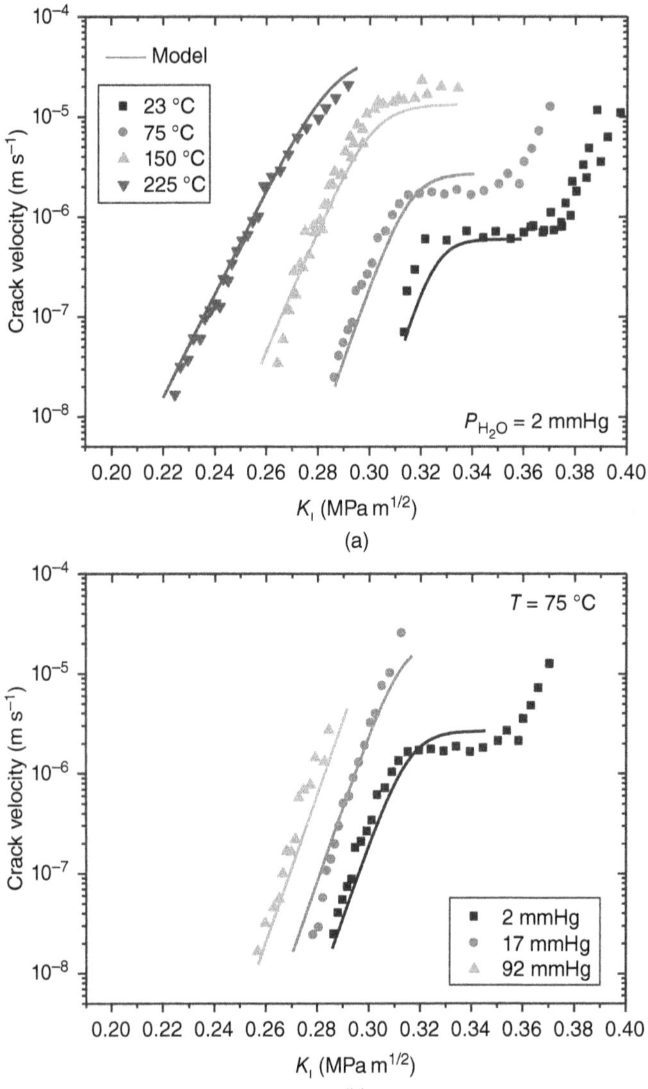

Figure 6.8 The crack-growth rate measured at various temperatures and constant water vapor pressures of (a) 2 mmHg and (b) 17 mmHg. The lines are fits to the data, using the reaction rate model, Equations (6.10)–(6.12). Source: Crichton et al. 1999 [17]. Reproduced with permission from John Wiley & Sons.

where A_o is a pre-exponential constant (m s^{-1}), V_a is the activation volume (m^3 mol^{-1}), ρ_c is the crack-tip radius (m), C_o is a pre-exponential constant (m s^{-1}), Q_I is the activation energy for hydrolytic reaction (kJ mol^{-1}), Q_{II} is the activation energy for transport (kJ mol^{-1}), p is partial water-vapor pressure, and p_o is the saturation water-vapor pressure. The reaction rate model has a fundamental foundation that allows accounts for water vapor, temperature, and different regimes and materials, and has been expanded to account for special phenomena such as crack-tip blunting, capillary condensation, change in glass water content, and anomalous temperature dependence [17, 18, 21].

As a slow crack grows, longer cracks appear. Depending on the geometry, the stress intensity may increase with crack length and approach K_{Ic}, leading to a catastrophic failure of the workpiece. A simple analytical expression for predicting workpiece lifetime using the reaction rate model has not been reported; Ritter and Meisel [23] used the power law for crack-growth relationship, however, to estimate workpiece time to failure t_f owing to slow-crack growth under an applied or residual stress of σ_o as follows:

$$t_f = \frac{2\sigma_o^{-N_c}\left(\dfrac{K_{Ic}}{\sigma_f}\right)^2}{A_o Y^2 (N_c - 2)}, \tag{6.13}$$

where σ_f is the failure strength of the workpiece in an inert environment (no slow-crack growth), and Y is the geometric constant of the flaw, $Y = \pi/2$ for surface flaws. If the stress and crack-growth constants are known for a given workpiece, Equation (6.13) may be used for determining whether a workpiece is in danger of a catastrophic failure.

6.4 Fracture Case Studies

In this section, some novel fractures that occurred during optical fabrication or end use are analyzed using fractography and other methods. While these cases largely involve optical workpieces fabricated for the NIF laser at Lawrence Livermore National Laboratory, the forensic methods and lessons learned may be applied to a variety of optical workpieces.

6.4.1 Temperature-induced Fracture

Optical fabrication often involves workpiece temperature fluctuations, for example, during elevated-temperature solvent cleaning, thermal annealing, thermal curing of an adhesive near or attached to the workpiece, or upon absorption of light. This exposure to a thermal gradient will result in thermal stresses in the workpiece; if the stress exceeds the critical stress for failure, as described in Equation (6.1), the workpiece may fracture.

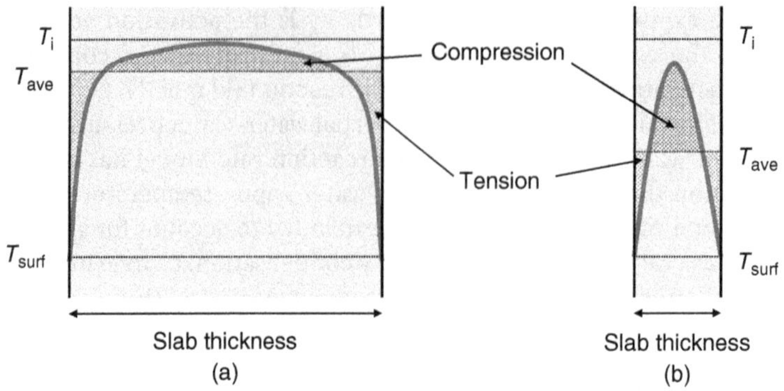

Figure 6.9 A schematic illustration of the temperature profile in a semiinfinite workpiece cooled from temperature T_i for a thick workpiece (a) and thin workpiece (b).

Finite-element analysis (FEA) is required for a precise determination of the temperature and stress distribution of complex thermal exposure recipes on complex geometries involving multiple materials. Nevertheless, some simple analytical expressions can provide quantitative estimates with which to make process decisions. Consider a common, simple case where a workpiece is considered as a semiinfinite slab (thickness $\ll$ length and width) and exposed to a thermal treatment. During heating, the surface is put in compression, and during cooling, the surface is put in tension. Because the largest flaws are typically on the surface of brittle materials, the tension on the surface during cooling puts the workpiece at greatest risk of fracture. The magnitude of the thermal stress σ_t for an infinite slab can be estimated as follows:

$$\sigma_t \approx \alpha_{t1} E_1 (T_{ave} - T_{surf}), \tag{6.14}$$

where α_{t1} is the workpiece thermal-expansion coefficient, E_1 is the workpiece elastic modulus, T_{ave} is the average temperature in the bulk, and T_{surf} is the surface temperature of some location on the workpiece [24].

Figure 6.9 qualitatively shows the thermal profile across the thickness of a workpiece and regions of compressive and tensile stress. In this figure, two workpieces, one with large thickness (a) and the other with less (b), are cooled from some initial temperature T_i. The average temperature in the thick workpiece will be higher than that in the thin workpiece, leading to much higher tensile stress for the same cooling rate in the thick workpiece. Note that the thermal stress in the glass is proportional to ΔT (i.e. $T_{ave} - T_{surf}$). The points in the solid that have the value of the average temperature are approximately the point of zero stress in the material. Temperatures above T_{ave} will be in compression and temperatures below T_{ave} will be in tension. The corresponding stress profile, as approximated by Equation (6.14), will largely mimic the temperature profile.

The ability of a given workpiece to withstand a given thermal cycle before catastrophically failing will depend on its

- Thermal-expansion coefficient, which determines the strain a workpiece will experience for a given temperature change.
- Elastic modulus, which determines the stress induced for a given strain in the material.
- Fracture toughness, which determines the workpiece's ability to resist fracture.
- The flaw size in the workpiece.

A thermal shock figure of merit (FOM), R_{ts}, for a given workpiece material can be defined as follows:

$$R_{ts} = \frac{(1 - \nu_1)}{E_1 \alpha_{t1}} \frac{K_{Ic}}{\sqrt{\pi a_c}}, \tag{6.15}$$

where ν_1 is the Poisson's ratio of the workpiece. The thermal-shock FOM is a useful rule of thumb for fracture criteria; in other words, during any thermal cycling, the workpiece should always have a $\Delta T < R_{ts}$ to prevent the catastrophic fracture. Table 6.4 compares the thermal-shock FOMs for selected optical materials. Note the large differences – for example, fused-silica glass has a very large value for $R_{ts} = 1000$ K, while laser phosphate glass has a much lower value of $R_{ts} = 34$ K and potassium dihydrogen phosphate (KDP) crystal has even a smaller value, $R_{ts} = 1.8$ °C. Hence, for the same size workpiece, the thermal cycling of KDP needs to be significantly slower than that of fused silica to prevent fracture.

6.4.1.1 Laser-Phosphate-glass Thermal Fracture

Consider an example of a phosphate-glass workpiece fractured by thermal cycling during treatment (see Figure 6.10). The workpiece was heated in an aqueous solution to 88 °C and then placed in an aqueous solution of 60 °C,

Table 6.4 Comparison of the thermal-shock FOM for selected optical materials.

Property	Units	KDP	Phosphate glass	Pyrex glass	Fused silica
Fracture toughness (K_{Ic})	MPa $m^{1/2}$	0.10	0.50	0.70	0.75
Thermal expansion (α_{t1})	$\times 10^{-7}$ K^{-1}	440	130	33	5
Elastic modulus (E_1)	GPa	63	47	64	70
Thermal-shock FOM for a 100-µm flaw (R_{ts})	K	1.8	34	150	1000

KDP = potassium dihydrogen phosphate single crystal.

Figure 6.10 A ground-phosphate-glass workpiece ($14.1 \times 4.1 \times 1.4\ cm^3$) that fractured from thermal cycling during cleaning.

where it was observed to fracture. Fractography analysis reveals some important characteristics of the fracture event. First, the overall fracture pattern includes numerous fractures initiating from the surface (see Figures 6.10 and 6.11a). Second, all the originating fractures are orthogonal to the workpiece surface, indicating that the mode-I tensile stress was perpendicular to the surface – a telltale sign of thermal stress. Third, many of the fractures penetrate a fixed distance into the bulk and then curve, indicating that the surface was in tension and the bulk in compression. Fourth, the fracture surface near one of the fracture origins (Figure 6.11a) confirms that the workpiece fractured in the presence of an aqueous environment, owing to the presence of Sierra scarp features (see Table 6.1 and Figure 6.3). Finally, because the workpiece had a ground surface, it contained large flaws ~200 μm in size, which reduces the critical stress-to-failure. See Equation (6.1) and Figure 6.11b.

The temperature distribution in the workpiece as a function of time can be calculated more precisely using Fourier's equation in 3D, using the following form:

$$\rho_1 C_{p1} \frac{\partial T(x,y,z,t)}{\partial t} = k_1 \left(\frac{\partial^2 T(x,y,z,t)}{\partial x^2} + \frac{\partial^2 T(x,y,z,t)}{\partial y^2} + \frac{\partial^2 T(x,y,z,t)}{\partial z^2} \right), \tag{6.16}$$

where x, y, and z are the Cartesian coordinates, ρ_1 is workpiece density, C_{p1} is workpiece heat capacity, k_1 is workpiece thermal conductivity, and t is time. In this analysis, only convection heat loss was considered. The boundary conditions then become

$$-k_1 \frac{\partial T(x_b,y,z,t)}{\partial x} = h_t(T(x_b,y,z,t) - T_{surf}), \tag{6.17}$$

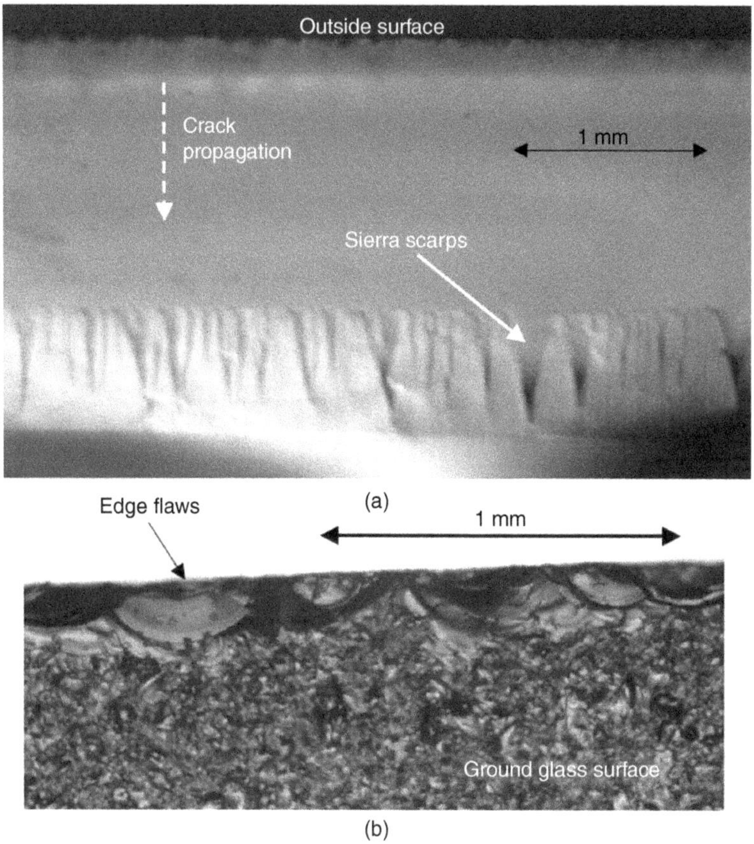

Figure 6.11 (a) An optical micrograph of the phosphate-glass-workpiece edge, illustrating large flaws on the ground surface (i.e. SSD) and (b) optical micrograph of a fracture surfaces.

$$-k_1 \frac{\partial T(x, y_b, z, t)}{\partial y} = h_t(T(x, y_b, z, t) - T_{surf}), \tag{6.18}$$

$$-k_1 \frac{\partial T(x, y, z_b, t)}{\partial z} = h_t(T(x, y, z_b, t) - T_{surf}), \tag{6.19}$$

where x_b, y_b, and z_b are dimensions of the workpiece, h_t is the heat-transfer coefficient (W m^2 K^{-1}), and T_{surf} is the surface temperature. These expressions can be solved using FEA analysis or available analytical solutions for a rectangular parallelepiped with convection (see for example, Ref. [25]).

Supported by the fractography analysis abovementioned, the thermal fractures for such geometries tend to originate from the 2D edges of the workpiece, because these edges have the largest combination of flaws and temperature differences (ΔT) within the workpiece. Based on Equation (6.14), the stress will be

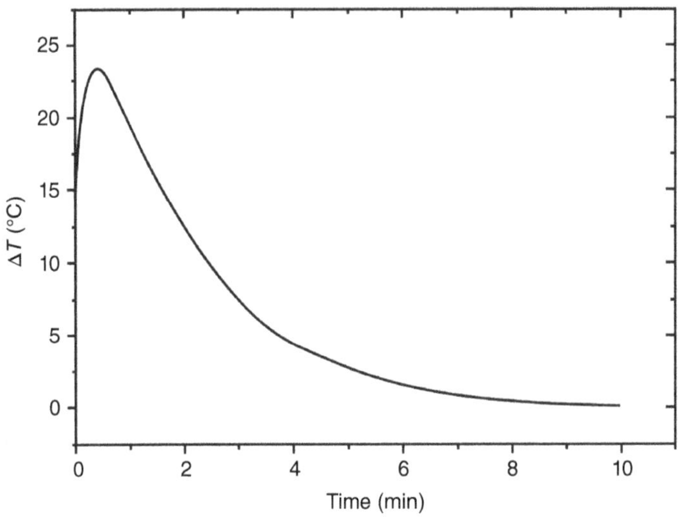

Figure 6.12 The calculated temperature difference between the center of the phosphate glass workpiece and the center of the long 2D surface edge as a function of time submerged in 60 °C water after initial temperature of 88 °C. Material properties of the phosphate glass are $k_1 = 0.58\,\mathrm{W\,mK^{-1}}$, $C_{p1} = 0.77\,\mathrm{J\,gm^{-1}\,K}$, $\rho_1 = 2.83\,\mathrm{gm\,cm^{-3}}$, $\alpha_{t1} = 127 \times 10^{-7}\,\mathrm{K^{-1}}$, $K_{Ic} = 0.51\,\mathrm{MPa\,m^{1/2}}$, $E_1 = 50.1\,\mathrm{GPa}$, and $v_1 = 0.26$.

proportional to ΔT or $T_{ave} - T_{surf}$. The calculated ΔT is plotted as a function of time during the thermal cycle in Figure 6.12. The maximum ΔT of 23 °C occurs 20 s after submerging in water, and, using Equation (6.14), the stress is ~20 MPa. This stress corresponds to a critical flaw size of ~200 μm, using Equation (6.1), and is consistent with the flaws present on a ground surface.

Note the heat-transfer coefficient h_t, which determines the rate of heat transfer from the surfaces of the workpiece, a key parameter, as seen in Equations (6.17)–(6.19). The value of the heat-transfer coefficient may vary dramatically, depending on the environment around the workpiece (see Table 6.5). The larger the heat-transfer coefficient, the larger the ΔT that will develop in the workpiece and the greater the stress.

6.4.1.2 KDP Crystal-Workpiece Thermal Fracture

Another source of heat transfer not commonly considered is surface evaporation. When a fluid such as water evaporates from a workpiece, energy, and hence heat, is drawn away from the surface by the heat of vaporization, resulting in an additional temperature-drop term in the thermal equations above. Here, a fracture in which evaporation is added to the thermal stress that contributed to a KDP crystal fracture is described.

As shown in Table 6.4, KDP has very poor thermal-shock resistance, due to its high-thermal expansion coefficient and low fracture toughness. Large KDP

Table 6.5 Approximate heat-transfer coefficients from a workpiece surface exposed to different environments.

Environment	h_t (W m^{-2} K^{-1})
(1) Air only (no flow)	10
(2) Water evaporation	60
(3) Air only (forced flow)	100
(4) Mineral oil	200
(5) Metal contact	20 000

Source: From VanSant 1983 [25].

crystals in ~0.5 m scale blocks are grown for use as nonlinear optics, e.g. for frequency conversion, in large-aperture fusion laser systems [26–28]. These crystals begin as a small single-seed crystal in a saturated salt solution (typically ~1000 L) and are grown as they slowly cool from an elevated temperature (typically 80 °C), from several months to over a year. At the end of one these crystal-growth runs, a full-grown crystal surrounded by a 62 °C salt solution needed to be further cooled and removed from the tank. First, the salt solution was pumped out of the growth tank and warm dry air (60 °C) was input at a rate of ~80 standard liters per minute into the growth tank. The air temperature was then cooled at a rate of 3 °C a day, down to room temperature (23 °C). During this time, a large single fracture was observed propagating through the center of the crystal (see Figure 6.13).

The fractography analysis of the crystal is shown in Figure 6.13b,c. The fracture originated from the bottom side of the pyramid face along a 2D edge, similar to that observed with the phosphate glass discussed in Section 6.4.1.1. The arrows indicate the direction of fracture propagation, as determined from standard fracture markings like Wallner lines and twist hackle (see Table 6.1). Most of the fracture marks near the origin point were not visible, because the salt solution penetrated the crack and partly dissolved the crystal fracture surface. Based on the presence of subtle Wallner lines near the origin, however, the fracture is thought to have originated from the circled area.

Three-dimensional transient-thermal analysis using FEA, based on Equations (6.16)–(6.19), was performed to calculate the ΔT during cooling using three simulation conditions, as summarized in Table 6.6. Based on the thermal shock FOM for KDP, a $\Delta T > 1.8$ °C is needed to initiate fracture of the crystal boule. Simulations 1 and 2, which relied on the crystal cooling rate and initial temperature drop from the dry input air without evaporative heat loss, led to a maximum ΔT that was less than the thermal shock FOM. Hence, fracture was not expected from these contributions alone. Simulation 3, which included evaporative heat loss, led to a significantly higher ΔT of

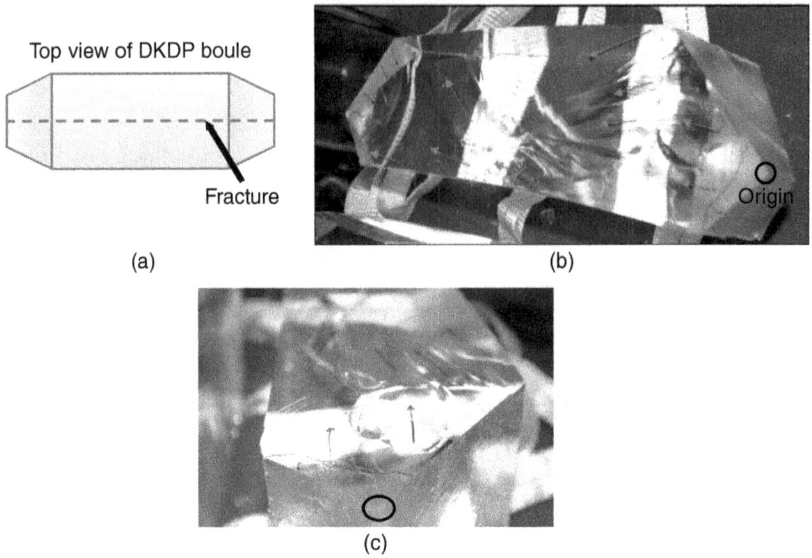

Figure 6.13 (a) A schematic of fractured DKDP crystal boule ($30.5 \times 19 \times 7.5$ cm^3); (b) photo of fracture surface of the same boule, illustrating the fracture origin (circled) and propagation directions (arrows); and (c) fracture-surface view with solid lines indicating regions where salt solution penetrated the fracture and dissolved the surface.

Table 6.6 Calculated maximum ΔT using FEA during cooling of KDP crystal boule under different heat-loss conditions.

Case	Hypothesis of failure mechanism	Simulation conditions	Calculated ΔT	Conclusion
1	Nominal cooling rate too fast	Cool crystal in air at 3 °C/day	0.18 °C	Should not fracture
2	Air-temperature drop upon solution removal	• Cool crystal in air at 3 °C/day • Initial air input at 25 °C	0.92 °C	Fracture unlikely
3	Evaporation off crystal surface	• Cool crystal in air at 3 °C/day • Initial air input at 25 °C • Heat loss from evaporation on all sides of crystal	5.7 °C	Should fracture

The calculations used $k_1 = 1.9$ W m K, $C_{p1} = 0.86$ J gm^{-1} K, $\rho_1 = 2.332$ gm cm^{-3}, $\alpha_1 = 9.48 \times 10^{-3}$ cm^2 s^{-1}, $K_{Ic} = 0.10$ MPa m$^{1/2}$, $E_1 = 63$ GPa, and $v_1 = 0.13$.

5.7 °C, well above the thermal shock FOM of 1.8 °C. Hence, the evaporative contribution to temperature distribution likely caused necessary thermal stress, leading to fracture. A relatively simple mitigation was to incorporate temperature-matched, water-saturated air (as opposed to dry air) into the growth tank as the salt solution was removed, followed by cooling at 3 °C per day.

6.4.1.3 Thermal Fracture of Multilayers

Another form of thermally induced fracture that can occur is on laminate or multilayer structures with materials that have differing thermal-expansion coefficients. One example occurred when assembling and transporting absorbing glass plates mounted on aluminum backing plates. These optical components are strategically placed in high-power laser systems such as NIF to absorb stray light. The component consists of a multilayer of absorbing glass (Pilkington SuperGrey) plates (~6.4 mm thick), a polyurethane adhesive, and an aluminum backing plate. After assembly, the glass was observed to fracture during transport upon exposure to direct sunlight. The glass was also observed to fracture upon thermal cycling the assembly from 21 to 71 °C back to 21 °C.

Figure 6.14 shows the fractured glass upon thermal cycling. The origin point was on the 2D edge overhanging the aluminum backing plate. The fracture propagated orthogonal to the front face of the glass and 2D-edge surface, indicating tensile stress parallel to the long width of the glass. As the fracture propagated through the center of the glass, it branched. Crack branching is typical for the longer fractures, whose branching distance is a function of stress (i.e. stored energy) [1]. The greater the stress, the greater the amount of

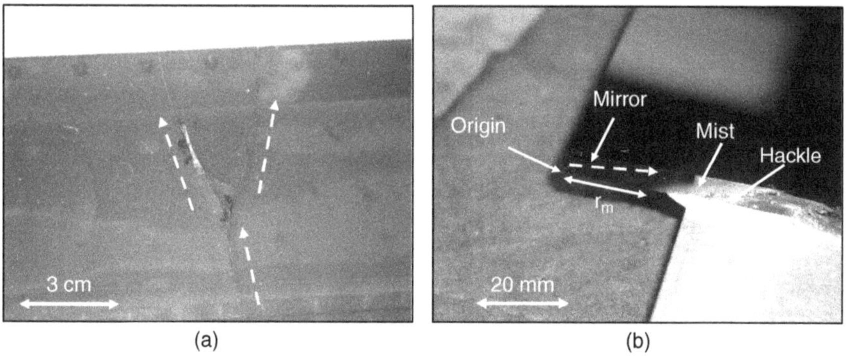

Figure 6.14 (a) A view of the top surface of glass baffle that fractured during thermal cycling (21–71 °C). Below the glass is the polyurethane adhesive and aluminum support. The dashed arrows indicate the direction of crack propagation. (b) A photograph of fracture surface of the baffle, where the dashed arrow indicates direction of crack propagation. Glass thickness is 6.4 mm.

crack branching and fracture area, and the shorter the distance from the origin before branching occurs. A nice example is seen in Figure 6.5b. Figure 6.14a shows the two branches of the absorbing fracture forming an angle of ~45° with respect to each other. This indicates that the principal stress occurred predominately along one axis of the component (see Figure 6.5a).

Figure 6.14b shows the fracture surface of the glass with the fracture origin and crack-propagation direction marked, as determined by fractography [1–3, 29]. The fracture originated on the 2D edge of the glass on the bottom side. The asymmetry of the mirror–mist–hackle regions on the fracture surface in Figure 6.14b also provides information on the type of stress and the stress distribution applied upon failure. The pattern shown indicates high tension on the bottom side of the glass (i.e. the side facing the adhesive and aluminum backing plate) and less tension or compression on the top side. Examples of various mirror–mist–hackle shapes and corresponding stress distributions are shown in Figure 6.15.

The effective stress at failure σ_f can be estimated from the size of the mirror region using Equation (6.2), where r_m is the mirror radius and A_m is the fracture mirror constant. For a typical window glass (Na_2O–CaO–SiO_2) that is similar to the SuperGrey glass composition, $A_m = 1.96\,\text{MPa m}^{1/2}$ [3]. Using Figure 6.14b, the mirror size was measured as 19 mm; hence, the stress at failure is calculated as ~15 MPa.

The stress distribution determined fractographically is consistent with the stress distribution expected from thermal stress caused by a thermal-expansion mismatch between the aluminum and the glass. Ignoring the effect of the polyurethane layer for simplicity, a conservative estimate of the tensile stress in the glass is given by

$$\sigma_t = E_1(\alpha_{Al} - \alpha_1)(T_f - T_i), \tag{6.20}$$

where E_1 is the elastic modulus of the workpiece glass (73.1 GPa), α_1 is the workpiece glass thermal-expansion coefficient ($8.6 \times 10^{-6}\,°C^{-1}$), α_{Al} is the aluminum thermal-expansion coefficient ($23.6 \times 10^{-6}\,°C^{-1}$), and T_i and T_f are the initial and final temperatures, respectively. The baffles are typically prepared under ambient conditions, hence $T_i = 23\,°C$. Since the thermal-expansion coefficient

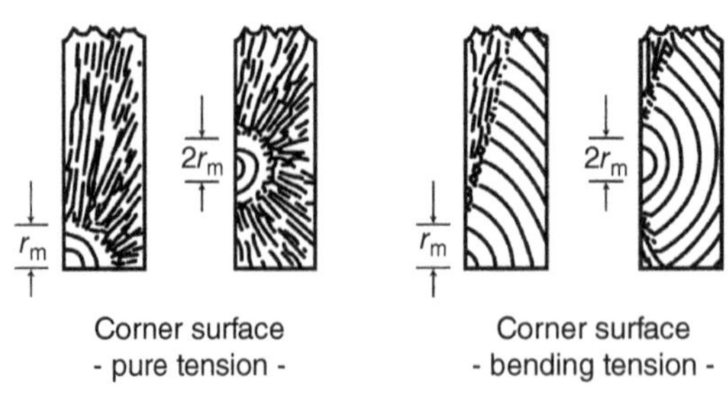

Figure 6.15 An illustration of the differences in fracture marking depending on origin point and type of stress applied (pure tension versus bending). Source: Source: From Fréchette 1990 [1].

of the aluminum is significantly higher than that of the glass, the glass will be in tension when the component is heated.

As stated above, this multilayer component was observed to fracture during transport when exposed to direct sunlight on a hot summer day. T_f was determined from a simple test by measuring the temperature of the laminate upon direct sunlight exposure. Originally at room temperature, the glass heated up to as high as 60 °C in approximately 15 min, owing to its large absorptivity. Using Equation (6.20), the calculated worst-case stress may approach 40.6 MPa.

The allowable temperature rise of a multilayer component before fracture can be estimated. The glass has ground edges, and large surface flaws are likely, possibly as large as 1000 μm. Using a flaw size of 1000 μm and Equation (6.1), the critical stress is ~17 MPa. Plugging this stress value into Equation (6.20) suggests that to prevent glass fracture, this component should not be exposed to temperatures greater than 16 °C above ambient, i.e. $T_f = 38\,°C$. A variety of other approaches can be used to create a greater fracture-resistance margin, including insulating and blocking the laminate from direct sunlight, reducing the flaw size in the ground glass by polishing the edges, and increasing the thickness or compliance of the low-modulus adhesive layer.

6.4.2 Blunt Loading with Friction

Another important and little-known failure mechanism is the mechanical loading of a blunt, often relatively soft material with significant friction (tangential) forces against a workpiece, leading to fracture initiation. It is common practice to minimize point loading of a hard, high modulus material against a brittle workpiece, because the load to initiate fracture decreases with decrease in the radius of curvature and with increase in the hardness of the indenter (see Equations ((3.1) and (3.2)). Hence, it is generally well accepted to minimize glass-to-metal contact in the various optical fabrication steps, especially for tools used to carry, transport, inspect, block, grind, or polish the workpiece. Using a blunt indenter with large radius of curvature and a softer material for contact with the workpiece is generally a good practice.

In this example, however, fracture initiation occurs even with a blunt, soft indenter. Nd-doped laser-phosphate-glass-slab workpieces used as amplifiers in high peak-power lasers are fabricated and mounted in specialized tooling called line replaceable units (LRUs) for installation in the laser (see Figure 6.17) [26, 30–32]. To minimize the distortion of the optical transmissive wavefront, the workpiece contacts the LRU at only a few points along the workpiece edge. Soft aluminum buttons, specifically Al 6061, with a relatively large radius of curvature (25 cm) make contact with the ground edge of the laser glass workpiece (Figure 6.16). Using Equation (3.1) and the value of Auerbach's constant for laser glass, $A = 29\,\mathrm{N\,mm^{-1}}$ (from Table 3.1), the critical load for fracture initiation is quite large, ~7200 N. The buttons should experience a load that corresponds

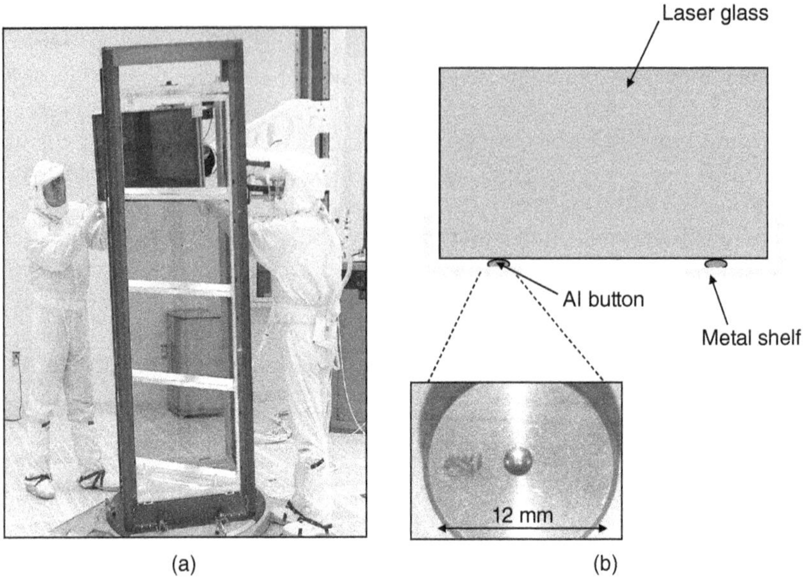

Figure 6.16 (a) A photo of laser-phosphate-glass slab being assembled in its mechanical mount and (b) a schematic illustration of the slab sitting on two aluminum buttons.

to only about half the weight of the slab (~220 N). In fact, the initiation load should be even higher, because of the softness of the aluminum button; testing under pure normal loads showed that fracture did not initiate until 40 000 N.

Despite the large load safety margin, this mounting configuration led to fracture initiation in practice and the growth of trailing indent fractures to ~1 mm deep (see Figure 6.17a). The analysis did not account for frictional or tangential forces occurring from normal loads applied to the buttons, which significantly drop the initiation load for fracture. The effect of a tangential load on stress distribution during loading was discussed in Section 3.1.1.3 in the context of scratch generation. Figure 3.9 compares the stress distribution of Hertzian loading with and without a tangential load. Note that with a tangential load, tensile stress at the leading edge is reduced and the tensile stress at the trailing edge increases. Hence, the initiation load for fracture decreases with an increase in tangential force at the trailing edge, leading to trailing indent fractures.

The magnitude of drop in the initiation load can be quite high (see Figure 6.17b). Experiments on the effect of metals of different hardnesses sliding across glass surfaces was examined by Ghering and Turnbull [34]. They found that sliding metals may initiate fracture at much lower load that in the purely normal loaded case. For iron, the initiation load dropped from 35 kgf without tangential to 0.23 kgf with a tangential load – a 100× decrease.

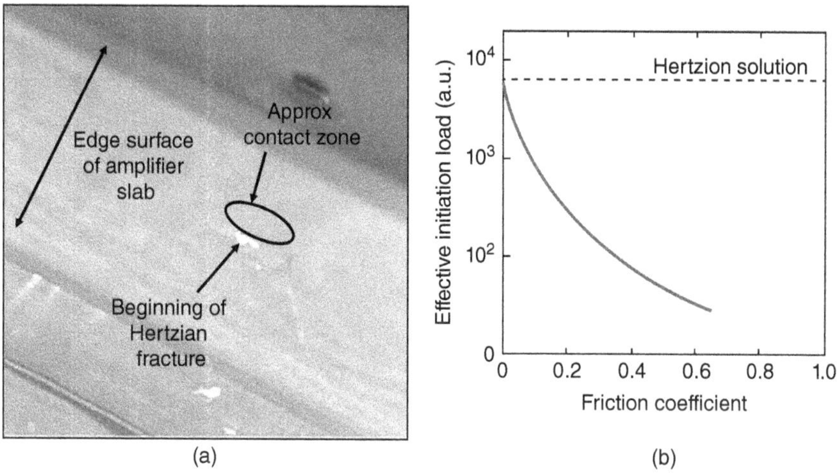

Figure 6.17 (a) A photo of ground edge of the laser slab, illustrating the fracture observed at the button contact zone and (b) the effect of friction on the initiation load for a blunt indenter. Source: Chiang and Evans 1983 [33]. Reproduced with permission from John Wiley & Sons.

Failure analysis of the fracture slab showed strong evidence of frictional or tangential forces. First, the fracture had a trailing-indent character, as opposed to an ideal Hertzian cone fracture (see Figure 6.17a). Second, aluminum was deposited on the ground glass surface of the slab workpiece at the contact zone, indicating friction-induced wear (see Figure 6.18A). No aluminum deposit was observed on the test coupons from offline testing, where off-axis loads are intentionally removed during testing; see Figure 6.18B.

6.4.3 Glass-to-metal Contact and Edge Chipping

Minimizing glass–metal contact during all stages of optical fabrication and end use to lower the risk of workpiece fracture is not a new practice. Nevertheless, in the following example, a glass workpiece fractured on mounting because of unforeseen glass–metal contact near the edge of the workpiece.

Square fused-silica glass windows (430 mm × 430 mm × 40 mm) are used as a vacuum barrier in high-energy, high-power lasers such as NIF [26, 35]. In mounting this window in its LRU, an aluminum cover is screwed onto a mount holding the vacuum window. A compliant gasket is placed between the aluminum frame and the face of the window to avoid glass-to-metal contact. In this failure, when the frame was screwed into the mount, the frame bent, as can be observed by the interface gap variations in Figure 6.19a. The bending was significant enough for the frame to make contact with face of the window and led to a large edge chip (see in Figure 6.19b).

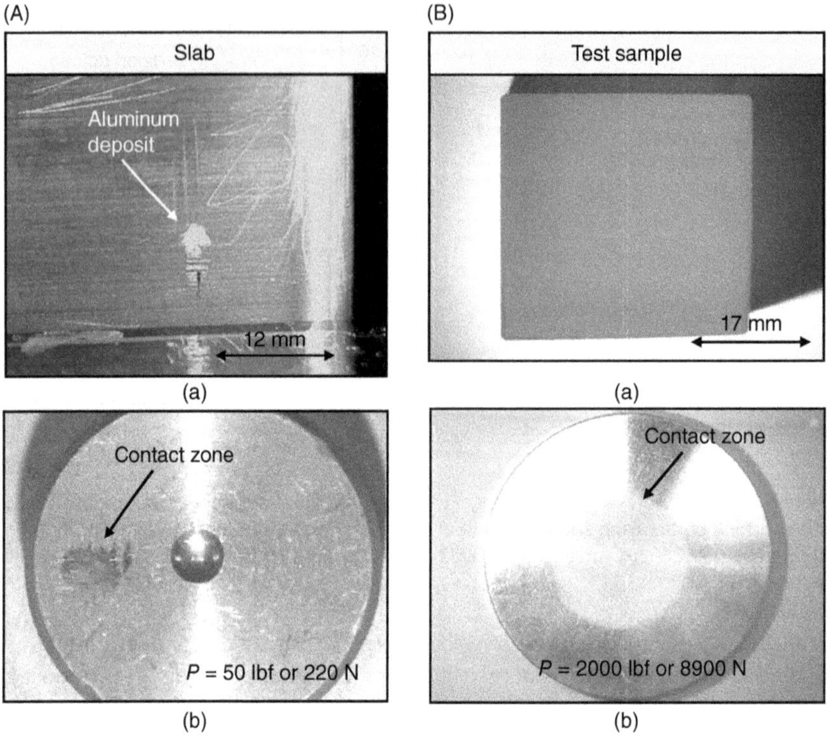

Figure 6.18 (A) The ground phosphate-laser-glass slab (a) loaded by an aluminum button (b) showing evidence of sliding friction and (B) the ground phosphate-laser-glass sample (a) normally loaded in a mechanical testing device by an aluminum button (b).

An aluminum deposit was present on the fused-silica window face near the chip origin on the edge, verifying that the aluminum frame made contact (see Figure 6.20a). The width of the deposit where the edge fracture occurred was ~300 μm and the location was ~500 μm away from the bevel, or 1680 μm from the window edge. The markings on the fracture surface, such as the mirror region, Wallner lines, and twist hackle, confirmed the direction of propagation and that the fracture originated from the window face at the point of the glass-to-metal contact (see Figure 6.20b). Treating the mirror region as a radial crack, an estimate of the load at failure is determined as ~240 lbf or ~1100 N using Equation (3.5), where $c_r = 1000\,\mu\text{m}$, $\chi_r = 0.022$ (from Table 3.1), and $K_{Ic} = 0.75\,\text{MPa m}^{1/2}$. Using the edge–toughness relationship, the critical load needed to cause the edge-chip failure is determined as $P_{ce} = 183\,\text{N}$ using Equation (3.7), where $d_e = 1680\,\mu\text{m}$ and $T_e = 109\,\text{N mm}^{-1}$ (from Table 3.1). Hence, the applied load that led to the edge chip was ~7× that required to cause the chip. Some simple mitigations to prevent failure can be applied: (i) avoid glass-to-metal contact by defining the proper torque on the frame screws

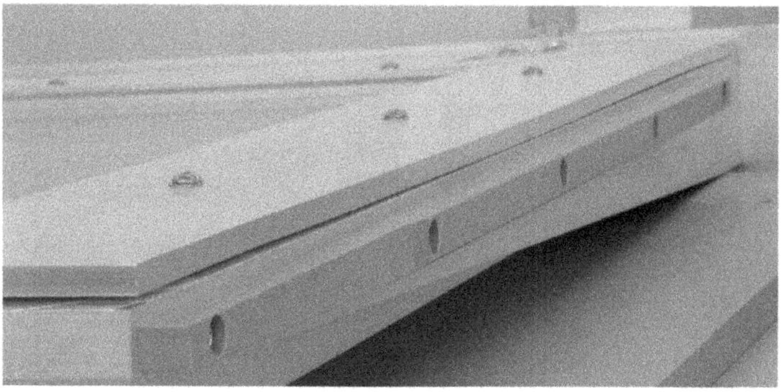

(a)

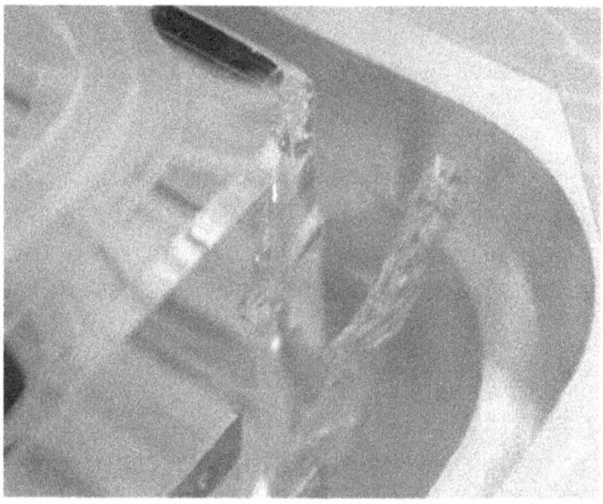

(b)

Figure 6.19 (a) A photograph of the edge of aluminum frame holding a fused-silica vacuum window. The images illustrate the bending of the aluminum cover frame when screws were tightened (see variation in gap between top frame and mount). (b) Resulting corner-edge chip on a fused-silica window, originating from the face of workpiece window.

and (ii) minimize bending by stiffening the frame with a different material or increasing its thickness.

6.4.4 Glue Chipping Fracture

In Section 3.1.1, various types of surface-crack formations from indentation were discussed – blunt versus sharp, static versus dynamic, on face versus near edge. In these cases, the normal load was the primary load. A lesser-known surface fracture type is described in this section: glue chipping, where the tangential load is the primary load. "Glue chipping" comes from an artisanal method to pattern glass surfaces by applying adhesive on a workpiece surface in a spatial pattern [36–38]. As the applied glue shrinks during drying and curing, a large local tensile stress is applied tangential to the surface, initiating conchoidal chips. The fractures run largely parallel to the glass surface,

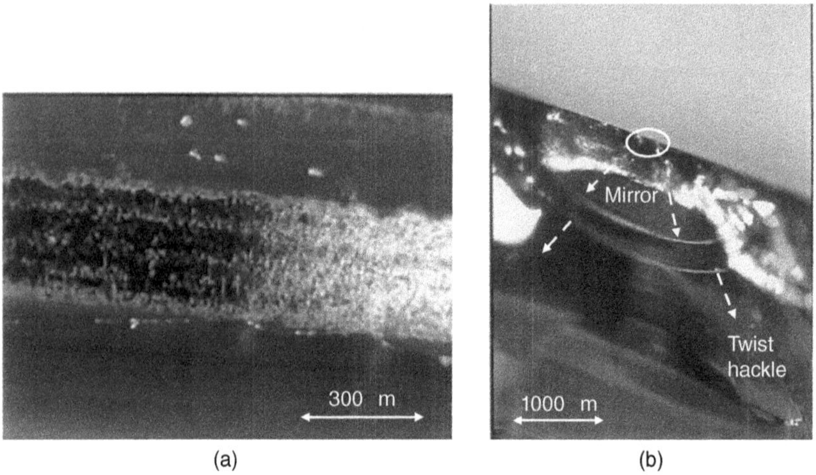

Figure 6.20 (b) An optical micrograph of top surface of an edge chip on a fused-silica window showing an aluminum deposit on the surface. The width of the aluminum deposit is approximately ~300 μm. (b) An optical micrograph of fracture surface of the edge chip. The top edge represents the face of the fused-silica window where glass-to-metal contact occurred.

ultimately removing a thin slice of glass from the surface. The initiation of the glue chip is analogous to that of a trailing-indent fracture during sliding indentation (see Section 3.1.1.3).

Glue-chipping failure may occur during optical fabrication, especially when a coating or barrier layer of some type is attached to the face of the workpiece. When differential shrinkage between the workpiece and coating occurs and adhesion is spatially nonuniform, a local point on the workpiece surface may experience a large local, concentrated tangential load against the surface, leading to glue chipping.

Figure 6.21 illustrates glue chipping in a glass tank with a fluoropolymer-based liner. This cylindrical 1000 L glass tank was used for growing large KDP crystals for nonlinear optics in large aperture, high-fluence lasers [26–28]. The liner was used to minimize secondary nucleation from surface imperfections on the glass surface and to reduce ppb-level impurities leaching from the glass, which can influence the final crystal habit, i.e. final shape [28]. The initial liner was not intentionally adhered to the glass tank; however, as the material shrank relative to the glass tank, it adhered locally, and a series of glue-chip fractures were formed (see Figure 6.21). One mitigation is to use a release agent at the interface between the liner and glass tank to eliminate the nonuniform adhesion.

A unique characteristic of this particular fracture is the repeating ripples, ~200 μm apart, on the fracture surface (see fracture surface images in

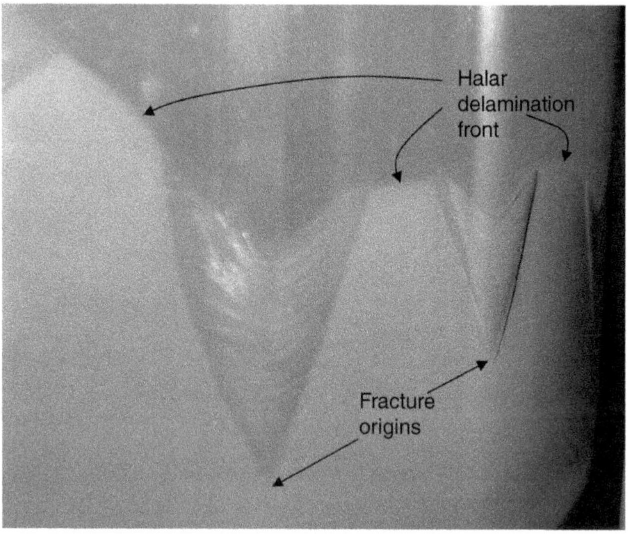

Figure 6.21 A photograph of glass tank, viewed from exterior, illustrating the formation of glue-chip fractures as the fluoropolymer liner delaminated from the glass inner surface. The delamination front of the fluoropolymer liner travels from bottom to top.

Figure 6.22). These features are stress-wave markings, indicative of some sort of oscillating force or vibration that modified the stress field during crack propagation (see also Table 6.1 and Figure 6.4). The source of this vibration is believed to be the repeated acceleration and deceleration of the rotating crystal during the crystal growth process.

6.4.5 Workpiece Failure from Differential Pressure

Another possible workpiece failure is fracture under vacuum- or gas-pressure loading. Examples include vacuum loading during mounting of workpiece for an optical-fabrication step or during end use as a pressure-differential barrier window. This type of workpiece failure may have large implications, such as the implosion of vacuum chambers or fracturing of pressurized aircraft windows.

The stress a workpiece is exposed to under such conditions is a function of the window geometry, pressure differential, and mechanical properties of the workpiece material. The failure stress is given by Griffith's failure criteria, Equation (6.1); thus controlling the flaw size or preventing the creation of flaws may prevent such failures. In some use applications, however, flaws may generate on the surface during use, e.g. in the case of fused-silica lenses used as barriers in vacuum spatial filters in high-power lasers, where the laser light may damage the lens [9, 39]. Despite methods employed to reduce the risk of laser damage, there remains a remote chance of creating a surface flaw, and thus unacceptable

(a) (b)

Figure 6.22 (a) An optical micrograph of the fracture surface of a glue chip on the KDP-crystal-growth glass tank. The lines at the left are millimeters apart. (b) A higher magnification of the same fracture surface showing the stress-wave markings.

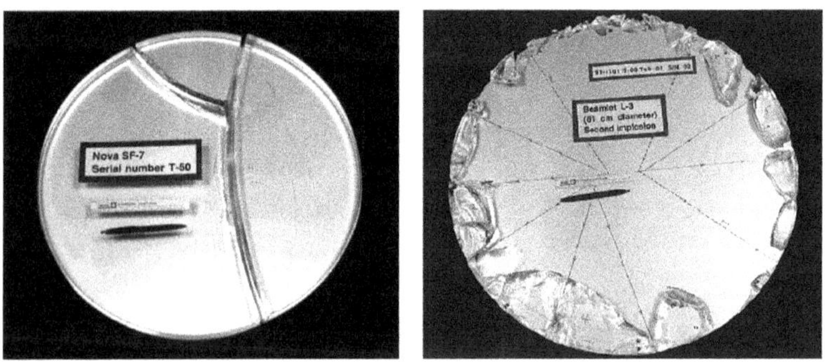

(a) (b)

Figure 6.23 (a) A fracture of 52 cm and (b) 61 cm diameter fused-silica lens that fractured during use as a vacuum barrier in an optical spatial filter in high-power lasers. Source: Campbell et al. 1996 [39]. Reproduced with permission from SPIE.

risk of an implosion and damage to the laser system. Two examples of fractured vacuum-barrier lenses are shown in Figure 6.23.

This remote risk may be mitigated by designing the window or lens such that it can break into at most two pieces (one full diameter fracture), allowing leakage, but not implosion and catastrophic damage. The lens in Figure 6.23a suffered only a vacuum leak; the lens in Figure 6.23b is an example of imploding into many pieces. The key factor that determines these two failure modes is the amount of stored energy in the window, i.e. the total stress from the pressure differential. The greater the stored energy or stress, the greater the fracture area and chance of implosion.

The peak tensile stress in the window, which is proportional to its stored energy, is located on the face center on the vacuum side. A flaw near this location is generally the origin of the fracture, as confirmed by fractography analysis of fractured glass plates from offline vacuum-load testing [9, 39]. The magnitude of the window's peak stress can be determined using FEA analysis or analytical relationships for simple geometries. For example, the peak stress for a square plate that is simply supported is given by

$$\sigma_{\mathrm{p}} = \frac{\beta \sigma_{\mathrm{o}} \ell_{\mathrm{o}}^{2}}{t_1^{2}}, \tag{6.21}$$

where σ_{o} is the applied-pressure differential, t_1 is workpiece thickness, β is the geometric constant (0.4345 for a square plate), and ℓ_{o} is the characteristic dimension (diameter or side length) of the workpiece [40]. Note that peak stress decreases as the square of the thickness and increases as the square of the size of the workpiece. Hence, smaller, thicker (or low-aspect-ratio) workpieces have lower peak stress and are less prone to multifracture, implosive failure.

Using a stored-energy-balance model, the fracture surface area A_{sf} is related to the volume of the lens V_{L} and the peak stress σ_{p} in the window as follows:

$$A_{\mathrm{sf}} = Z_{\mathrm{sf}} \sigma_{\mathrm{p}}^{2} V_{\mathrm{L}}. \tag{6.22}$$

Z_{sf} is a constant determined experimentally, where a series of round and square glass plates of various thicknesses were fractured by vacuum loading and the resulting surface fracture area was measured (see Figure 6.24). For fused-silica glass and conventional window glass, Z_{sf} was determined as 8.4×10^{-5} cm^2 psi^{-2} L^{-1} [9, 39].

Using the information above, a failsafe design criterion that does not allow for more than one full diameter fracture (e.g. $A_{\mathrm{sf}} \leq 2\, t_1\, \ell_{\mathrm{o}}$ for round plates) can be derived as follows:

$$\sigma_{\mathrm{p}} \leq \frac{K_{\mathrm{f}}}{\sqrt{\ell_o}}, \tag{6.23}$$

where K_{f} is an empirically determined constant: 3.79 MPa m$^{1/2}$ for round optics and 3.32 MPa m$^{1/2}$ for square optics [9, 39]. From a Griffith's law standpoint, this design criterion can be considered as a critical flaw size that is larger than the thickness of the workpiece, as described in Equation (6.1). The design peak stresses for round and square optics as a function of the characteristic dimension is shown in Figure 6.25.

To illustrate the utility of this design criterion, take the design of a NIF-laser square vacuum-barrier window with a characteristic length of $\ell_{\mathrm{o}} = 42.6$ cm with a critical peak design stress of 750 psi (Figure 6.25). The design thickness of this vacuum barrier was adjusted to ~40 mm, using Equation (6.21) to

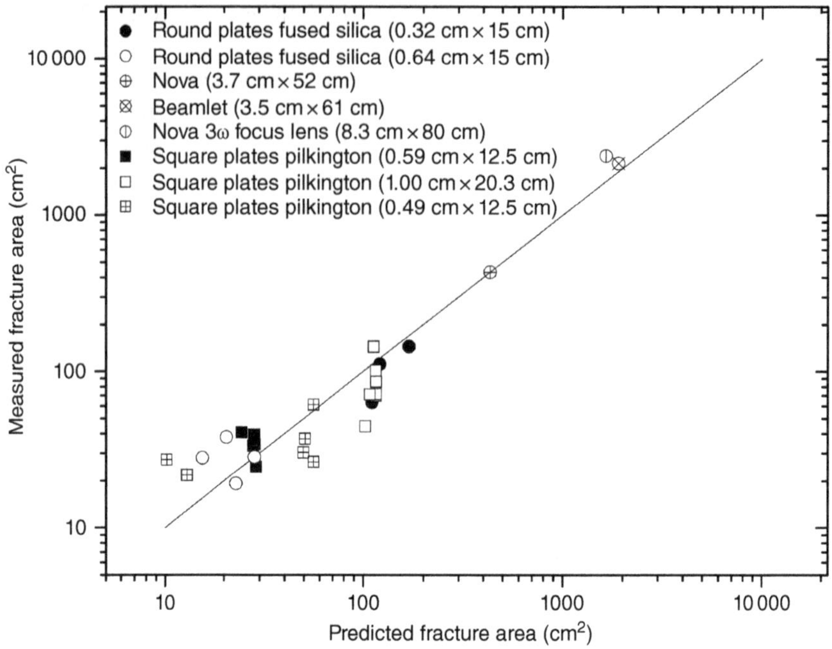

Figure 6.24 Measured fracture area from vacuum loading a variety of glass round and square windows compared with the calculated fracture area using Equation (6.22). Source: Suratwala et al. 1998 [9]. Reproduced with permission from SPIE.

ensure that the peak stress (in this case 500 psi) was less than the peak design stress (750 psi). Following this design constraint helps ensure that a window cannot break into more than two pieces and implode.

6.4.6 Chemical Interactions and Surface Cracking

Chemical interactions can not only enhance crack propagation, as with the slow-crack growth described in Section 6.3, they can modify some materials, leading to the generation of surface tensile stress and, hence, surface fracturing. In some ways, this phenomenon is the opposite of the chemical tempering commonly used for smartphone glass windows, where larger alkali atoms are substituted for smaller cations, in the glass structure, a process called ion exchange – leading to a large local surface compression in the glass [7, 41]. In the discussion below, phosphate glass and DKDP crystals exemplify this behavior.

6.4.6.1 Surface Cracking of Phosphate Glass

Laser phosphate glass doped with Nd ions is a novel optical material used as amplifiers in high-fluence laser systems [32]. This material is an attractive

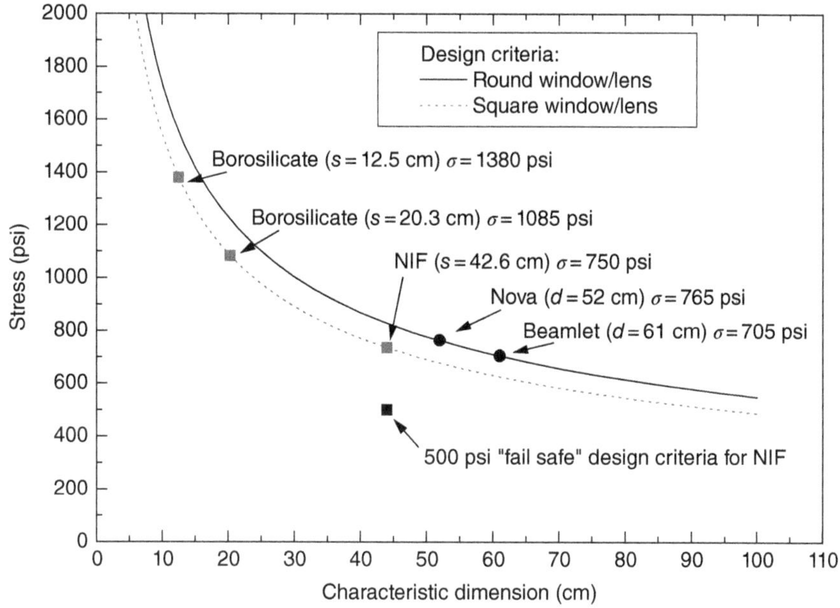

Figure 6.25 Design peak stress determined for round (dashed line) and square (solid line) window-glass workpieces as a function of the window characteristic dimension ℓ_0. Source: Suratwala et al. 1998 [9]. Reproduced with permission from SPIE.

gain-media host but has unique chemical characteristics that makes fabrication challenging, including (i) a highly corrosive nature during melting, requiring the use of Pt melting vessels, (ii) a hydroscopic nature, leading to embedded OH that is detrimental to laser performance and challenging to remove, (iii) reactivity with ambient moisture, leading to surface chemical weathering and optical-transmission degradation due to scatter, and (iv) water diffusion during annealing, leading to surface tension and surface fracturing [31, 42–44]. An example of the latter is shown in Figure 6.26, where a test block of phosphate glass was annealed at ~500 °C under ambient humidity and cooled to room temperature. (An annealing process is required for most glasses to remove residual stresses and improve the optical homogeneity.) [45].

The mechanism for this surface-crack generation stems from the diffusion of H_2O from the environment into the glass surface during annealing, which changes the thermal-expansion properties of the surface layer, leading to surface tension [43]. The incorporation of H_2O/OH into the glass surface was confirmed by depth profiling the OH absorptivity at 3.333 μm, using infrared spectroscopy (see Figure 6.27). After diffusion, the H_2O can react with the glass via hydrolysis of the linear phosphate chains P–O–P, changing the glass's properties, namely its transition temperature. This hydrolytic reaction is

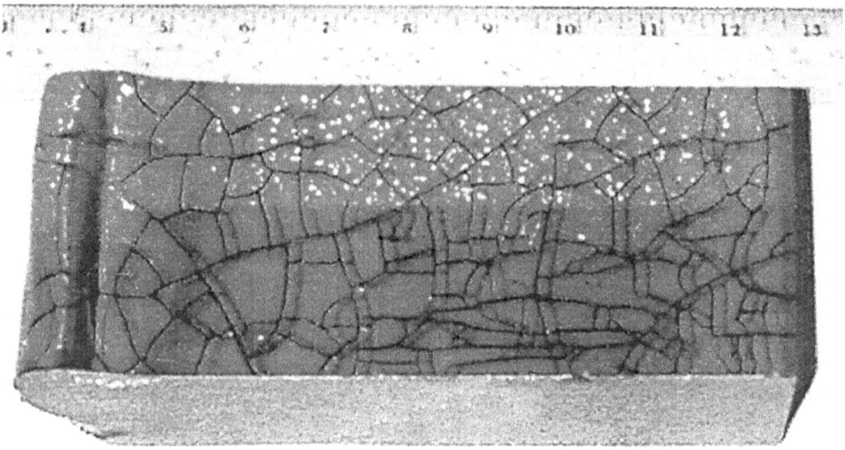

Figure 6.26 The surface cracking observed on the surface of laser phosphate glass after annealing (traced in black).

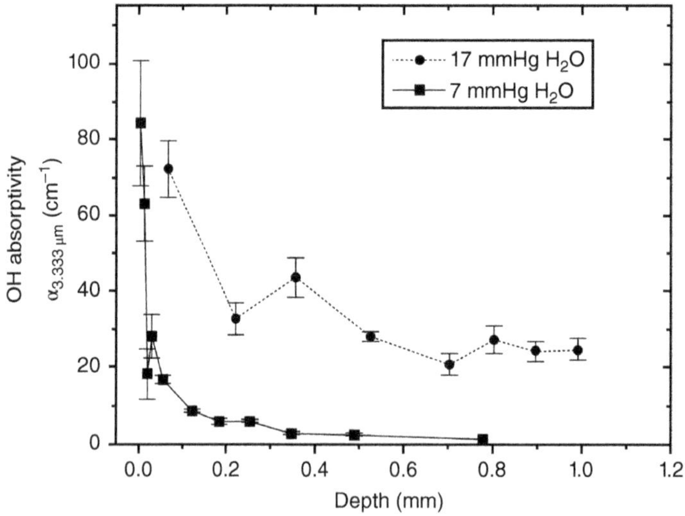

Figure 6.27 Measured OH concentration as a function of depth from the top surface for two laser phosphate glasses (LG-770) after annealing near the glass transition temperature at two water-vapor pressures. Source: Hayden et al. 2000 [43]. Reproduced with permission from Elsevier.

similar to that proposed for the chemical-tooth mechanism for polishing in Equation (5.15). An increase in the OH content of the glass from about 80 to 5000 ppmw causes a decrease in the glass transition temperature from 468 to 405 °C and an increase in the linear thermal-expansion coefficient from 12.9 to $14.2 \times 10^{-6}\,°C^{-1}$.

Hayden et al. [43] experimentally simulated and quantified the generation of surface tension by fusion-bonding at 500 °C a bilayer of two phosphate glass samples of the same composition, but with OH contents that differed by up to 50×. Using the measured thermal-expansion behavior of the two glasses, the generation of surface strain during the cooling of a bilayer phosphate glass was calculated (see Figure 6.28). Surface stress can be estimated from the determined elastic strain δ_e as follows:

$$\sigma_s = \frac{\delta_e E_1}{1 - \nu_1}\left(\frac{2t_1}{t_s} + \frac{E_1}{E_s}\right)^{-1}, \tag{6.24}$$

where E_1 and E_s are Young's moduli of the core and surface material, respectively, ν_1 is a composite Poisson's ratio, and t_s and t_1 are the thicknesses of the surface and core glass, respectively. Using $E_s = E_1 = 47$ GPa and $\nu_1 = 0.26$, and the measured strain $\delta_e = 700$ ppm, the calculated surface tensile stress is ~42 MPa. This value was fairly consistent with a measured value of ~20 MPa using stress birefringence [43].

6.4.6.2 Surface Cracking of the DKDP Crystals

Deuterated potassium-dihydrogen-phosphate single crystals (KD_2PO_4 or DKDP is important for nonlinear optical materials used for frequency conversion and polarization rotation in high-fluence lasers [28]. This material can also exhibit surface cracking, observed as high-density parallel cracks that run parallel to the extraordinary axis of the crystal and referred to as crazing or e-cracking (see Figure 6.29a). The surface fractures are typically only 10–200 μm deep. Fractography analysis of the surface of an e-crack shows that the fracture initiates uniformly from the top surface, likely from the edge, under a very uniform mode-I tension orthogonal to the crystal face (see Figure 6.29b). As evidenced by the roughly evenly spaced arrest lines parallel to the top surface, the fracture propagates in 20–30 μm increments, as if stress is built up and relieved, repeatedly.

Huser et al. [46] used the Raman spectroscopy to surface profile the Deuterium concentration of a variety of DKDP crystals (see Figure 6.30). They showed that crazed DKDP crystals had significantly more deuterium depletion than those not crazed. The deuterium depletion rate varied from crystal to crystal.

The depletion of deuterium from the DKDP crystal is a result of the deuterium–hydrogen exchange in the crystal, where the source of the hydrogen is from H_2O in the atmosphere. After the exchange, the surface crystal layer

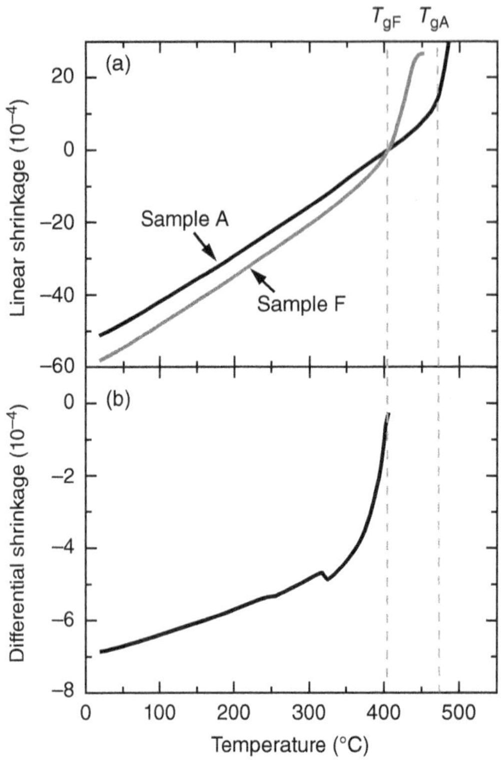

Figure 6.28 (a) Measured thermal expansion using dilatometry of phosphate glass A (low OH) and phosphate glass F (high OH), shifted to intersect at the glass transition temperature of the high-OH glass. (b) Differential shrinkage during cooling of a laminate (fusion sample) composed of glasses A and F. The differential shrinkage is calculated from the difference between the two curves in (a). Source: Hayden et al. 2000 [43]. Reproduced with permission from Elsevier.

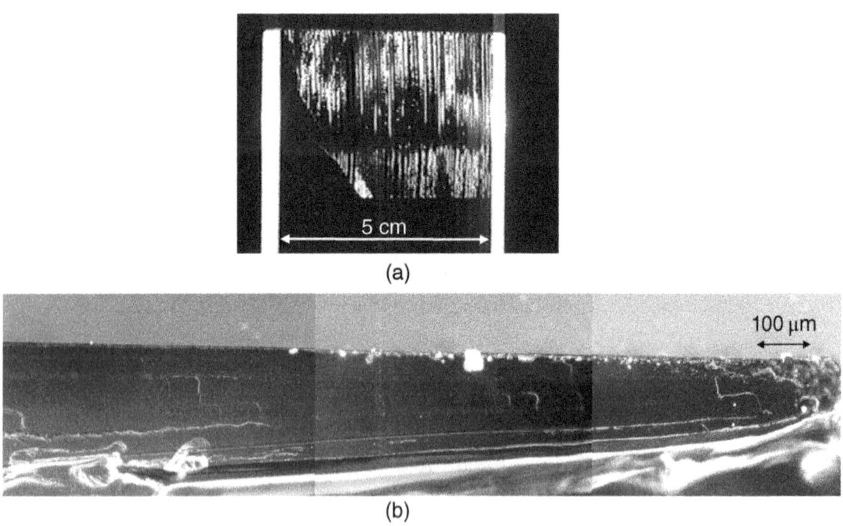

Figure 6.29 (a) A photo of e-cracking observed on the DKDP crystal after growth and storage in air for a few months and (b) the fracture surface of a single e-crack.

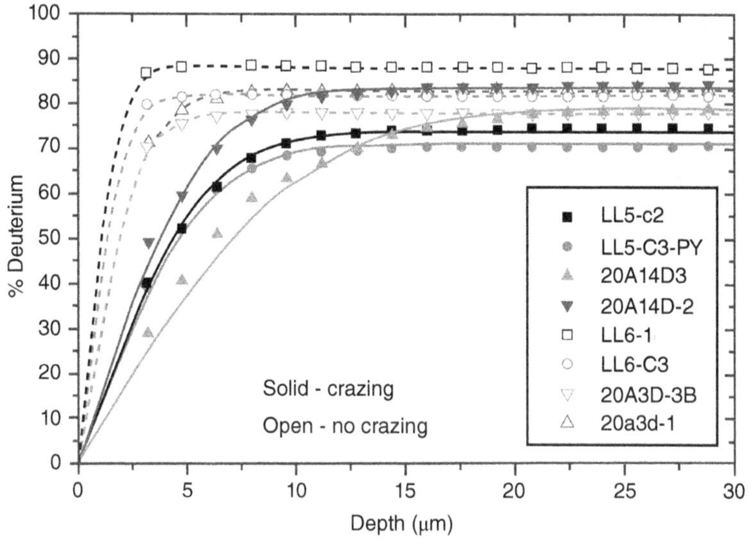

Figure 6.30 Deuterium concentration as a function of depth in the surface of various DKDP crystals. Open data points are samples that crazed and solid data points are samples that were not observed to craze. Data was measured by the Raman spectroscopy.

becomes KDP-like, or KH_2PO_4. The change in the crystal composition results in a large tensile-strain development on the surface (~0.2%) due to changes in the lattice spacings ($a_{KDP} = 7.4521$ Å and $a_{DKDP} = 7.4690$ Å). The mechanism for the chemical change and stress development has been proposed as a sequence of steps involving: (i) adsorption/desorption of H_2O, D_2O, or HDO on and off the surface; (ii) the exchange of the hydrogen in H_2O with the deuterium on the crystal surface; and (iii) diffusion of deuterium/hydrogen in the bulk.

The rate-limiting step was determined to be the diffusion of deuterium/hydrogen in the bulk. The effective diffusion coefficient for samples that craze was $>0.5 \times 10^{-14}$ cm^2 s^{-1} (typically $1\text{–}5 \times 10^{-14}$ cm^2 s^{-1}). In contrast, the diffusion coefficients for samples that did not craze were $<0.1 \times 10^{-14}$ cm^2 s^{-1}. Hence, materials with high diffusion coefficients are expected to craze within months, and those with low diffusion coefficients are expected to craze in years. It was later determined that changing the temperature at which the crystal was initially grown by lowering the temperature below 40 °C helped reduce the diffusion coefficient of H within the crystal and prevent e-cracking from occurring. Altering the growth temperature served as a primary mitigation for the e-cracking. It is hypothesized that growth temperature influences D or H vacancy concentration, which affects its diffusivity. Another possible mitigation is to create a D_2O environment around the DKDP crystals during storage and laser installation, to prevent D/H exchange; however, this method is difficult to implement in practice.

References

1 Frechette, V.D. (1990). *Failure Analysis of Brittle Materials*. Westerville, OH: American Ceramic Society.
2 Varner, J.R. and Frechette, V.D. (2007). *Fractography of Glasses and Ceramics V*. Hoboken, NJ: Wiley.
3 Quinn, G.D. (2007). *Fractography of Ceramics and Glasses*. Washington, DC: National Institute of Standards and Technology.
4 Freiman, S. and Mecholsky, J.J. Jr., (2012). *The Fracture of Brittle Materials: Testing and Analysis*. Wiley.
5 Bradt, R.C. and Tressler, R.E. (1994). *Fractography of Glass*. Springer.
6 Anderson, T.L. (1995). *Fracture Mechanics: Fundamentals and Applications*, 2e, 688. New York: CRC Press.
7 Varshneya, A.K. (2013). *Fundamentals of Inorganic Glasses*. Elsevier.
8 Griffith, A.A. (1921). The phenomena of rupture and flow in solids. *Philos. Trans. R. Soc. London, Ser. A* 221: 163–198.
9 Suratwala, T.I., Campbell, J.H., Steele, W.A., and Steele, R.A. (1998). Fail safe design for square vacuum-barrier windows. Proceedings of SPIE, Volume 3492, pp. 740–9.
10 Suratwala, T., Miller, P., Feit, M., and Menapace, J. (2008). Scratch forensics. *Opt. Photonics News* 20 (9): 12–15.
11 Lawn, B.R. (1998). Indentation of ceramics with spheres: a century after hertz. *J. Am. Ceram. Soc.* 81 (8): 1977–1994.
12 Suratwala, T., Steele, R., Feit, M.D. et al. (2008). Effect of rogue particles on the sub-surface damage of fused silica during grinding/polishing. *J. Non-Cryst. Solids* 354 (18): 2023–2037.
13 Suratwala, T., Wong, L., Miller, P. et al. (2006). Sub-surface mechanical damage distributions during grinding of fused silica. *J. Non-Cryst. Solids* 352 (52–54): 5601–5617.
14 Swain, M.V. ed. (1979). Microfracture about scratches in brittle solids. *Proc. R. Soc. London, Ser. A* 366 (1727): 575–597.
15 Lawn, B.R. (1993). *Fracture of Brittle Solids*, 2e, xix, 378. Cambridge, New York: Cambridge University Press.
16 Wiederhorn, S. (1967). Influence of water vapor on crack propagation in soda-lime glass. *J. Am. Ceram. Soc.* 50 (8): 407–414.
17 Crichton, S., Tomozawa, M., Hayden, J. et al. (1999). Sub-critical crack growth in a phosphate laser glass. *J. Am. Ceram. Soc.* 82 (11): 3097–3104.
18 Suratwala, T.I., Steele, R.A., Wilke, G.D. et al. (2000). Effects of OH content, water vapor pressure, and temperature on slow crack growth behavior in phosphate laser glass. *J. Non-Cryst. Solids* 263&264: 213–227.
19 Wiederhorn, S. (1978). Mechanisms of subcritical crack growth in glass. In: *Fracture Mechanics in Ceramics* (ed. R.C. Brandt, D.P.H. Hasselman and F.F. Lange), 549–580. New York: Plenum Press.

20 Freiman, S. (1984). Effects of chemical environments on slow crack growth in glasses and ceramics. *J. Geophys. Res. Solid Earth* 89 (B6): 4072–4076.
21 Suratwala, T. and Steele, R. (2003). Anomalous temperature dependence of sub-critical crack growth in silica glass. *J. Non-Cryst. Solids* 316 (1): 174–182.
22 Wiederhorn, S. (1972). A chemical interpretation of static fatigue. *J. Am. Ceram. Soc.* 55 (2): 81–85.
23 Ritter, J. and Meisel, J. (1976). Strength and failure predictions for glass and ceramics. *J. Am. Ceram. Soc.* 59 (11–12): 478–481.
24 Kingery, W.D., Bowen, H.K., and Uhlmann, D.R. (1976). *Introduction to Ceramics*, 2e, 1032. New York: Wiley.
25 VanSant, J. (1983). *Conduction Heat Transfer Solutions.* Lawrence Livermore National Laboratory. Contract No.: UCRL-52863.
26 Baisden, P.A., Atherton, L.J., Hawley, R.A. et al. (2016). Large optics for the National Ignition Facility. *Fusion Sci. Technol.* 69 (1): 295–351.
27 Zaitseva, N.P., De Yoreo, J.J., DeHaven, M.R. et al. (1997). Rapid growth of large-scale (40–55 cm) KH_2PO_4 crystals. *J. Cryst. Growth* 180 (2): 255–262.
28 Hawley-Fedder, R.A., Robey, H.F. III,, Biesiada, T.A., et al. ed. (2000). Rapid growth of very large KDP and KD* P crystals in support of the National Ignition Facility. International Symposium on Optical Science and Technology. International Society for Optics and Photonics.
29 Frechette, V.D. (1990). Examples of glass plate fracture without branching. In: *Failure Analysis of Brittle Materials.* Westerville, OH: American Ceramic Society/Advances in Ceramics, 44–97.
30 Suratwala, T., Campbell, J., Miller, P. et al. ed. (2004). Phosphate laser glass for NIF: production status, slab selection, and recent technical advances. Proceedings of SPIE, Volume 5341.
31 Campbell, J.H., Suratwala, T.I., Thorsness, C.B. et al. (2000). Continuous melting of Nd-doped phosphate laser glasses. *J. Non-Cryst. Solids* 263&264: 342–357.
32 Campbell, J.H. and Suratwala, T.I. (2000). Nd-doped phosphate glasses for high-energy/high-peak-power lasers. *J. Non-Cryst. Solids* 263&264: 318–341.
33 Chiang, S.S. and Evans, A.G. (1983). Influence of a tangential force on the fracture of two contacting elastic bodies. *J. Am. Ceram. Soc.* 66 (1): 4–10.
34 Ghering, L. and Turnbull, J. (1940). Scratching of glass by metals. *Bull. Am. Ceram. Soc.* 19 (8): 290.
35 Spaeth, M.L., Manes, K.R., Kalantar, D.H. et al. (2016). Description of the NIF laser. *Fusion Sci. Technol.* 69 (1): 25–145.
36 Ridge, J.D. (2000). Insulated glass unit window assembly including decorative thermoplastic sheet and method for forming. Google Patents.
37 Schrunk, T.R. (1982). Decorative glass chipping method. Google Patents.
38 Messer, J.A. (1984). Glue chipped glass and method. Google Patents.

39 Campbell, J., Hurst, P., Heggins, D. et al. (1996). Laser induced damage and fracture in fused silica vacuum windows. Proceedings of SPIE, Volume 2986, pp. 106–25.

40 Roark, R. and Young, W. (1982). *Formulas for Stresses and Strain*, 386. New York: McGraw-Hill.

41 Barefoot, K.L., Dejneka, M.J., Gomez, S. et al. (2013). Crack and scratch resistant glass and enclosures made therefrom. Google Patents.

42 Thorsness, C., Suratwala, T.I., Steele, R.A. et al. ed. (2000). Dehydroxylation of phosphate laser glass. International Symposium on Optical Science and Technology. International Society for Optics and Photonics.

43 Hayden, J.S., Marker, A.J. III,, Suratwala, T.I., and Campbell, J.H. (2000). Surface tensile layer generation during thermal annealing of phosphate glass. *J. Non-Cryst. Solids* 263&264: 228–239.

44 Tischendorf, B.C. (2005). *Interactions Between Water and Phosphate Glasses*. Rolla, MO: University of Missouri.

45 Doremus, R.H. (1973). *Glass Science*. Wiley.

46 Huser, T., Hollars, C.W., Siekhaus, W.J. et al. (2004). Characterization of proton exchange layer profiles in KD_2PO_4 crystals by micro-Raman spectroscopy. *Appl. Spectrosc.* 58 (3): 349–351.

7

Novel Process and Characterization Techniques

There are many conventional optical-fabrication process and characterization techniques, and they are well documented in the literature [1, 2]. These include the following:

- Interferometry techniques and test plating to measure *surface figure*, wedge, and parallelism
- Needle profilometry, white-light interferometry, and atomic force microscopy (AFM) to measure *surface roughness*
- Optical, confocal, and dark-field microscopy and visual inspection to measure *surface quality*
- Baume density floats, pH sensors, and particle-size analyzers to characterize polishing slurries
- Hardness testers to characterize lap materials, particularly pitch

In this chapter, some novel techniques (and perhaps obscure techniques) that provide additional insight into the optical-fabrication process are described. Chapters 2–5 focused on the phenomena that govern the basic properties of a workpiece during and after optical fabrication, namely *surface figure, surface quality, surface roughness,* and *material removal rate.* The novel process and characterization techniques explored in this chapter are described for an optical-fabrication engineer or an optician who might wish to use them.

The chapter contains three major sections:

- Section 7.1 discusses process techniques to improve or isolate a process variable during fabrication
- Section 7.2 provides workpiece characterization techniques to enhance understanding of workpiece structure, chemistry, and properties
- Section 7.3 describes polishing or grinding-system characterization techniques for better control or understanding of the subsystems of a given polishing or grinding process.

Materials Science and Technology of Optical Fabrication, First Edition. Tayyab I. Suratwala.

7.1 Process Techniques

7.1.1 Stiff Versus Compliant Blocking

Just as the stiffness of the tool or lap affects the degree of conformal versus nonconformal removal (as shown in Figure 2.20), so does the workpiece. Depending on the relative stiffness of the workpiece and how it is blocked, the degree of conformal removal can be controlled. Blocking is a method to contact and hold a workpiece against an applied weight or a blocking plate. Consider two extreme methods of blocking a high-AR workpiece in terms of workpiece bending (see Section 2.5.6.3), as shown in Figure 7.1. The freestanding workpiece shown is flat on the one side and concave on the side to be processed. When blocked with a stiff interface to a stiff weight, the workpiece plus block will behave as a stiff element – that is, have a high effective composite elastic modulus – and the workpiece will bend minimally under load. Hence, the contact of the workpiece with the lap will initially result in high pressure on the edges of the workpiece and low pressure at the center. This will result in a nonconformal removal and convergence of the *surface figure* to the shape of the lap. In contrast, when the same workpiece is blocked using a compliant interface (such as low modulus foam or rubber), the workpiece will conform to the shape of the lap upon loading. If the lap is flat, conformal removal (i.e. uniform removal) from the workpiece will occur. After the polishing process and separation from the block, the workpiece will largely revert to its freestanding shape.

Stiff blocking can be challenging for high-AR workpieces. For example, making direct contact between the workpiece and the block, without an interface layer, requires the workpiece back and block surfaces to be perfectly matched

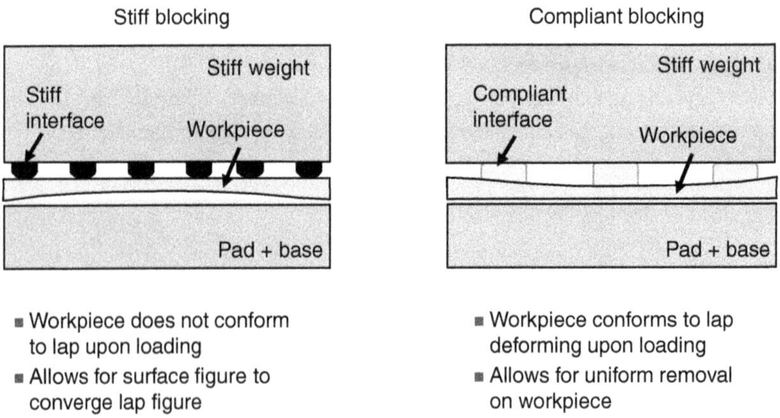

Figure 7.1 A comparison between stiff blocking and compliant blocking of a workpiece for grinding or polishing.

in surface shape, so as to not to deform the workpiece during blocking. In addition, direct contact requires that the interface surfaces be free of rogue particles,, which can lead to digs and scratches. Achieving such cleanliness may be difficult to fulfill reliably and repeatedly – hence an interface layer is often used. Blocking with an interface layer requires a minimal level of workpiece deformation during blocking, enough interface strength and adhesion to survive the shear forces of polishing and not delaminate, and minimal creep or flow under load during polishing to minimize workpiece bending.

Pitch has been used not only as a medium to polish on (i.e. a lap) but also as a blocking agent to hold optics and other workpieces for polishing. There are several methods for using pitch as a blocking medium [1, 2]. One uses a thin layer of pitch across the whole blocking plate to block single or multiple parallel workpieces. Since pitch is applied over the entire back surface, the difference in thermal expansion between workpiece and pitch during cooling will lead to residual stresses that can ultimately distort the workpiece. The second method blocks a lens or lenses to a curved blocking tool with a relatively thick, full-aperture pitch button on the back surface of each lens. The third method, pitch-button blocking (PBB) [3, 4], attaches small pitch buttons to the individual workpiece to block it without unduly distorting it, as often happens to high AR workpieces.

The process for PBB, outlined in Figure 7.2, involves the application of pitch droplets of a controlled volume to a block and attaching it to the workpiece before heating and cooling it [4]. The heating and cooling flow the pitch so that it can conform to the shape of the workpiece in a stress-free state and, when cooled, harden with minimal distortion to the workpiece. Figure 7.3 shows two PBB fused-silica workpieces.

A tradeoff arises between increasing the adhesive strength of the block interface, which is related to the areal fraction of buttons A_f and distortion during blocking, which is driven by the differential expansion between the workpiece and the block. After performing PBB using various button geometries, pitch materials, and workpiece materials, Feit et al. [4] showed a correlation between workpiece deformation during blocking, represented by ΔPV or change in the PV (peak-to-valley) *surface figure* before and after blocking, with button spacing and area fraction (see Figure 7.4a,b). When the spacing between buttons is greater than 10–15 mm, workpiece deformation is small (Figure 7.4a), and the magnitude of workpiece deformation increases linearly with the area fraction of the buttons (Figure 7.4b). A thermoelastic model using an effective pitch thermal-expansion coefficient to account for stress relaxations was used to explain the observed deflections quantitatively as a function of workpiece material and PBB geometry, as indicated by the solid lines in Figure 7.4b. The details of the model are described elsewhere [4].

Using this model and data, a set of PBB design rules was developed for workpieces of various sizes and materials. The rules define the size, spacing,

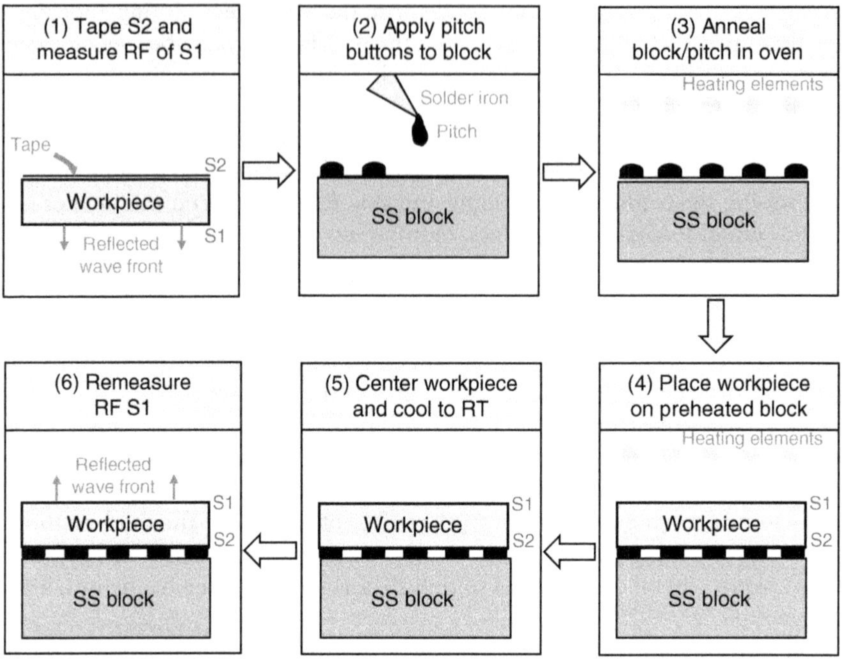

Figure 7.2 The process steps for blocking a high AR workpiece using PBB. (S1 = side 1, S2 = side 2, SS = stainless steel, RF = reflected wavefront, RT = room temperature). Source: Feit et al. 2012 [4]. Reproduced with permission from The Optical Society.

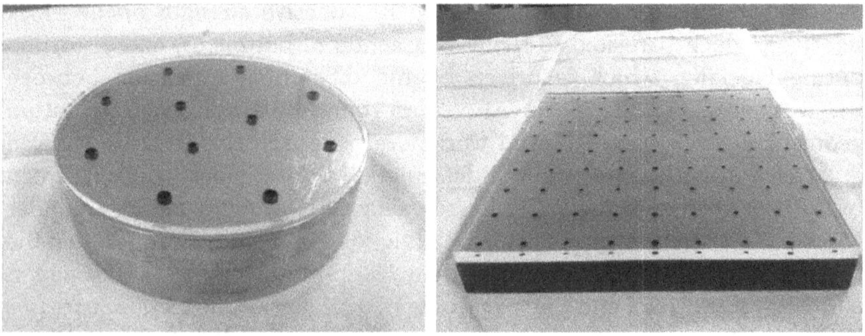

(a) (b)

Figure 7.3 The photographs of optimized PBB pattern used for (a) a 100 mm diameter round, fused-silica workpiece and (b) a 265 mm square fused-silica workpiece. Source: Feit et al. 2012 [4]. Reproduced with permission from The Optical Society.

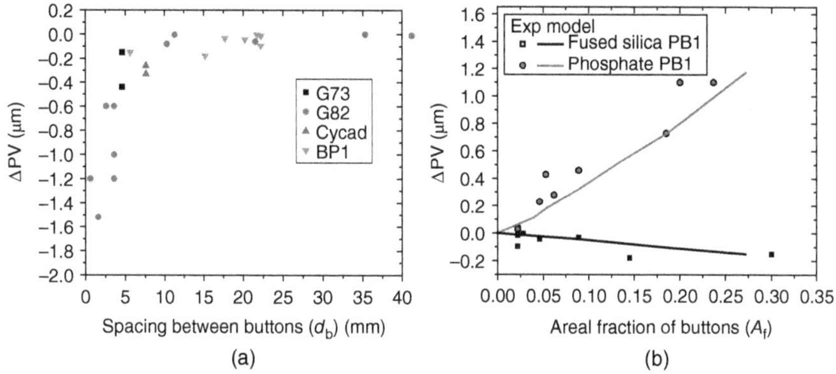

Figure 7.4 A change in surface figure (ΔPV) before and after blocking as a function of (a) button spacing and (b) area fraction of buttons using PBB. G73, G82, Cycad, and BP1 are various pitch materials. Source: Feit et al. 2012 [4]. Reproduced with permission from The Optical Society.

and number of pitch buttons for a given acceptable blocking deflection using the following relationships [4]:

$$r_p = \frac{d_b}{\sqrt{\frac{\pi}{A_f}} - 2}, \tag{7.1}$$

$$N_b = \frac{\pi r_o^2}{(2r_p + d_b)^2}, \tag{7.2}$$

$$\Delta PV = A_f C_{pb}, \tag{7.3}$$

where r_p is the radius of the pitch button, r_o is the radius of the workpiece, d_b is the spacing between buttons, N_b is number of buttons needed on the workpiece, ΔPV is the expected workpiece deflection, A_f is the areal fraction of buttons, and C_{pb} is the linear rate of increase in deflection with area fraction. From Figure 7.4b, $C_{pb} = +4.3\ \mu m$ for phosphate glass and $C_{pb} = -0.56\ \mu m$ for fused-silica glass.

To illustrate the utility of the design rules, consider a 50 mm radius workpiece, where $A_f = 0.05$ has sufficient interfacial strength to survive polishing. Using Equation (7.3), the ΔPV upon blocking would be −0.03 μm for fused silica and 0.22 μm for phosphate glass, as long as $d_b > 15$ mm. For some of the added margin to minimize crosstalk between pitch buttons, d_b is set at 20 mm. Equations (7.1) and (7.2) are then used to determine the ideal button radius ($r_p = 3.3$ mm) and the number of buttons ($N_b = 11$).

In a recent study, the concept of convergent polishing (see Section 8.4) was introduced and demonstrated, where a workpiece, regardless of its initial *surface figure* will converge to the lap shape in a single iteration [5, 6]. This technique requires that all sources of material-removal nonuniformity be removed, except for the workpiece-lap mismatch owing to the workpiece surface shape. For a high AR workpiece, bending upon loading on the polisher can prevent such a technique from working. Hence, PBB is an attractive blocking method for convergent polishing.

To test if a low areal fraction of pitch at the interface ($A_f = 0.05$) has enough strength to withstand a polishing run, the convergent polishing technique was conducted on a 50 mm radius fused-silica workpiece using the PBB design. The PBB-blocked workpiece survived polishing over tens of hours, confirming sufficient interface strength [4].

Figure 2.26 compares the optimized PBB-process workpiece *surface figures* versus a foam blocking process, before and after polishing [6]. Note that the PBB workpiece *surface figure* converged to flat while the foam-blocked workpiece did not change *surface figure* after polishing. This indicates that the foam-blocked workpiece bends during polishing, resulting in uniform spatial material removal (conformal removal), while the PBB workpiece was stiff, leading to nonuniform material removal (nonconformal removal) because of workpiece–lap mismatch [6].

7.1.2 Strip Etch and Bulk Etch

As discussed in Section 3.1.4, subsurface damage (SSD) on a workpiece is often hidden because of the index-matched Beilby layer and because cracks may have closed. A strip-etch process is used to chemically etch the workpiece slightly to expose SDD. Figures 3.39 and 7.5 show how such microfractures may become more visible after an etch of ~1 μm from a polished fused-silica workpiece. Using a strip-etch process on the workpiece after final polishing or between polishing steps can be useful for performing a *surface quality* metrology of

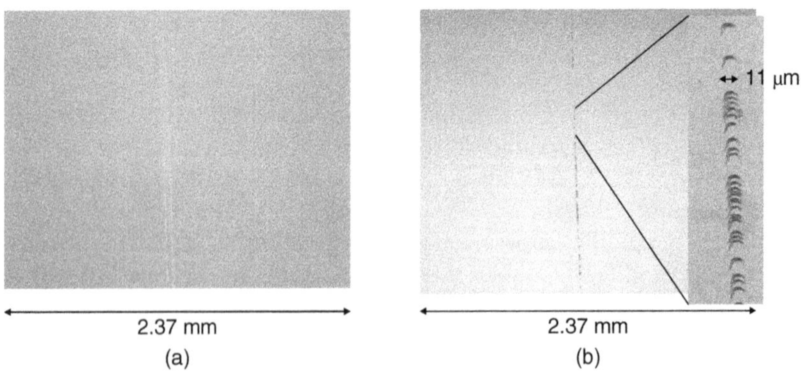

Figure 7.5 An optical micrograph of vertical scratch on fused-silica glass (a) before and (b) after 1 μm strip etch using BOE.

the workpiece for hidden scratches and digs. Combining the poststrip etch inspection with scratch forensics, as discussed in Section 6.2, the cause of these fractures can be determined along with an insight on the removal.

Bulk etching is a relatively long etching process (typically removing 50–150 μm) performed on a workpiece after grinding and reducing the net amount of material to be removed in subsequent polishing steps. The bulk-etch process may reveal very deep fractures that occurred during grinding, informing the decision of whether the workpiece should move forward to polishing or go back to grinding. The bulk-etch process results in a surface with etch cusps (see Figure 3.40), making it easier to assess whether all the SSD on a workpiece has been polished through, because etch cusps are much easier to detect on the surface than closed microfractures. Because a bulk-etch system is generally less costly than a polishing system, it is potentially an attractive process step between grinding and polishing. Based on the etching results on the ground surfaces, bulk etching can reduce polishing time by as much as two to four times. Details of how ground surfaces behave with etch time and the amount of removal needed are discussed in Section 3.1.4.1.

7.1.3 Pad Wear Management with Septum or Conditioner

For achieving the desired *surface figure* during polishing, controlling the shape of the lap is critical, because it affects the workpiece–lap mismatch (see Section 2.5.6). During pitch polishing a conditioner, typically a large block of glass, is altered either in load or position to adjust the shape of the lap, which is viscoplastic. During pad polishing, workpiece contact may create nonuniform pad wear, as discussed in Section 2.5.6.2. One workaround is to use a small-tool, kinematic-controlled diamond conditioner that strokes back and forth across the pad to compensate for nonuniform wear [7]. However, the high wear rates incurred can dramatically reduce the pad's lifetime and the kinematics for spatially uniform pad wear is difficult to optimize.

An alternative method is to use a specially shaped septum – a sacrificial component that also creates wear on the pad – to counteract the spatially nonuniform pad wear caused by the workpiece [5, 6]. In designing a septum, the aim is that the total rate of radial pad wear from the workpiece and septum is combined to be kept constant (see Figure 2.23a), as given by:

$$\frac{\mathrm{d}h_{\mathrm{lap}}(r)}{\mathrm{d}t} = C_{\mathrm{L}} = f_{\mathrm{o}}(r)k_{\mathrm{lap}}\mu V_{\mathrm{r}}\sigma_{\mathrm{o}} + f_{\mathrm{s}}(r)k_{\mathrm{lap}}\mu V_{\mathrm{rs}}\sigma_{\mathrm{sep}}, \tag{7.4}$$

where $f_{\mathrm{o}}(r)$ is the fraction of the lap circumference loaded by the workpiece as a function of radial distance (r) from the lap center, $f_{\mathrm{s}}(r)$ is the fraction of the lap circumference loaded by the septum as a function of radial distance (r) from the lap center, C_{L} is a chosen constant for the removal rate on the pad, V_{rs} is the time-averaged relative velocity on the septum, which is given by $2\pi R_{\mathrm{L}}r$, and σ_{sep} is applied pressure on the septum. For conditions of matched rotation ($R_L = R_o$) and matched workpiece and septum pressures $\sigma_o = \sigma_{sep}$, the septum shape can be determined from

$$f_s(r) = \frac{C_L}{2\pi k_{lap}\mu R_o r\sigma_o} - \arcsin\left(\frac{x_L(r)}{r}\right)\frac{s}{\pi r}, \tag{7.5}$$

where $x_L(r)$ is x component of a point on the leading edge of the workpiece (see Figure 2.22). Using Equation (7.5), the calculated width of the septum, that is, its circumferential width, as a function of distance from lap center – that is, $2\pi r f_s(r)$) for the specific case where $r_o = 50$ mm and $s = 75$ mm – is shown in Figure 7.6a. Note that there are two limitations in the design of the septum as designated by the dashed lines in Figure 7.6a. The first is the minimum radial width of the septum, which is needed for structural integrity and to allow a minimal distance for pad viscoelastic relaxation (see Ref. [8] for details). The second is a limit near the lap center where the circumference of the pad must be large enough to provide the required counter-wear by the septum. The latter limit can be expanded by applying a higher load on the septum or by increasing the separation distance s between the lap and workpiece center [6].

Figure 7.6b illustrates the resulting shape of the septum (septum R) using Equation (7.5). The photos in Figure 7.7 show two polishing setups (for the round and square workpieces) that have successfully demonstrated compensation of pad wear using septum designs [6]. Equation (7.5) applies to a round workpiece only; the septum design for a square workpiece was determined numerically using Equation (7.4).

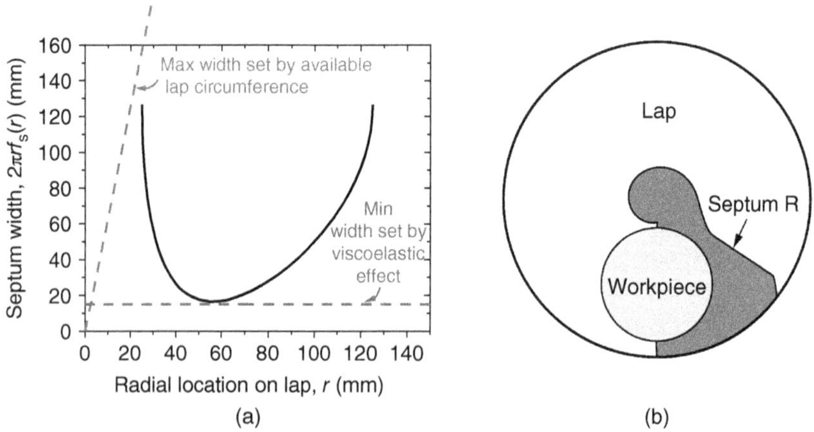

Figure 7.6 (a) The calculated septum radial width for a round workpiece using Equation (7.5) ($\mu = 0.7$, $\sigma_o = 2068$ Pa (0.3 psi), $R_o = 20$ rpm, $s = 75$ mm, $r_o = 50$ mm, $r_{lap} = 150$ mm); (b) the calculated shape of septum R for a round workpiece from that shown in (a). Source: Suratwala et al. 2012 [6]. Reproduced with permission from John Wiley & Sons.

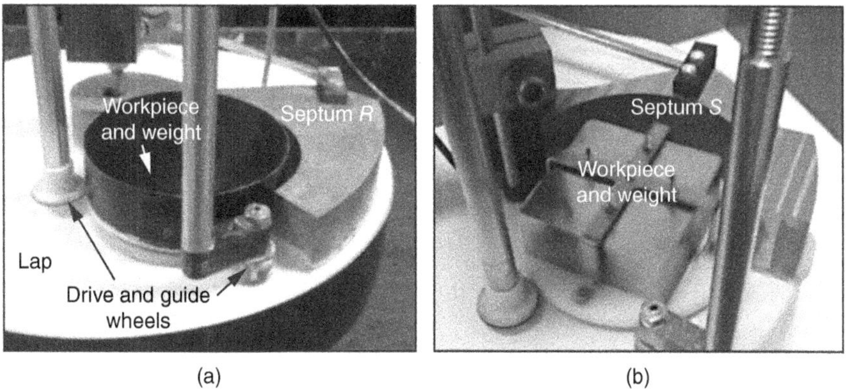

Figure 7.7 Photos of convergent polishing setups for (a) round and (b) square workpieces with accompanying pad-wear-compensating septum. Source: Suratwala et al. 2012 [6]. Reproduced with permission from John Wiley & Sons.

The effectiveness of the septum in reducing pad wear was determined by measuring the degree of spatial nonuniformity in material removal from the workpiece. Figure 7.8 shows the workpiece *surface figure* radial lineout at various polishing times for a 50 mm round fused-silica workpiece using septum R. Comparing the equivalent data from the cases without a septum in Figure 2.24, a significant improvement in spatial uniformity is seen, owing to more-uniform pad wear with a septum. Using the septum, the within-workpiece nonuniformity (WIWNU) was significantly reduced to <1.2% (Figure 7.8). In other words, only ~1 μm deviation was observed after polishing 83 μm of material from the surface. In comparison, the WIWNU for the cases without the septum was 92% and 27%, respectively (Figure 2.24) [6].

Figure 7.8 A radial-surface profile of fused-silica workpiece after polishing for various durations using septum R. Source: Suratwala et al. 2012 [6]. Reproduced with permission from John Wiley & Sons.

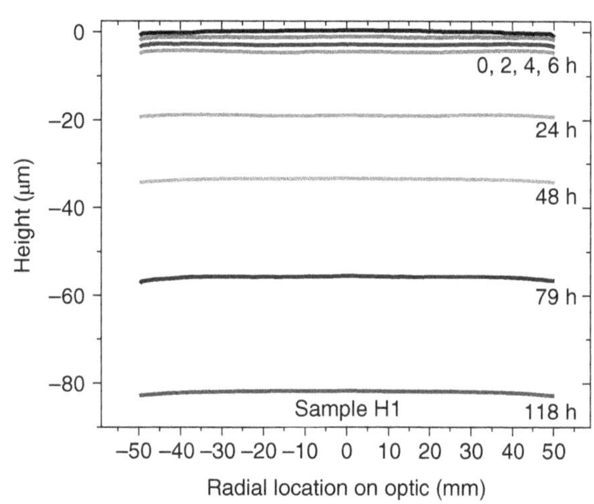

The other benefits of pad-wear-compensating septums include the following:

- More-uniform slurry distribution
- Reduced viscoelastic-pad stress nonuniformities from prepad compression at the leading edge of the workpiece (see Section 2.5.5.1)
- Reduced friction-induced temperature nonuniformities (Section 2.5.6.5)
- Reduced local nonlinear material deposit (see Section 2.5.6.8).

All these phenomena contribute to improved workpiece spatial material-removal uniformity.

7.1.4 Hermetically Sealed, High-humidity Polishing Chamber

As discussed in Sections 3.1.3 and 4.3, the management of rogue particles and the tail end of the slurry particle-size distribution (PSD) are paramount in reducing surface microfractures and achieving low *surface roughness.* There can be many sources of rogue particles, whether in the external environment (such as cross contamination among process steps or general workplace dirtiness) or internally generated from abrading components, dried slurry, or agglomerated slurry particles. In particular, dried slurry particles may collect in the system between polishing runs, especially in crevices, corners, and hard-to-reach places. For example, many dried ceria-slurry particles form hard agglomerates that lead to workpiece scratching [9] and adhere to surfaces, making it very difficult to clean and recover the polishing system.

A novel strategy to manage many rogue-particle sources is to polish in a hermetically sealed, high-humidity chamber to prevent the entry of external rogue particles. This is an effective method if good practices are enforced when the chamber is open between polishing runs and during maintenance. By incorporating a humidifier within the polishing chamber, 100% humidity can be maintained, making it much less likely for dry, hard slurry agglomerates to form. An example of a polisher with a hermetically sealed, high-humidity chamber is given in Figure 8.14.

Besides rogue-particle management, a hermetically sealed, high-humidity chamber has other advantages:

- A sealed system can achieve greater temperature stability and uniformity, helping minimize thermally induced material-removal nonuniformities (see Section 2.5.6.5).
- The polyurethane pad is prone to changes in properties with humidity variations and drying (see Section 2.5.6.6). The stable humidity allows for a greater control of the lap shape and hence a more deterministic control of the *surface figure.*
- Slurry stability is improved. Many polishing systems require frequent slurry concentration (i.e. Baume) and pH adjustments, due to water evaporation

or loss of dried slurry in the system. The hermetic enclosure greatly reduces slurry evaporation for a more stable slurry requiring less maintenance.
- A sealed polishing system contributes to overall facility cleanliness by keeping slurry contained.

7.1.5 Engineered Filtration System

Another strategy for managing rogue particles and controlling the tails of the slurry PSD is to recirculate the slurry through a particle-filtration system, continuously removing any rogue particles generated. While a number of polishing systems use filtration systems, it can be difficult to achieve reliable, economical, satisfactory level of performance. Some of the challenges include the following:

- Frequent clogging and possible bypassing of the filters, requiring frequent filter changes.
- Slurry settling and drying in various nooks and crannies in the filtration-system plumbing, creating a secondary source of rogue particles.
- Impurity, chemical or particulate contamination from materials used in the filtration system.
- Large thermal heating by slurry pump, leading to thermal changes in the polishing system.
- Difficult maintenance of slurry adjustments and replacements.

A number of slurry filtration systems have been engineered for the chemical–mechanical planarization (CMP) industry to overcome these challenges [10–12]. An example of the engineered filtration system tailored for optical polishing is shown in Figure 7.9. This system can feed slurry to one or

Figure 7.9 An engineered filtration system for polishing slurries.

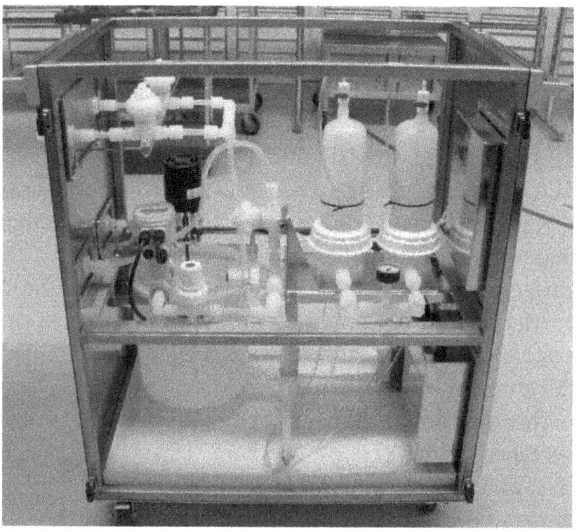

more polishing systems with a slurry feed rate up to 5 gallons per minute. Key features of such a system include the following:

1. Flow lines that minimize crannies and dead space (i.e. zero-velocity regions).
2. Easy maintenance by a design for filter changes, slurry draining, and system cleaning.
3. Slurry lines that are composed of fluorinated plastic, using Furon flare fittings, perfluoroalkoxy (PFA) tubing, and slurry pump to minimize metallic cation contamination, which may disrupt slurry colloidal stability and reduce slurry particle adhesion to the inner surfaces.
4. Upside-down use of CMP-type filters for ease of maintenance and operation (see Figure 7.9).
5. High-velocity slurry flow to reduce settling.
6. Translucent slurry lines for visual inspection of system cleanliness.
7. Pump mounted below the fluid level to allow pump priming by gravity fill.
8. Instrumentation to measure pressure drop, pH, and flow rate.
9. Easy access sampling port for external measurement of slurry properties such as Baume.

Not only can a filtration system catch large rogue particles, it can also modify the tail end of the slurry PSD. As discussed in Section 4.3, the slope of the tail end may strongly influence the resulting roughness, where increasing the slope yields lower roughness (see Figures 4.10 and 4.11). Using the single-particle optical-sensing (SPOS) technique for measuring the tail end, the engineered filtration can further improve the PSD by employing various filter sizes (see Figure 7.10).

7.1.6 Slurry Chemical Stabilization

Yet another method to improve roughness and prevent rogue particle scratching is to use chemistry to improve and maintain the slurry PSD. Much of this technology is housed in commercial, and usually proprietary, processes for developing narrow PSD slurries and maintaining them during storage and use. The latter can be quite a task due to settling and agglomeration.

A novel method has been developed to treat as-born narrow PSD slurries to maintain their stability in storage and use [13, 14]. This involves using a class of surfactants that form a cloud of micelles around the slurry particles, preventing particles from agglomerating and keeping active sites on the particle surface available for chemical polishing (see Figure 7.11). This method is referred to as the charged-micelle halo mechanism.

This method has been demonstrated on a ceria-based slurry with an average particle size of ~200 nm (Hastilite PO). Despite an original particle size of 200 nm, the slurry may be largely agglomerated upon storage, as illustrated by measuring the settling time of the slurry (see Figures 7.10 and

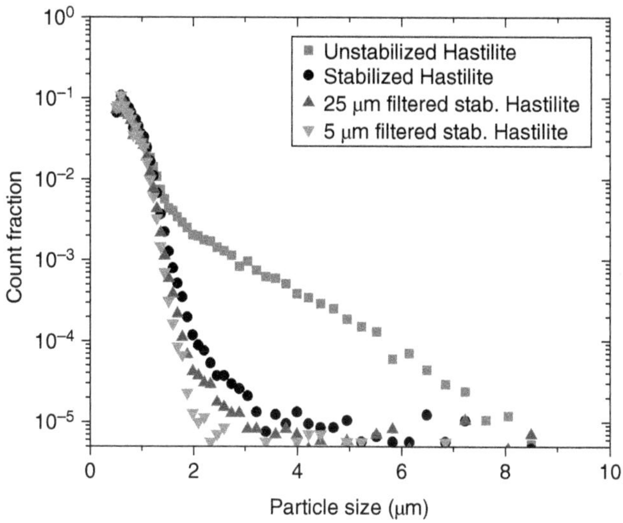

Figure 7.10 Measured slurry PSD using SPOS on a ceria polishing slurry (Hastilite PO), as a function of chemical stabilization and various levels of filtration through the filtration system in Figure 7.9. Source: Dylla-Spears et al. 2014 [13]. Reproduced with permission from Elsevier.

Figure 7.11 A schematic illustration of how similarly charged surfactants to the particle surface can form a cloud of micelles around polishing particles, resulting in improved stability. Source: Dylla-Spears et al. 2014 [13]. Reproduced with permission from Elsevier.

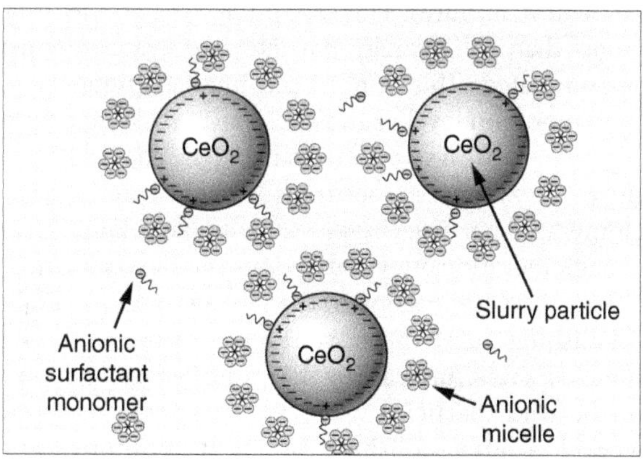

7.12a) – untreated ceria slurry settles in just a few minutes. Using Stokes's law [16], an estimate of the effective agglomerate particle size (d) can be determined as follows:

$$d^2 = \frac{18\eta_s}{tg(\rho_3 - \rho_f)d_{set}}, \tag{7.6}$$

where t is settling time, η_s is viscosity of the fluid (in this case water), ρ_3 is ceria-particle density, ρ_f is water density, g is the gravitational constant

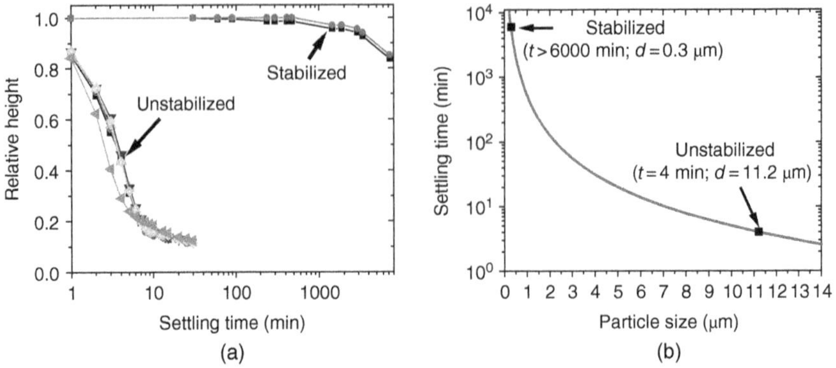

Figure 7.12 (a) Relative settling height as a function of time of unstabilized (untreated) and stabilized (treated) ceria slurry when slurry is placed in a standalone beaker without agitation or mixing; (b) calculated settling time of slurry as a function of an average particle or agglomerate size using Equation (7.6). Source: Dylla-Spears et al. 2017 [15]. Reproduced with permission from Elsevier.

(9.80 m s^{-2}), and d_{set} is settling distance. The calculated settling times for this system as a function of particle size are shown in Figure 7.12b, using $\rho_3 = 7.1\,\text{g cm}^{-3}$, $\rho_f = 1\,\text{g cm}^{-3}$, $\eta_s = 0.01$ P, and $d_{set} = 10$ cm. Settling occurs in just a few minutes, corresponding to particles that are ~11 μm in size, far larger than the average particle size, which confirms the presence of large agglomerates. This simple settling technique is useful in determining the agglomeration level or the effective particle size for a given slurry.

After chemical stabilization using the charged-micelle halo method, the same slurry did not start to settle visually until after 4 days, consistent with an unagglomerated particle size of ~200 nm (see Figure 7.12) [13]. The large reduction in agglomeration after stabilization is also confirmed using SPOS, which shows a large increase in slope of the tail end of the PSD (Figure 7.10).

Polishing fused-silica workpieces with these slurries showed equivalent removal rates, indicating that micelles did not interfere with the reactivity of the ceria particles with the workpiece [13]. The AFM roughness of the workpiece after polishing significantly improved, as shown in Figure 7.13, consistent with the improved tail of the slurry PSD of the treated slurry. See Section 4.3 for details of this correlation.

There are two other significant benefits of stabilized ceria slurry. First, the stabilized slurry is filterable and hence longer lasting. The large agglomerates of the unstabilized slurry quickly clog the filtration systems (typically in just minutes), making it impractical and expensive to use for filtration. By contrast, stabilized slurry can be filtered easily and indefinitely without the addition of glass products. Figure 7.14 compares filters employed with these slurries. On the left, unstabilized ceria slurry produced a large buildup at the entry of a

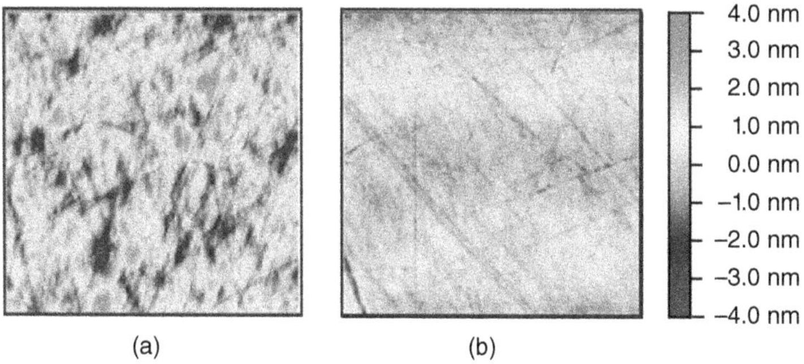

Figure 7.13 AFM images (50 μm × 50 μm area) on fused-silica surface after polishing with 4.4 wt% Hastilite PO ceria slurry, before stabilization (a) and after (b). Source: After [13].

Figure 7.14 Photos of used CMP polishing-slurry filters (a) using 50 μm filter with untreated ceria slurry and (b) using a 5 μm filter with treated slurry.

50 μm CMP filter; on the right, a more challenging 5 μm CMP filter used with stabilized ceria slurry showed little buildup at the filter entry.

The second benefit is the reversibility and cleanability of the slurry after drying. As discussed in Section 7.1.4, drying ceria slurry is known to form hard agglomerates that adhere strongly to the surface, making it difficult to clean and maintain the polishing system. Unlike unstabilized slurry, stabilized slurry does not form hard agglomerates as it dries, and when dispersed back into water, it largely returns to its original PSD. This is illustrated in Figure 7.15, where the stabilized and unstabilized slurries were dried, viewed with SEM,

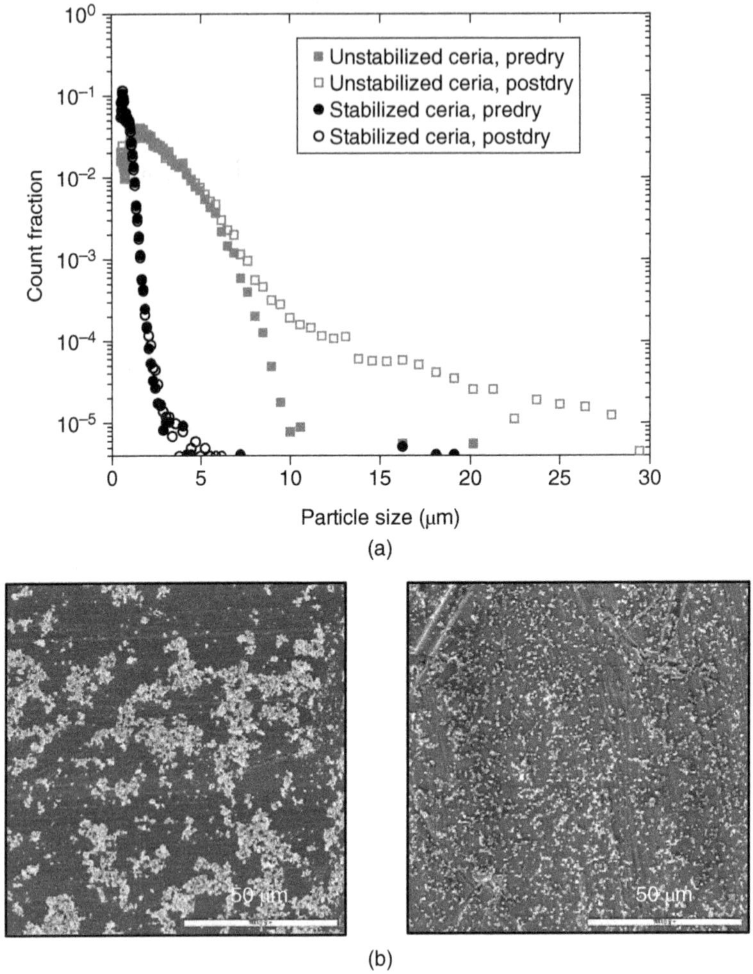

Figure 7.15 (a) PSD distribution of unstabilized and stabilized ceria slurry before and after drying and resuspending. Source: After [13]; (b) SEM images of the unstabilized (left) and stabilized (right) ceria slurries after drying.

and resuspended into the solution, upon which their PSDs were again measured. The agglomeration of the dried unstabilized slurry compared with the stabilized slurry is apparent in the SEM micrographs presented in Figure 7.15.

7.1.7 Slurry Lifetime and Slurry Recycling

An important factor in the economics of polishing is the cost of consumables such as polishing slurry. This concern has been problematic in the case of ceria, since market conditions over the last decade have led to spikes in slurry cost. Two new methods have been developed to increase slurry lifetime.

Using the combined techniques of a hermetically sealed, high-humidity chamber, an engineered filtration system, chemical stabilization of the slurry, routine slurry maintenance for pH control (see Sections 7.1.4–7.1.6), and a minimum required slurry concentration, i.e. Baume, (see Figure 5.10), the useful life of a slurry batch and filter can be significantly increased such that it is limited only by the rate of collection of glass products. A chemical analysis of the cross section of a used CMP filter after polishing multiple fused-silica workpieces showed the preferential deposit of glass products (i.e. silica) relative to the ceria slurry particles at a Si:Ce molar ratio of 2.9. Hence, the filter is acting as a getter for the glass-removal products, and the amount of workpiece polishing determines the filter lifetime. Therefore, the slurry can be used subject to periodic filter replacements and minor slurry additions to account for losses in the filter.

The precise lifetime of the slurry depends on many factors, including slurry type and volume, workpiece size, *material removal rate*, filter-pore size, and filter surface area. With the techniques above, a single batch of slurry was demonstrated to last several months, using 3 gallons of Baume-5 stabilized ceria slurry with daily polishing of 265 mm square fused-silica workpieces at a polishing rate of 4 $\mu m\,h^{-1}$ with filter replacements (25 μm CUNO CMP filters, 12″) nearly every 2 weeks. The ultimate lifetime of a slurry batch was dictated by the slow buildup of glass products, which was observed during polishing as rapid pad glazing, or buildup of glass products on the pad surface, even after a filter change.

Another method to reduce the cost of polishing consumables is to recycle the polishing slurry [17, 18]. The used slurry may be chemically processed to remove the glass products or "swarf." This process has been commercialized for ceria polishing slurries and shown to be 30% cheaper than new polishing slurry [17].

7.1.8 Ultrasonic Pad Cleaning

Another phenomenon described in Section 2.5.6.8 is pad glazing – the buildup of glass products in the slurry and embedment into the pad. Glazing may lead to degradation in the *surface figure*, midspatial frequencies, and overall *material removal rate* [19]. Figure 7.16 shows a used polyurethane pad with glazing along several radial bands. Glazing is more common when using slurry-recirculation systems and develops more rapidly without filtration due to the greater rate of glass-product buildup in the slurry (see discussion in Section 7.1.7). Optical micrographs show that glazed regions arise from the material deposits that lands on the surface of the pad (see Figure 7.17). A chemical analysis of the pad surface confirms the preferential deposit of glass products relative to slurry

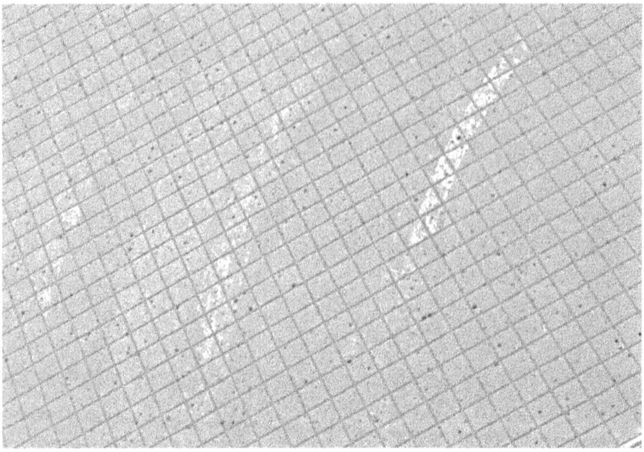

Figure 7.16 The top surface of a polyurethane-foam pad (MHN) after prolonged polishing of fused-silica glass and rinsing of remaining slurry. Lap center is to the right of the photo; grooves on the pad are 1 cm apart.

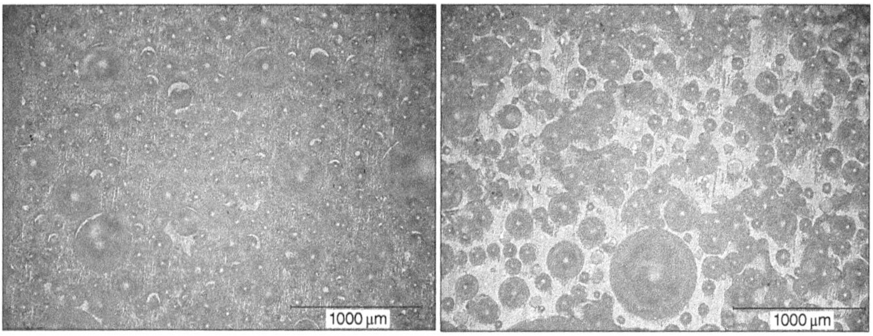

(a) (b)

Figure 7.17 Optical micrographs of the (a) nonglazed region and (b) glazed region of a used polyurethane-foam pad.

particles. The molar ratio of Si:Ce was 2.4 – significantly higher than the ratio expected if Si glass products were evenly distributed in the slurry, as given in Table 7.1.

The most common method of reducing pad glazing is periodic conditioning or dressing of the pad using a diamond conditioner. The challenge with such a treatment is that it results in significant pad wear, resulting in a modification of the lap shape (and hence the *surface figure* of the workpiece) and in a significant reduction of the pad lifetime [20, 21]. Some proposed alternatives are acid or base chemical treatment to remove the pad glazing [22] and removal

Table 7.1 Elemental analysis of polyurethane-foam pad (MHN) surface of a new pad, a nonglazed portion of a used pad, and a glazed region of a used pad.

	Mole (%)		
Element	**New pad**	**Nonglazed pad**	**Glazed pad**
C	81	84	84
N	12	6	5
O	8	9	10
Ce	ND	0.4	0.3
Si	ND	ND	0.7
Cl	ND	0.6	0.3

ND = not detected.

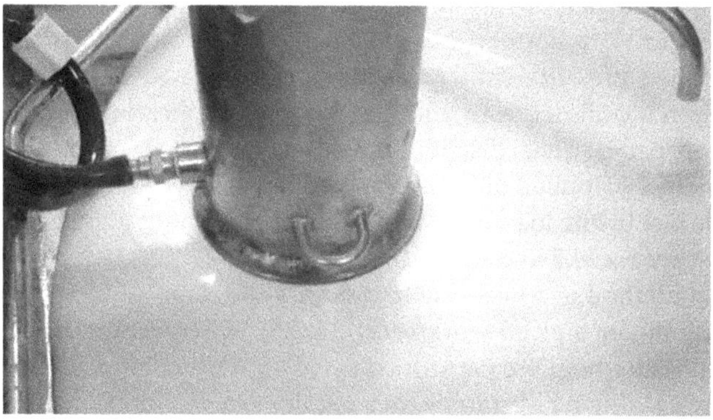

Figure 7.18 A photo of an ultrasonic pad-treatment system (Advanced Sonics Processing System, model URC4).

of the glazing using an ultrasonic water rinse, a method developed along with convergent polishing technology [5, 23].

In situ ultrasonic water treatment is shown in Figure 7.18. Between polishing runs, the ultrasonic treatment is performed by placing a transducer face down in close contact with the pad surface, in the presence of the deionized (DI) water rinse. This dislodges any embedded slurry from the pad with little or no modification to the pad itself and has been shown to effectively remove glaze regions and increase the longevity of the pad. For example, using a polyurethane-foam pad (MHN) with periodic ultrasonic treatment, >2000 h of polishing has been demonstrated on a single pad, with minimal wear and stable workpiece *surface figure*.

7.2 Workpiece Characterization Techniques

7.2.1 Single-particle Removal Function Using Nanoscratching

The removal function of a polishing particle – how much material a single particle removes from the workpiece surface as a function of load – is important in determining the resulting workpiece *surface roughness* and *material removal rate* (see Chapters 4 and 5). For an example of how the removal function can be used, refer to the measured-removal functions in Figure 4.8. The depth of removal was determined to change abruptly depending on the load regime. For example, for fused silica at low loads, the removal depth is ~0.04 nm, and at higher loads, 0.55 nm (~10× higher). Now consider slurries with certain PSDs that lead to load-per-particle distributions as shown in the figure. If the objective is low roughness, it is better to use a load-per-particle distribution that leads to low removal depth (in the molecular or chemical regime). However, if the objective is to obtain a high *material removal rate*, a slurry with a load-per-particle distribution that leads to a higher removal depth (nanoplastic regime) is better. This goal can also be accomplished by increasing the applied pressure sufficiently to shift the load-per-particle distribution from one removal depth to another. Another application of this concept is if the load-per-particle distribution is well within a particular removal-depth regime – the applied pressure can be increased to increase the *material removal rate* just by the loading of more particles, without increasing *surface roughness* (see Section 4.8 and Figure 4.37).

The nanoscratching method is a nice way to quantify the removal function in the nanoplastic regime for a given workpiece [24, 25]. The creation of the nanoscratch and its measurement are performed using an AFM equipped with a special stiff diamond tip. The AFM starts with a regular tapping mode during the initial approach to the sample workpiece surface. Once the tip is in contact with the surface to engage a normal scan, the AFM probe's cantilever oscillation is turned off and the tip is forced into the surface until a predetermined load is reached; the tip is then moved laterally (specifically, 10 μm at a rate of 0.3 mm s^{-1} with a tip rotation of 12°). Multiple pass scratches may be generated by repeatedly scratching the same location under constant load. After the scratch is created, the same AFM tip can be used in tapping mode to determine the topography of the nanoscratch (see Figure 4.5a,b). The normal load P of a scratch is determined by the spring constant K_{sp} of the AFM probe used and the tip displacement Δz, according to Hook's law:

$$P = K_{sp}\Delta z. \tag{7.7}$$

The typical load per particle during polishing in the nanoplastic regime is 1–200 μN. Under such loads, optical-workpiece materials measured to date have effective nanoscratch depths ranging from 0.2 to ~4 nm (see Figures 4.5,

5.14, and 5.15). The hardness of the workpiece material is the key factor that determines the depth of nanoplastic removal.

7.2.2 Subsurface Damage Measurement Using a Taper Wedge

Statistically determining SSD depth distribution for a given grinding process and workpiece material is a powerful characterization that aids in the design of a more economical or SSD-free fabrication process (see Figure 1.2). Also, periodic measurements of the SSD depth distribution can be used as process control metrics or diagnose processes when yield is lower than desired or when required due to *surface quality.*

Section 3.1.2.1 describes a method to determine SSD distribution called the taper-wedge technique. The basic process is shown in Figure 3.17 [26]. First, a sample workpiece material is treated with the grinding process of interest. Next, a linear wedge is polished into the workpiece using magnetorheological finishing (MRF) with a profile as shown in Figure 7.19. Since MRF is known to create little or no SSD, only fractures generated from the grinding process of interest can be evaluated. To reveal SSD for ease of characterization by optical microscopy, the workpiece is chemically etched ~1 μm. For fused-silica glass, buffered oxide etch (BOE) is often used as the etchant. The sample is then imaged using optical microscopy at various locations along the wedge

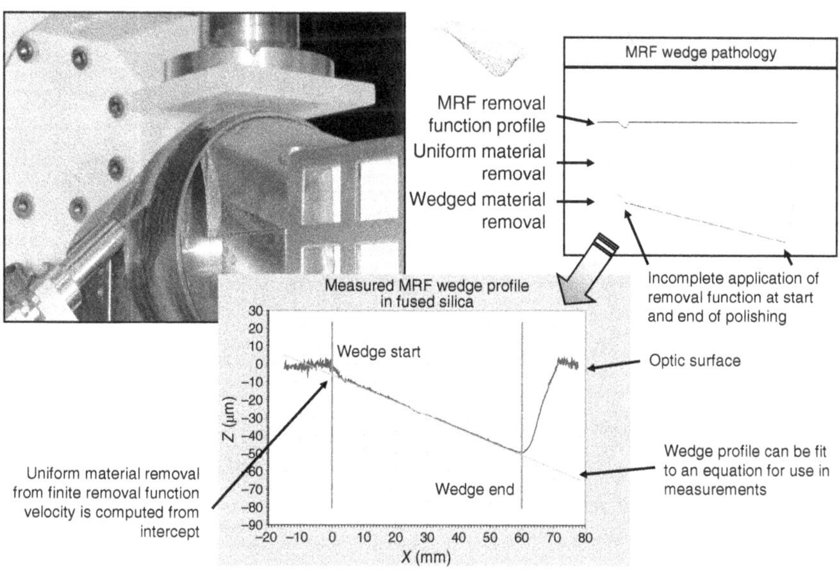

Figure 7.19 A description of polished wedge imparted on ground-glass sample using MRF. Source: Reproduced with permission of Menapace 2005 [27].

(and hence at different depths into the workpiece) to determine the characteristics of the SSD, including crack areal density, length distribution, and, most important, depth distribution.

Figure 3.20a shows the results of measured crack-depth distributions for a variety of grinding processes on fused-silica glass. The crack areal density is described in terms of obscuration (the areal fraction of cracks observed in the microscope images). The y-axis is on a log scale; crack densities drop exponentially and have a maximum crack depth of c_{max}. c_{max} typically represents 1 out of 10^6 cracks. Sufficient area of the workpiece sample (>9 cm^2) must be wedged to achieve this level of statistical resolution, and accurate measurement of the wedge profile (depth versus position) is needed to ensure accurate determination of the depth. Needle profilometry or optical coordinate-measuring devices work well for this.

As discussed in Sections 3.1.2 and 6.2, data obtained from the taper-wedge technique has enabled many useful scientific insights and practical rules of thumb regarding the characteristics of grinding processes. In terms of optical-fabrication process design, once the value of c_{max} is determined, the ideal removal amount required for a given process can be determined from an SSD point of view. For example, the 15 μm loose-abrasive-grind process in Figure 3.20 has a measured SSD depth or c_{max} of ~25 μm. Hence, the minimum amount to remove in the next step should be at least 25 μm to ensure removal of all the SSD. The historic rule of thumb for determining removal amounts has been to remove three times the abrasive size of the particles used in the previous process, which would imply removing 45 μm. Knowledge of the precise SSD depth allows optimizing of the removal amount, potentially saving downstream processing time, increasing throughput, and reducing fabrication cost.

Also, take another example where the yield due to scratch-dig has declined significantly in the overall process. Remeasuring the SSD depth of the 15 μm loose abrasive process again reveals a c_{max} of 50 μm, instead of 25 μm. This large change is likely the source of the yield problem. One way to mitigate the problem is to modify the grinding process, for example, by identifying the source of the rogue particles that likely lead to deeper SSD. Another method is to double the amount of removal in the next process step, but this is usually undesirable, because of higher costs.

7.2.3 Stress Measurement Using the Twyman Effect

As discussed in Section 2.5.6.4, grinding results in the generation of significant residual compressive stress on the workpiece surface. As a result, the workpiece may deflect, especially if it has a high aspect ratio (length/thickness). This deflection reduces the ability to control the *surface figure* during polishing, especially for workpieces polished on one side only, such as mirrors. The Twyman test is useful to determine the residual stress for a given grinding process. In this procedure, the surface figure of a thin sample workpiece material, initially polished on both sides, is measured on one side, using interferometry or other means. The opposing surface is treated with the grinding process of interest, and the polished surface is then remeasured for its *surface figure*. The bending of the workpiece occurs under the residual stress imparted, and the stress can be quantitatively determined by Equation (2.41), using the PV change in the *surface figure*. Figure 2.28 illustrates this for a grinding process with 9 μm loose abrasive and another with 30 μm loose abrasive. The compressive stress on the ground side makes the workpiece bend so as to make the opposing surface more concave, and a more aggressive grind leads to greater residual stress and greater *surface figure* changes. The stresses measured for various grinding processes are given in Figure 2.27.

7.2.4 Beilby Layer Characterization Using SIMS

As discussed in detail in Section 3.3, the surface layer of the workpiece during polishing can be modified chemically and structurally. This Beilby layer is usually very thin, typically only tens of nanometers thick. Water in the form of OH, polishing compounds, and impurities in the slurry are common species that build up in the Beilby layer. These impurities can modify surface properties such as the refractive index, chemical reactivity, and wetting.

A useful way to determine the chemical characteristics of the Beilby layer on the workpiece is to measure the concentration of atomic species as a function of depth, using secondary ion mass spectroscopy (SIMS) [28]. After imparting a focused-ion beam on a sample workpiece surface in vacuum, the surface layer is slowly sputtered away. The number of sputtered ions and their atomic or molecular mass are analyzed using a mass spectrometer. Most optical-fabrication workbenches do not have the appropriate tools for this well-established technique, but measuring services are offered by a number of facilities. The measurement requires the use of a small, clean sample (10–20 mm in size) treated with the polishing process of interest. Figure 3.57 shows example measurements of the incorporation of Ce (polishing slurry) and K (a pH-controlling species often added to polishing slurries) in fused-silica glass.

For most end-use applications, the Beilby layer has little impact on performance, but knowledge of the Beilby layer may be necessary for niche applications such as laser optics, where chemical impurities can lead to laser damage, chemical sensors, where the substrate is chemically active and surface chemical impurities may alter performance, and precision reflectors, where even a very small refractive-index change on the surface may change reflectivity. The polishing process, particularly the *material removal rate*, has been shown to influence the extent of change in the Beilby layer (see Figure 3.60), and chemical etching is a proven method for removing the Beilby layer postpolishing (see Section 9.3).

7.2.5 Surface Densification Using Indentation and Annealing

Another type of surface modification is the possible local densification of the workpiece surface, especially in the Beilby layer. As with chemical modification, local densification may alter the properties of the surface. Local densification may arise from the mechanically loaded particles, if the yield stress is exceeded.

In a novel method used by Yoshida et al. [29], the degree of possible densification can be quantified for a given workpiece material. A workpiece sample is first prepared by polishing and subsequent chemically etching to remove densification that may have resulted from the polishing process. The surface is then treated by static indentation, such as Vickers, or by nanoscratching (see Sections 4.1 and 7.2.1) at the load range of interest. Next, the surface topography of the surface modification is measured by AFM or white-light interferometry. The change in the surface can be due to plastic deformation or densification. Finally, the sample is heat treated, and the surface topography remeasured. For glasses, typically a temperature of $0.9T_g$ is used, where T_g is the glass transition temperature.

Figure 7.20 illustrates this characterization method for both nanoscratching and static indentation on fused-silica glass [24]. At the top, the surface topography of these features is measured before and after annealing; the bottom

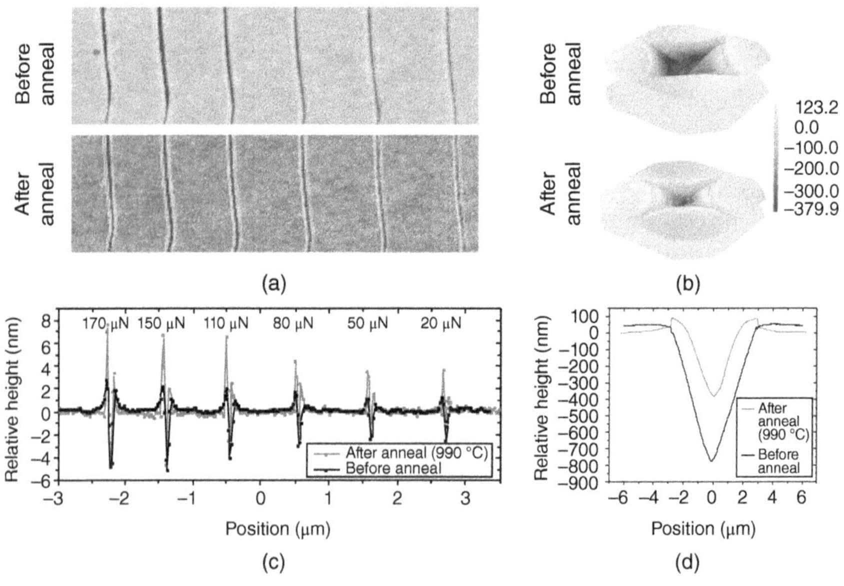

Figure 7.20 (a) AFM images of series of nanoscratches on fused silica (S1) before and after annealing at $0.9T_g$; (b) AFM images of a 0.5 N Vickers indent before and after annealing at $0.9T_g$; (c, d), average 1D lineout of the scratches for the samples above. Source: Shen et al. 2016 [24]. Reproduced with permission from John Wiley & Sons.

shows the corresponding lineout of the surfaces. With fused-silica glass after annealing, the surface topography is relaxed, removing the densified volume. The change in volume of the surface modification after annealing, relative to the total surface-modification volume just after the indentation or nanoscratch, is used to quantify the amount of densification occurring. In Figure 7.20, ~50% of surface modification is due to densification.

7.2.6 Crack Initiation and Growth Constants Using Static Indentation

Section 3.1 describes microfractures (i.e. SSD) generated during grinding processes and some general rules for predicting SSD depth using radial and trailing-indent initiation and growth relationships, and grinding rates, using lapping hardness and lateral crack-growth relationships. These relationships stem from knowing the quantitative behavior of crack growth for a given workpiece material as a function of load (see Equations (3.1)–(3.6)). Though crack initiation and growth constants can be estimated or scaled using the macroscopic material properties of the workpiece, it is more accurate to measure them directly, using static indentation. Figure 7.21 provides optical micrographs of cracks resulting as a function of applied load on various optical materials, using Hertzian and Vickers indentation. The measured crack characteristics of initiation load, length, and depth for each load and material can be used to determine the crack constants in Equations (3.1)–(3.6). Figure 3.3 shows the resulting measured crack dimensions as a function of scaled load. The slope of each data set is used to determine the crack constants summarized in Table 3.1.

7.3 Polishing- or Grinding-system Characterization Techniques

7.3.1 Tail End of Slurry PSD Using SPOS

As discussed in Sections 3.1.2.4 and 4.3, the PSDs of abrasive particles during grinding and polishing strongly influence the SSD distribution and *surface roughness* of the workpiece. It is shown that the larger particles of the PSD actively participate in workpiece interactions. Hence, quantitatively understanding and characterizing the tail end of the distribution allows for quantitative assessments of the resulting workpiece properties. For example, Figures 4.10–4.12 show the slope of the tail end of the PSD distribution as directly linked to the resulting *surface roughness* on fused-silica glass workpieces.

A number of techniques are used to measure PSD, including sedimentation, optical counting, laser diffraction, and scattering [30]. For most techniques, the

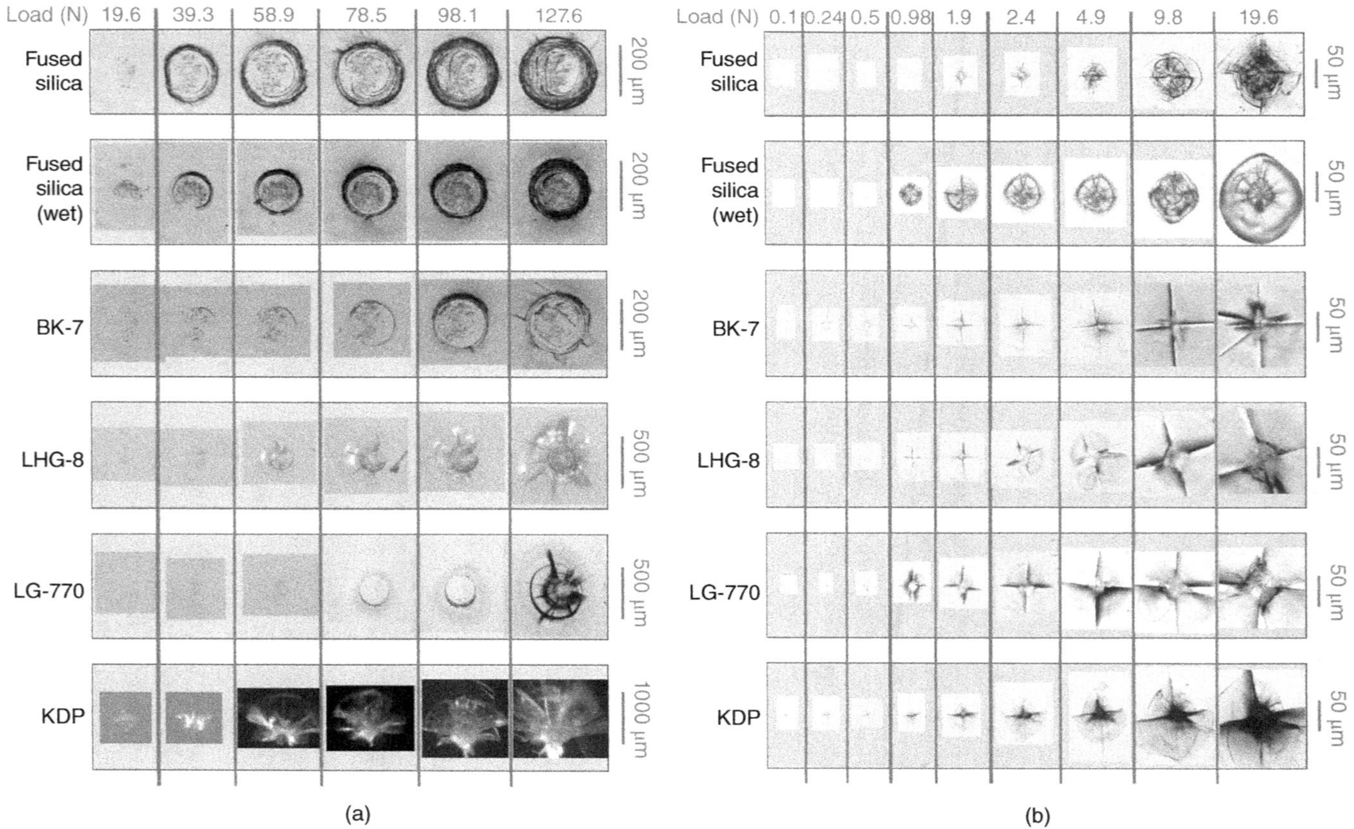

Figure 7.21 Optical micrographs of a series of (a) blunt Hertzian indents and (b) sharp Vickers indents placed on a variety of optical materials as a function of applied load.

emphasis is on obtaining the general shape of the whole particle-size distribution. For example, static light-scattering techniques simultaneously measure the scatter of ensemble of particles and use Mie scattering theory to determine the best-fit PSD. This technique does well for small particles down to 40 nm and for obtaining general shape.

Another technique, called single particle optical sensing (SPOS), greatly dilutes the particles and measures the scatter of one particle at a time. This technique provides much greater sensitivity to the largest particles at the tail end of the distribution, but is not well suited for smaller particles. Comparison of the measured PSDs of various polishing slurries using static light scattering and SPOS is shown in Figure 4.10. The tails of the distribution of many of the slurries are indistinguishable under static light scattering, but shown as very different with SPOS. The SPOS method for measuring grinding and polishing slurries is clearly the preferred choice for relating PSD characteristics with grinding and polishing behavior on the workpiece.

7.3.2 Pad Topography Using Confocal Microscopy

As discussed in Sections 4.4 and 4.7.2, the topography of the lap surface and buildup of the embedded slurry influence both the *material removal rate* and the resulting fine- and mid- spatial length scale *surface roughness* of the workpiece. Conventional optical-microscopy measurements do not provide sufficient quantitative height topographical information concerning the lap surface because of the large depth of focus. Another method, confocal laser-scanning microscopy, uses a spatial pinhole at the confocal plane, which eliminates unfocused light. This results in a very narrow depth of focus, and the reconstruction of the 3D surface can be performed by scanning the surface over various heights under high resolution [31]. Some images using confocal microscopy to image a pad surface are shown in Figures 4.14 and 4.19.

7.3.3 Slurry Stability Using Zeta Potential

As discussed in Section 4.5, the stability of polishing slurries is driven by their colloidal chemistry, where the effective surface charge of the particles and surrounding surfaces influences whether the particles will repel or attract other particles and surrounding surfaces. The zeta potential is a measure of this effective surface charge; it is the electric potential in the interfacial double layer at the location of the slipping plane, relative to a point in the bulk fluid away from the interface [32]. The general rule is that colloidal particles with a zeta potential greater than 30 mV or less than −30 mV are stable among themselves and will generally not agglomerate. When the absolute value of the zeta potential is less than 30 mV, colloidal particles are prone to agglomeration.

There are a number of parameters that determine the value of the zeta potential, including particle composition and concentration, pH, and counter-ion concentration [32, 33]. An example of the general behavior of zeta potential as a function of pH is shown in Figure 4.24. Usually, surface charge decreases (i.e. becomes more negative) with an increase in pH. The pH at which the zeta potential is zero is the isoelectric point (IEP) or point of zero charge (PZC). Usually, the pH of the slurry must be adjusted away from the IEP charge to avoid agglomeration.

During polishing, it is important to monitor and adjust the pH and slurry concentration or Baume as a function of polishing time, especially for recirculated slurries. This often allows maintaining the stability of the colloidal polishing slurry. As described in Section 4.5, polishing can introduce counter ions from the glass-removal products or deliberate pH adjustments, which can change the pH and zeta potential, as seen in Figure 4.25.

At the polishing interface, interactions are more complex, because stability is needed not just with colloidal particles among themselves but also with the workpiece and lap surface. Hence, the zeta potential of the workpiece and lap surface must also be accounted for. In an example using ceria slurry on a polyurethane lap to polish both fused-silica and phosphate-glass workpieces, the measured zeta potentials of each of the surfaces, including the effects of additional glass removal products, are used in a three-body force model to evaluate the conditions in which slurry particles collect on the lap surface (see Section 4.5.4 and Figure 4.26).

The propensity of slurry particles to adhere to a workpiece surface can lead to problems, as discussed in Section 4.6. This attraction can lead to reduced material removal because the relative velocity of the particle is reduced, there is difficulty in cleaning the workpiece after polishing, and there is increased apparent *surface roughness* of the final workpiece. In Figure 7.22, an example of slurry particles adhering to a workpiece surface, the surface topography, is

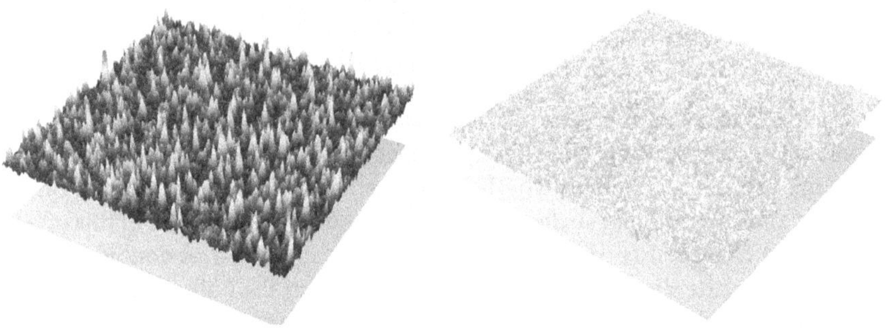

(a) (b)

Figure 7.22 AFM image of fused-silica workpiece after polishing with colloidal-silica slurry (S27) adjusted to (a) pH = 2 and (b) pH = 9. The images are 5 μm × 5 μm in area with a z scale of −2.5 to 2.5 nm.

measured by AFM of two fused-silica workpieces polished with colloidal-silica slurry at pH = 2 or pH = 9. Since the IEP of fused silica is 2, polishing at this pH results in a low zeta potential and hence a higher likelihood of attraction between the colloidal-silica slurry particles and the fused-silica workpiece. In contrast, polishing at pH = 9 should result in a large negative zeta potential for both the workpiece and the slurry particles, resulting in a repulsion between the two. The sample polished at pH = 2 shows significant deposit as compared to that polished at pH = 9, consistent with this mechanism.

Based on the situations described above, directly measuring the zeta potential as a function of the pH of the slurry, workpiece, and lap can be used in determining the best pH range with which to polish and to minimize detrimental interactions with the workpiece and lap. There are several methods to measure zeta potential. The basic principle used is electrophoresis, where an electric field is applied to a dispersion of particles and the velocity of the charged particles is measured. The measured velocity is used to calculated the effective zeta potential using known models [32, 34, 35]. Measurements can be conducted at low-frequency applied fields, which is better for measuring solutions that have been diluted, or at high-frequency, dynamic applied fields, better for measuring concentrated solutions. Commercial instruments for zeta-potential measurements and services that perform these measurements are available.

7.3.4 Temperature Distribution During Polishing Using IR Imaging

A useful diagnostic for characterizing a polishing system is measuring the temperature distribution, for example, with an infrared (IR) thermal imaging camera. IR imaging provides detailed spatial temperature information in a single image and does not require installation of discrete thermocouples and wiring, which can be challenging, especially on the moving components. The downside of IR imaging, however, is that only the surface temperature can be measured, and, for accurate absolute temperature measurements, the emissivity of the material surface must be determined.

As discussed in Section 2.5.6.5, spatial and temporal temperature variations may influence the polishing system, resulting in increased levels of nonuniform material removal via a number of mechanisms, including the following:

- Increase in slurry reactivity with the workpiece at regions of temperature rise
- Change in workpiece–lap mismatch, owing to differential thermal expansion at regions of temperature rise and resulting in deformation of and within the workpiece and lap
- Change in workpiece–lap mismatch from changes in lap properties, such as the elastic modulus, with temperature.

Frictional heating, at the workpiece–lap interface, and slurry heating, by the recirculating pump, are common heat sources leading to temperature rise within the polishing system. Example images of measured temperature variations in a polishing system are shown in Figure 2.31. Since an IR imaging system can measure surface temperatures only, a useful trick to measure the workpiece–lap interface temperature distribution is to rapidly and safely remove a workpiece after polishing and image the polished face of the workpiece (see bottom of Figure 2.31). This works because the thermal transport within the workpiece is generally much slower than in the slurry. Because net frictional heating changes as a function of lap radius, a radial temperature distribution is usually observed (see Section 2.5.6.5).

For the finishing of optics, tight *surface figure* control is usually required – therefore, temperature distribution needs to be managed. A rule-of-thumb requirement is to make the temperature distribution across the lap <1 °C and across the workpiece <0.3 °C. Some example strategies to reduce temperature variations include

- Increasing slurry flow
- Slowing down kinematics (e.g. rotation rates), leading to less frictional heating
- Using a slurry-recirculation pump that leads to minimal heating
- Using a compensating septum that reduces spatial temperature variations (see Figure 2.32b and Section 7.1.3).

7.3.5 Slurry Spatial Distribution and Viscoelastic Lap Response Using a Nonrotating Workpiece

Slurry spatial distribution, as discussed in Sections 2.5.6.7 and 2.5.6.8, and the viscoelastic lap response (Section 2.5.5.1) can influence the uniformity of material removal on the workpiece at different spatial scale lengths. A novel method developed to measure and quantify these effects is to measure the change in *surface figure* of the workpiece without rotating the workpiece during polishing. For continuous-polisher (CP) type kinematics (see Figure 2.6) using a nonrotated workpiece ($R_o = 0$ rpm), the rate of material removal based on kinematics alone should result in a linear wedge across the workpiece. The lower *material removal rate* will occur on the workpiece location nearest the lap center, and the highest *material removal rate* will occur farthest from the center. Measured deviations in material removal from the expected linear wedge can be used to quantify these material-removal spatial nonuniformities.

An example of using nonrotated workpiece polishing for measuring fine-scale slurry distribution on the pad is shown in Figures 2.36 and 2.37, where ripples are observed on the workpiece surface. The ripples are attributed to a distribution of slurry islands nominally spaced ~150 μm apart.

The magnitude of the ripples is quantified by PV roughness and was found to change significantly with pad type and topography (see Figure 2.37b). A slightly longer-range slurry distribution can also be evaluated for localized nonlinear deposit, as illustrated in Figure 2.40b.

Another example of using this technique for measuring material-removal spatial nonuniformities due to viscoelastic relaxation of the lap is shown in Figure 2.18. Enhanced material removal occurs at the leading edge of the workpiece because of high stresses at the interface before the viscoelastic pad relaxes with continued loading. With a rotated workpiece, this nonuniformity is observed as the *surface figure* of the workpiece's becoming convex (see Figure 2.24b).

7.3.6 Slurry Reactivity Versus Distance Using Different Pad Grooves

As a fixed amount of polishing slurry is trapped in the small interface gap between the workpiece and lap during polishing, its reactivity or material-removal effectiveness can decrease as the slurry travels farther in the interface. This effect is more pronounced for large workpieces. The decrease in removal effectiveness is likely due to the buildup of workpiece removal products in the slurry, reducing the condensation reaction rate as in Equation (5.14). The observed result is a spatially dependent Preston coefficient (see Section 2.2).

A novel method to diagnose and quantify this effect is to change the degree of slurry replenishment at the interface by altering the groove spacing on the lap or using a lap without grooves. The grooves act as a reservoir of fresh slurry to replenish the interface slurry. By polishing on a lap without grooves, the inherent reactivity of the slurry with distance can be determined. This method also requires use of the convergent polishing method to separate the effects of slurry reactivity with distance from other material-removal nonuniformities (see Chapter 2 and Section 8.4). The slurry travel path in the interface starts at the leading edge of the workpiece, as defined as point (x_L, y_L) in Figure 2.16, and is dictated by the kinematics.

In this method, the measured rate of change in the *surface figure* can be inferred as related to the change in slurry reactivity with distance across the workpiece. An example of this technique is given in Figure 2.3. The workpiece polished on a pad without grooves has a significantly different *surface figure* evolution than that on a pad with 1 cm spaced grooves. Using the SurF model described in Section 2.6, these results can be explained by a decrease in slurry reactivity of ~50% after traveling 8 cm through the workpiece–lap interface in the case of the pad without grooves. Understanding this behavior for a given polishing system can aid in optimizing the groove spacing and pattern on the lap.

References

1 Karow, H.H. (1992). *Fabrication Methods for Precision Optics* (ed. J.W. Goodman), 1–751. New York: Wiley.
2 Twyman, F. (1952). *Prism and Lens Making; A Textbook for Optical Glass-workers*, 2e, viii, 629. London: Hilger & Watts.
3 Scott, R. (1965). *Applied Optics and Optical Engineering*, vol. 3 (ed. R. Kingslake), 87. New York and London: Academic Press.
4 Feit, M.D., DesJardin, R.P., Steele, W.A., and Suratwala, T.I. (2012). Optimized pitch button blocking for polishing high-aspect-ratio optics. *Appl. Opt.* 51 (35): 8350–8359.
5 Suratwala, T., Steele, R., Feit, M. et al. (2012). Method and System for Convergent Polishing. WO 2012129244 A1.
6 Suratwala, T., Steele, R., Feit, M. et al. (2012). Convergent pad polishing of amorphous silica. *Int. J. Appl. Glass Sci.* 3 (1): 14–28.
7 Liang, H. and Craven, D.R. (2005). *Tribology in Chemical–Mechanical Planarization*, 185. Boca Raton, FL: Taylor & Francis.
8 Suratwala, T.I., Feit, M.D., and Steele, W.A. (2010). Toward deterministic material removal and surface figure during fused silica pad polishing. *J. Am. Ceram. Soc.* 93 (5): 1326–1340.
9 Suratwala, T., Steele, R., Feit, M.D. et al. (2008). Effect of rogue particles on the sub-surface damage of fused silica during grinding/polishing. *J. Non-Cryst. Solids* 354 (18): 2023–2037.
10 Adams, J.A., Krulik, G.A., and Harwood, C.R. (1997). Slurry recycling in CMP apparatus. Google Patents.
11 Yueh, W. (1998). Slurry recycling system for chemical–mechanical polishing apparatus. Google Patents.
12 Obeng, Y.S. and Schultz, L.D. (2000). Apparatus and method for continuous delivery and conditioning of a polishing slurry. Google Patents.
13 Dylla-Spears, R., Wong, L., Miller, P.E. et al. (2014). Charged micelle halo mechanism for agglomeration reduction in metal oxide particle based polishing slurries. *Colloids Surf., A* 447: 32–43.
14 Dylla-Spears, R., Feit, M., Miller, P.E. et al. (2015). Method for preventing agglormeration of charged colloids without loss of surface activity. US Patent 20,150,275,048.
15 Dylla-Spears, R., Wong, L., Shen, N. et al. (2017). Adsorption of silica colloids onto like-charged silica surfaces of different roughness. *Colloids Surf., A* 520: 85–96.
16 Robinson, R. and Stokes, R. (1959). *Electrolyte Solutions*. London: Butterworths Scientific Publications.
17 Naselaris, M. and Hobbs, Z. (2014). Improving slurry recycling makes green process greener. *Photonics Spectra* 48 (8): 69–73.

18 Kim, J.-Y., Kim, U.-S., Byeon, M.-S. et al. (2011). Recovery of cerium from glass polishing slurry. *J. Rare Earths* 29 (11): 1075–1078.
19 McGrath, J. and Davis, C. (2004). Polishing pad surface characterisation in chemical mechanical planarisation. *J. Mater. Process. Technol.* 153: 666–673.
20 Hooper, B., Byrne, G., and Galligan, S. (2002). Pad conditioning in chemical mechanical polishing. *J. Mater. Process. Technol.* 123 (1): 107–113.
21 Y-Y, Z. and Davis, E.C. (1999). Variation of polish pad shape during pad dressing. *Mater. Sci. Eng., B* 68 (2): 91–98.
22 Sun, L., Li, S., and Redeker, F.C. (2003). Elimination of pad glazing for Al CMP. Google Patents.
23 Suratwala, T., Steele, R., Feit, M. et al. (2014). Convergent polishing: a simple, rapid, full aperture polishing process of high quality optical flats & spheres. *JoVe-J. Vis. Exp.* (94): doi: 10.3791/51965.
24 Shen, N., Suratwala, T., Steele, W. et al. (2016). Nanoscratching of optical glass surfaces near the elastic–plastic load boundary to mimic the mechanics of polishing particles. *J. Am. Ceram. Soc.* 99 (5): 1477–1484.
25 Shen, N., Feigenbaum, E., Suratwala, T. et al. (2018). Single Particle Nanoplastic Removal Function of Optical Materials. TBD.
26 Suratwala, T., Wong, L., Miller, P. et al. (2006). Sub-surface mechanical damage distributions during grinding of fused silica. *J. Non-Cryst. Solids* 352 (52–54): 5601–5617.
27 Menapace, J.A., Davis, P.J., Steele, W.A. et al. (2005). MRF applications: measurement of process-dependent subsurface damage in optical materials using the MRF wedge technique. Proceedings of SPIE, Volume 5991.
28 Skoog, D.A. and Leary, J.J. (1992). *Principles of Instrumental Analysis*, 4e, 1–700. Orlando, FL: Saunders College Publishing.
29 Yoshida, S., Sanglebœuf, J.-C., and Rouxel, T. (2005). Quantitative evaluation of indentation-induced densification in glass. *J. Mater. Res.* 20 (12): 3404–3412.
30 Gee, G.W. and Bauder, J.W. (1986). Particle-size analysis. In: *Methods of Soil Analysis: Part 1–Physical and Mineralogical Methods (methodsofsoilan1)* (ed. E. Klute), 383–411. Madison, WI: American Society of Agronomy/Soil Science Society of America.
31 Pawley, J.B. (2006). Fundamental limits in confocal microscopy. In: *Handbook of Biological Confocal Microscopy*, 20–42. Springer-Verlag.
32 Hunter, R.J. (2013). *Zeta Potential in Colloid Science: Principles and Applications*. Academic Press.
33 Israelachvili, J.N. (2011). *Intermolecular and Surface Forces*, 3e. Academic Press.
34 Lyklema, J. (2005). *Fundamentals of Interface and Colloid Science: Soft Colloids*. Academic Press.
35 Russel, W.B., Saville, D.A., and Schowalter, W.R. (1989). *Colloidal Dispersions*. Cambridge University Press.

8

Novel Polishing Methods

In this chapter, some novel polishing methods are briefly discussed and related to principles and phenomena outlined in Chapters 2–5. Details of these polishing methods and capabilities are well described in the broad literature, and hence not covered here.

8.1 Magnetorheological Finishing (MRF)

Magnetorheological finishing (MRF) is a novel polishing process for deterministic *surface figure* correction. It uses a subaperture lap created by a flowing, magnetic field-stiffened magnetorheological (MR) fluid ribbon loaded against the workpiece surface. The MRF process was devised in the late 1980s by Kordonski and Jacobs [1–3]; its historical evolution is nicely summarized by Harris [4]. A schematic of the MRF contact between a workpiece and lap is shown in Figure 8.1. The workpiece is kept at a fixed distance from the subaperture tool (i.e. the moving MR fluid in a magnetic field), whose width typically ranges from the millimeter to the centimeter scale, depending on the magnetic wheel size. Because of its relatively low, defined viscosity, the tool naturally conforms locally to the shape of the workpiece. The kinematics of the ribbon contact and its dwell time on the workpiece determine the *surface figure* modification for a given polishing run. To achieve the desired final *surface figure*, the initial (i.e. prepolish) *surface figure* is measured and registered to the MRF polishing system to determine the optimum kinematics and dwell time at each surface location.

MRF technology has had a high impact on the optical fabrication industry because of the following advantages:

- Workpieces with improved *surface figure*.
- The ability to fabricate novel free-form optics.

Materials Science and Technology of Optical Fabrication, First Edition. Tayyab I. Suratwala.

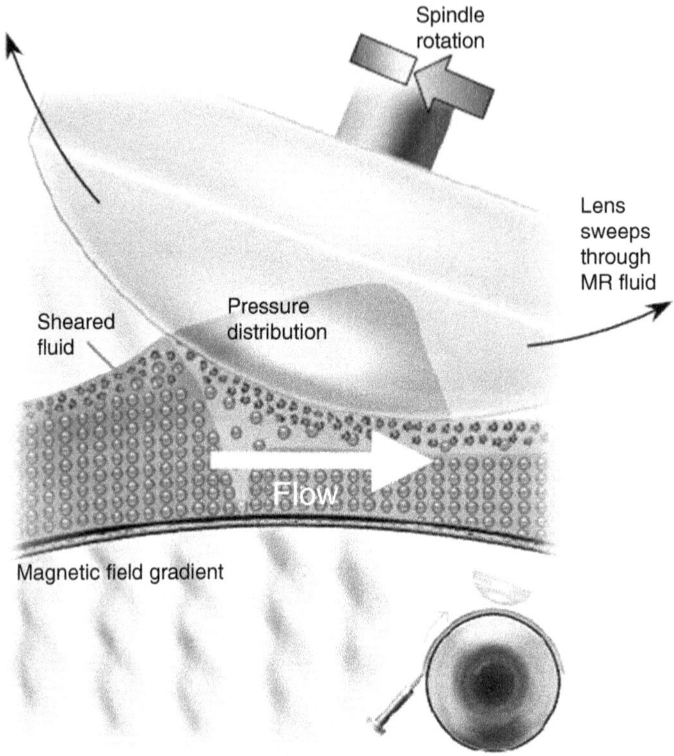

Figure 8.1 A schematic of MRF at the workpiece–lap interface, illustrating the hydrodynamic contact between magnetized MR fluid ribbon and workpiece. Source: Shorey et al. 2004 [5]. Reproduced with permission from SPIE.

- Workpieces with complex *surface figures* unachievable by conventional polishing techniques.
- Workpieces with improved laser damage resistance.

These capabilities stem from several fundamental characteristics of the MRF process.

First, the MRF removal function is very repeatable, which allows very deterministic material removal and, hence, greater control of the *surface figure* of the workpiece. There is essentially no workpiece–lap mismatch at the tool contact due to the low viscosity or elastic modulus of the MR fluid under magnetic field (50–1000 Pa s or 0.5–1.0 MPa, respectively) compared to conventional pitch or pad (10^9 Pa s or 50–100 MPa) [6, 7]. In other words, within the contact zone between workpiece and lap, removal is not driven by the shape mismatch, but by factors such as ribbon width, wheel rotation rate, MR slurry composition, and workpiece material. For a given MR slurry and workpiece material, this leads to a very well-defined removal function, both in terms of volumetric removal and shape (see Figure 2.8) [5]. However, the overall removal rate (up to ~1 μm day^{-1} on a ~0.5 m workpiece) is generally much slower than conventional full-aperture polishing (typically several micrometer per hour). Hence, in many optical fabrication processes, conventional polishing is first used to get as close as possible to the desired *surface figure*. If additional *surface figure* improvement is needed, MRF permits a more stringent *surface figure* requirement.

Second, because MRF is a subaperture tool with low normal loads, it can (i) orientate itself on the workpieces with large surface slopes (e.g. aspheric optics with large departures [8] or thermal correcting optics [9]; (ii) create novel, complex *surface figure* shapes (e.g. continuous phase-plates type optics [10]; and (iii) improve the *surface figure* on thin-membrane optical elements [9, 11]. Examples are shown in Figure 8.2.

Third, MRF has the ability to make sub-surface mechanical damage (SSD) free workpieces with low *surface roughness*. Because the normal load is low due to the low viscosity and elastic modulus of the MR fluid even under a magnetic field, removal is dominated by a high relative velocity (V_r) due to the shear of many polishing particles. Hence, the MRF process operates in hydrodynamic mode, just as with float polishing and some chemical mechanical planarization (CMP) processes (see Figure 2.5). Calculations of the normal load using a hydrodynamic model are ~0.1 μN [12, 13] with MRF, significantly lower than the load estimates for conventional polishing (typically >50 μN). In addition, the normal load per particle using MRF is below that needed to cause nanoplastic deformation on most glasses (see Figure 4.8), contributing to lower roughness surfaces. The relatively low elastic modulus lap of MRF leads to a weak load-per-particle dependence as a function of particle size. Therefore, since the normal load per particle is a primary driver for crack initiation and growth, as is seen in Equations (3.1)–(3.6), the low normal loads make MRF more immune to rogue particles causing SSD. This allows SSD-free or very low SSD workpieces, which is particularly attractive for the fabrication of high laser-damage threshold optics [14].

A typical MR fluid contains polishing particles such as cerium oxide, alumina, or nanodiamonds (6%), stabilizers such as sodium carbonate or glycerol (3%), magnetic carbonyl iron particles (36%), and water (55%) [15]. A scanning electron microscope (SEM) image of a dried MR fluid is shown in Figure 8.3, illustrating the larger carbonyl-iron particles (4.5 μm) and the much smaller CeO_2 particles [13]. In the presence of a magnetic field, most of the carbonyl-iron particles consolidate to form a stiff layer against the rotating wheel. A thin film of water, nanodiamond abrasives, and some residual carbonyl-iron particles are thought to move above this supporting layer, against the part, to remove material (see Figure 8.1) [16]. As proposed by Cook, water introduces chemistry into the removal process, and the presence

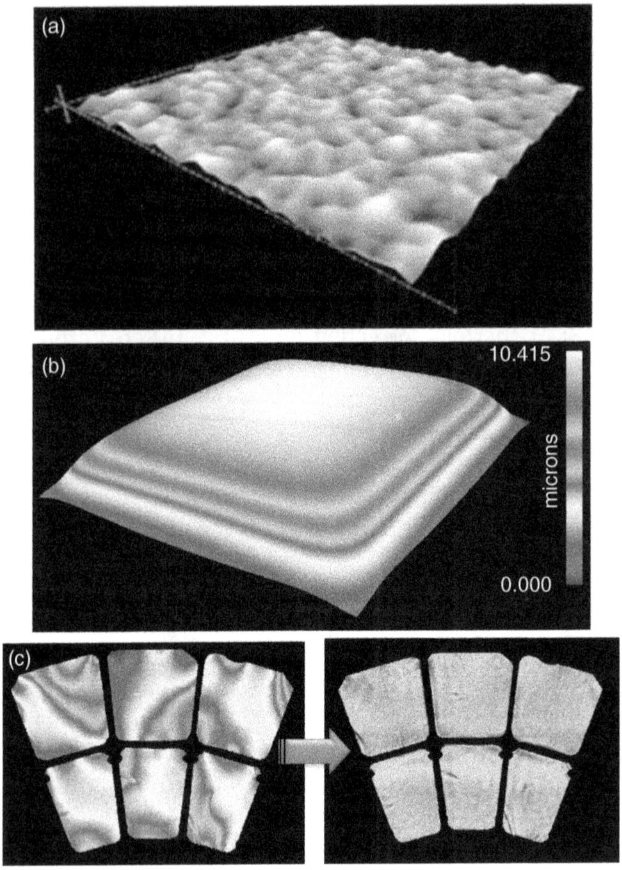

Figure 8.2 Examples of the types of optics fabricated using MRF, having complex surface figures and overall shape: (a) continuous phase-plate optic on 430 mm × 430 mm fused silica with a PV of 8.6 μm; (b) thermal corrector optic for HAPLS laser with a PV of 10.4 μm; and (c) subsection of 5 m diameter primary objective membrane of MOIRE space telescope. Source: Reproduced with permission from Menapace et al. 2016 [9] and Menapace et al. 2006 [11].

of water at the interface layer is likely critical [17]. Stabilizers are required for aqueous fluids because they help reduce corrosion and sedimentation.

Several studies have evaluated the mechanism of material removal with MRF [13, 16, 18–21]. DeGroote et al. [16] and Miao et al. [19] performed a systematic experimental study of the factors affecting *material removal rate* using nanodiamond containing MR fluid. These factors include slurry characteristics, such as particle size and concentration of various species; polishing parameters, such as relative velocity; and workpiece characteristics, including mechanical properties, bond strength, and chemical durability. The results and the various

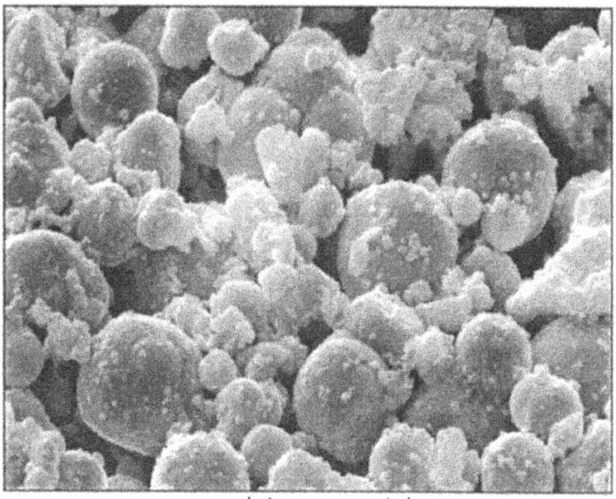

Figure 8.3 SEM image of dried MR fluid illustrating larger carbonyl-iron particles and smaller CeO_2 particles. Source: Shorey et al. 2001 [13]. Reproduced with permission from The Optical Society.

correlations were combined into the following unified *material removal rate* expression, reformulated according to the conventions of this book as follows:

$$\frac{dh}{dt} = k_p \tau V_r, \tag{8.1}$$

$$k_p \propto D_r \frac{E_1}{K_{Ic} H_1^2} (B_{nd} \varphi_{nd}^{-1/3} C_{nd}^{1/3} + B_{CI} \varphi_{CI}^{-4/3} C_{CI}) D_s(pH_{MRF})^{0.3} e^{-E_{sbs}/bRT}, \tag{8.2}$$

where τ is the shear stress, V_r is relative velocity, k_p is the Preston coefficient, D_r is the ribbon penetration depth, B_{nd} and B_{CI} are constants ($\mu m^{4/3}$), φ_{nd} is the nanodiamond particle size, φ_{CI} is the carbonyl-iron particle size, C_{nd} is the nanodiamond concentration, C_{CI} is the carbonyl-iron particle concentration, $D_s(pH_{MRF})$ is percent weight loss of the glass in the MR fluid supernatant as a function of MR fluid pH, E_{sbs} is the single-bond strength of the glass-network formers, and b is a unitless coefficient determined empirically to be 1000. It is important to note that the relative velocity is smaller than the wheel velocity, because there is a slipping of the ribbon near the workpiece–ribbon interface. This effect is greater at higher wheel velocities. Also, instead of considering the normal applied pressure, as used in the mechanism for removal during conventional slurry-based polishing, per Equation (1.3), this *material removal rate* is dominated by shear stress. Note that shear stress is indirectly driven by normal applied stress and interface friction.

Embedded within Equations (8.1) and (8.2) are useful scaling relationships to predict *material removal rate* trends with MRF. First, the overall *material removal rate* increases linearly with the relative velocity, ribbon penetration depth, and lapping hardness ($E_1/K_{Ic}H_1^2$) of the workpiece. As discussed in Section 5.1, lapping hardness is a measure of mechanical removal, namely by lateral cracks, for various workpiece materials [22]. In the case of MRF, the lapping-hardness correlation may just be a hardness correlation for nanoplastic removal. DeGroote found that this mechanical figure of merit (FOM) showed the best correlation with *material removal rate* using nanohardness values when the workpiece was measured in a water environment, suggesting the near surface is softened and more prone to shear nanoplastic removal [16]. Second, the *material removal rate* scales linearly with the term in parentheses in Equation (8.2), describing the effect of concentration and particle size of the nanodiamond and carbonyl iron. Figure 8.4a shows this correlation for a variety of glasses. Finally, the chemical durability and bond strength of the workpiece are proposed to influence the *material removal rate* – see the right end of Equation (8.2). Figure 8.4b illustrates this with measured *material removal rates* scaling with $D_s(\mathrm{pH}_{\mathrm{MRF}})^{0.3}$. Since Equation (8.2) accounts for both mechanical and chemical properties of the workpiece (i.e. lapping hardness and chemical durability), the belief is that both nanoplastic and chemical reaction removal mechanisms are both contributing to material removal (see Figure 1.9).

The MRF tool contact, and hence removal function, has a very characteristic shape, as shown in Figure 2.8. Kordonski et al. [7, 23] have modeled the contact between the MR fluid and workpiece. MR fluid under a magnetic field can be described as a Bingham plastic material, which deforms like a solid when applied stress is below the yield point and like a Newtonian fluid when stress exceeds the yield stress. There is a gradient in the magnetic field, normal to ribbons, that result in the accumulation of abrasive nonmagnetic particles on the exterior of the ribbon, furthest from the wheel or closest to the workpiece, and of carbonyl-iron particles on the interior of the ribbon. The ribbon acts as a moving wall contacting the workpiece and is dragged through the interface between the workpiece and the wheel (Figure 8.1). The resulting contact zone forms the top view shape of the removal spot (see Figure 2.8). When the moving wall of the ribbon contacts the workpiece, it experiences high shear stress, or strain, at the top of the wall. As the high shear stress exceeds the yield stress of the Bingham plastic, this part of the ribbon becomes more fluid-like. The shear stress was modeled across the contact spot using computational fluid dynamics of the hydrodynamic flow through the interface. The calculated particle-shear force correlates well with the measured removal function (see Figure 8.5). Lambropoulos and coworkers [13, 18, 20, 21] directly measured the drag and normal forces during MRF, which allowed determining the shear and normal stresses at the workpiece–tool contact spot. Typical values of the normal and shear stresses were between 5 and 10 N. At low magnetic fields, the measured shear and normal stress at the workpiece–ribbon interface increased with magnetic field due to increased stiffening of the MR fluid, or increase in the MR-fluid yield stress. At higher magnetic fields, the normal and shear stresses

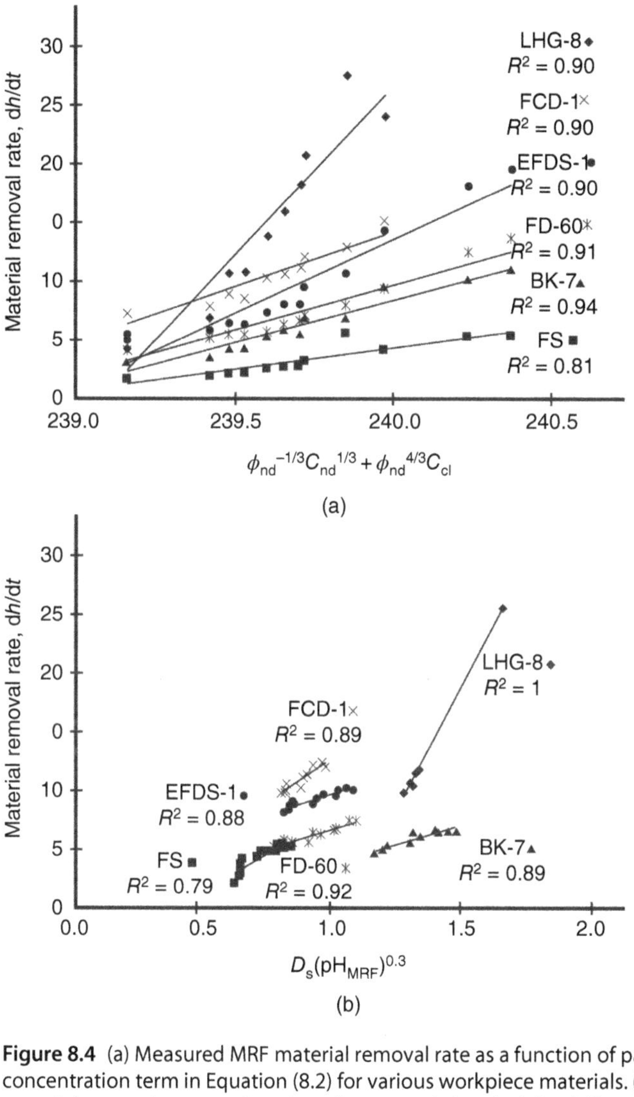

Figure 8.4 (a) Measured MRF material removal rate as a function of particle size and concentration term in Equation (8.2) for various workpiece materials. (b) Measured MRF material removal rate as a function of measured chemical durability term in Equation (8.2) for various workpiece materials. Source: DeGroote et al. 2007 [16]. Reproduced with permission from The Optical Society.

stabilize likely due to a lack of further stiffening of the MR fluid. The *material removal rate* also followed this behavior with the magnetic field [20].

Despite the low normal loads, a high local thickness *material removal rate* is achieved with MRF compared with conventional optical polishing. For example, for fused-silica glass, the peak removal rate is ~5 μm min^{-1}

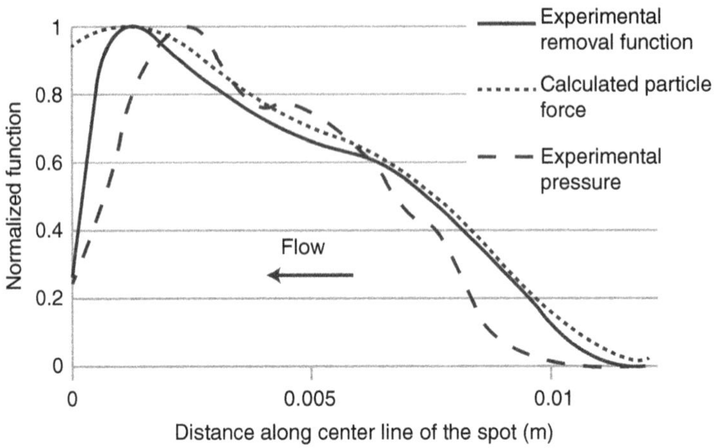

Figure 8.5 A comparison of the calculated material-removal rate profile, with measured removal rate and measured pressure distribution for typical MRF conditions. Source: Kordonski and Gorodkin 2011 [7]. Reproduced with permission from The Optical Society.

(or ~300 μm h^{-1}) with MRF and ~1 μm h^{-1} using conventional ceria polishing. This is partly due to the high velocity of the ribbon, which leads to a greater number of particles contacting the surface. For typical MRF processing, $V_r \sim 300$ cm s^{-1}, and for typical conventional polishing, $V_r \sim 15$ cm s^{-1}.

The high local thickness *material removal rate* with MRF has been shown to influence the surface characteristics of the workpiece, namely the Beilby layer. For the Beilby layer model discussed in Section 3.3.2, Ce penetration into the workpiece increases with polishing *material removal rate*. Figure 8.6 compares Ce and Fe penetration into fused silica polished conventionally and with MRF polishing as measures by secondary-ion mass spectroscopy (SIMS). Ce and Fe penetration were significantly higher after polishing with MRF. Specifically, Ce concentration was ~10× higher using MRF, as compared with conventional pad polishing. These results are qualitatively consistent with the Beilby layer model described by Equation (3.70).

8.2 Float Polishing

Similar to MRF, float polishing is a finishing technique that operates in hydrodynamic mode. In contrast with subaperture MRF polishing, float polishing is a full-aperture technique in which ultralow roughness surfaces on flats, with little surface damage, can be achieved. Developed in the late 1970s, float polishing was largely used for polishing magnetic heads (polycrystalline ferrites Mn—Zn and Ni—Zn) for videocassette recorders [24–26].

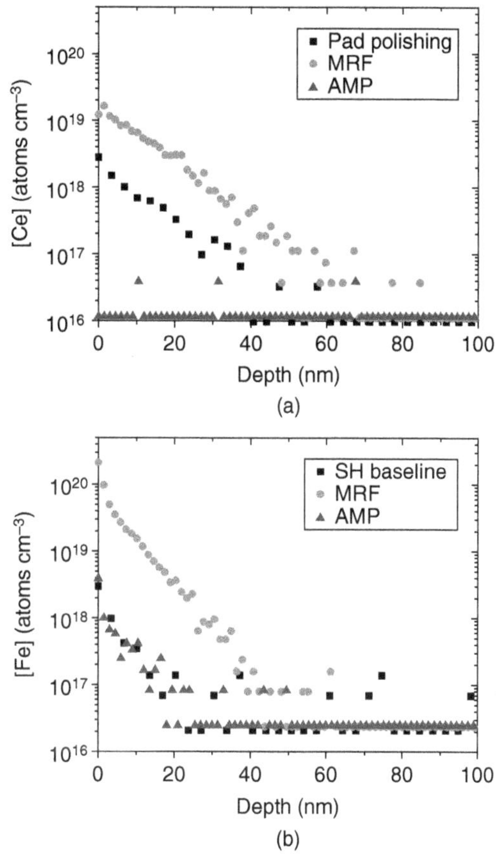

Figure 8.6 A penetration of slurry components (a) Ce and (b) Fe into the surface of a fused-silica workpiece after polishing using full-aperture pad polishing (CeO_2 on polyurethane pad), MRF (CeO_2-based slurry), and MRF followed by an HF-based treatment, AMP (advanced mitigation process).

This process provided very good flatness with minimal edge roll-off (<100 Å), leading to better performance than conventional polishing at the time. Later, float polishing was demonstrated on a variety of optical materials, including sapphire, Zerodur, BK7, and CaF_2 [25–27]. With an appropriate flat lap, *surface figures* near $\lambda/20$ with *surface roughness* in the 1–2 Å RMS range can be achieved. A demonstration of how this process results in little damage to the surface is illustrated for CaF_2 with transmission electron microscope (TEM) cross-section images shown in Figure 8.7 [27].

A schematic of a typical float-polishing setup is shown in Figure 8.8. A relatively flat workpiece is loaded at high pressures (typically tens of psi) and at high relative velocities (80–300 cm s^{-1}) in the presence of a 100 nm sized silica slurry. The lap is a high-precision metal flat (typically diamond turned) containing a unique groove pattern. A typical groove pattern contains two sets of grooves, one set 100 μm apart and 100 μm deep and the other 500 μm apart and 500 μm deep. During polishing, a micrometer level fluid gap is typically achieved.

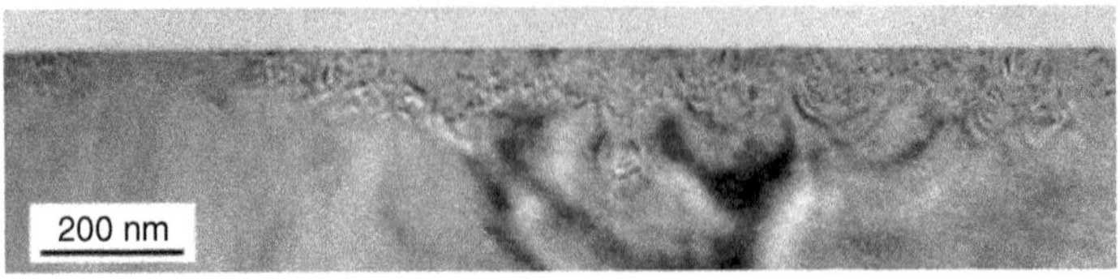

(a) Optically polished surface

(b) Float-polished surface

Figure 8.7 A cross-section of the TEM image of CaF_2 after the top surface was (a) conventionally polished and (b) float polished. Source: Namba et al. 2004 [27]. Reproduced with permission from Elsevier.

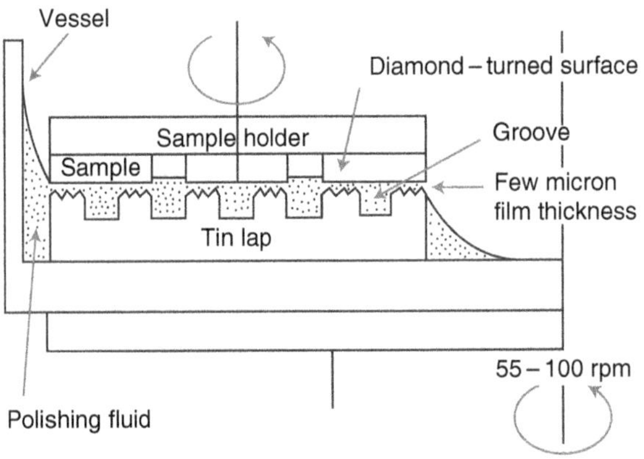

Figure 8.8 A schematic of a float polishing setup. Source: Namba et al. 1987 [26]. Reproduced with permission from Elsevier.

The combination of high relative velocities and grooves on the lap aids in getting into a hydrodynamic mode. Referring to Figure 2.5, hydrodynamic regime is aided by increasing $\eta_s V_r/\sigma_o$ FOM, or in this case, by an increase in relative velocity (V_r). However, on a flat, ungrooved lap, high applied pressures, σ_o, help increase the *material removal rate* but reduce the driving force toward achieving hydrodynamic mode. A study by Soares et al. [28] showed that the grooves on the lap provide interface pressure gradients that aid in achieving a hydrodynamic interface (i.e. causing the workpiece to float). They showed that the interface pressure, σ, when under laminar flow, varied across the groove as follows:

$$\sigma = \sigma_o - 1.2\,\eta_s \frac{v_r x}{\alpha_g{}^2}, \tag{8.3}$$

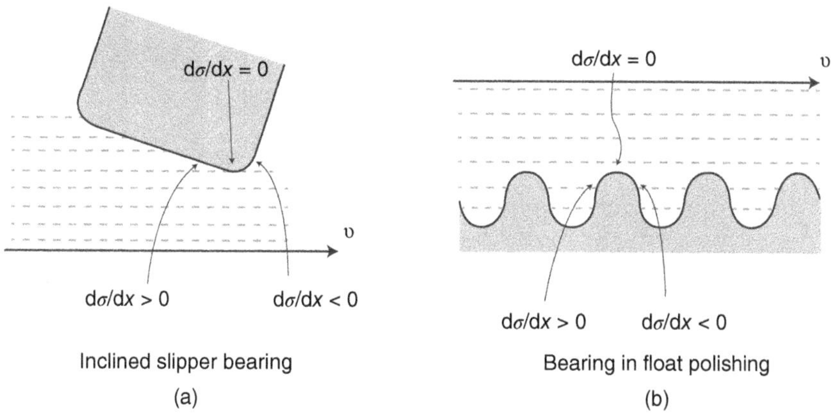

Figure 8.9 A comparison of pressure gradients due to hydrodynamic forces generated using (a) a tilted workpiece and (b) grooves at the workpiece–lap interface. Source: Soares et al. 1994 [28]. Reproduced with permission from The Optical Society.

where σ_o is applied pressure, η_s is slurry viscosity, v_r is relative velocity, x is the distance from the center of the groove, and α_g is groove height. The slurry, which is essentially an incompressible fluid, travels through the interface gaps and, hence, is the source of the pressure changes derived from Reynold's equation (see Section 2.5.3). Figure 8.9 compares the pressure changes created using a tilted plate versus grooves at an interface.

8.3 Ion Beam Figuring (IBF)

Ion beam figuring (IBF) is an ultraprecision, subaperture surface-finishing method that started in the mid-1980s [29]. The IBF method has a unique removal mechanism wherein the workpiece is placed in a vacuum chamber and an ion source accelerates a stream of ions (typically Ar ions, sometimes accompanied by other reactive species such as F ions) that remove material from the workpiece either by sputtering from atomic bombardment and dislodgement or by chemical reaction. These methods are also referred to as ion-beam sputtering or ion-beam etching. Ion sources range in size, having an ion flux of a Gaussian shape, which leads a removal spot on the workpiece of a similar shape. Because of the soft spatial decay of the Gaussian removal function, this method generally leads to an improved midspatial figure, as compared to other subaperture methods with a sharper spatial decay in the removal function (such as MRF). As with other subaperture methods, figure correction of the workpiece is achieved via kinematics (i.e. dwell times at each location) that is predetermined based on the initial *surface figure*. The size of the removal function or ion source determines which spatial scale lengths can be corrected and how long it takes to process a workpiece.

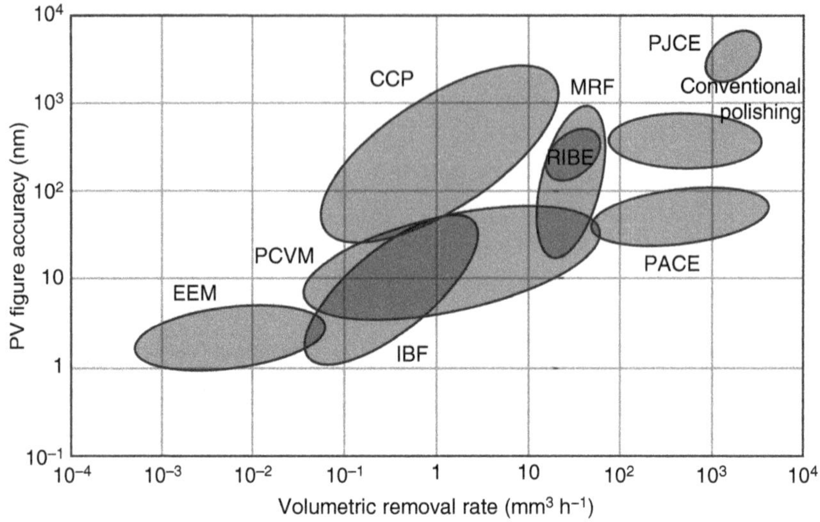

Figure 8.10 A comparison of the surface figure achievable with the corresponding volumetric removal rate using various finishing methods. EEM: elastic emission machining, PCVM: plasma chemical-vaporization machining, IBF: ion-beam figuring, CCP: computer-controlled polishing, RIBE: reactive ion-beam etching, PACE: plasma-assisted chemical etching, PJCE: plasma-jet chemical etching, MRF: magnetorheological finishing, FJP: fluid-jet polishing. Source: Mori et al. 2005 [34]. Reproduced with permission from Elsevier.

IBF has some unique characteristics compared with other polishing techniques [30–33]. First, because removal occurs at the atomic level, the removal function is highly predictable and deterministic, offering greater control on the *surface figure* and roughness. However, for the same reason, *material removal rates* are typically slower: 0.1–1 $mm^3\ h^{-1}$ for IBF, 10–60 $mm^3\ h^{-1}$ for MRF, and ~1000 $mm^3\ h^{-1}$ for conventional polishing. A nice comparison of the volumetric *material removal rate* of various finishing methods is shown in Figure 8.10, which illustrates that greater *surface figure* accuracy is generally achieved with methods that have smaller volumetric *material removal rates* [34]. Second, IBF is a noncontact technique. This has the advantage of avoiding complicating factors such as nonuniform spatial material removal (e.g. lap wear, viscoelastic effects, workpiece bending), which contribute to *surface figure* control (see Section 2.5), and the impact of rogue particles that lead to SSD and *surface quality* degradation (see Section 3.1).

IBF is well suited for performing the final finishing on workpieces that have very high *surface figure*, midspatial frequencies, and high spatial frequency requirements such as extreme ultraviolet (EUV) lithography and soft X-ray optics [35]. However, its utility is limited for niche optics types, materials, and specifications for various reasons. First, high vacuum systems are expensive, especially for large workpieces, leading to high upfront costs. Second, since *material removal rates* are relatively slow, long process times may be required. Third, the ion bombardment of the workpiece can lead to significant energy deposition. Part of this energy is used for removing material from the workpiece and part leads to a local temperature rise. Gailly et al. [33] performed infrared thermal imaging during the IBF process and showed a significant temperature rise of 50–100 °C in the ion-deposition region. Workpiece heating can lead to damage to the workpiece itself or distortion in the removal function, due to workpiece thermal expansion or changes in surface reactivity, complicating the ability to control *surface figure*. Fourth, the workpiece surface is prone to contamination, both from sputtered material from the workpiece or from surrounding materials such as grid, screen, and chamber wall depositing onto the workpiece surface [32, 36]. For high-laser-damage optics, for which redeposited materials are known absorbing precursors (see Chapter 9), this contamination is particularly detrimental. Methods have been employed to reduce redeposition, such as using an ionic concentrator, where a magnet can drive impurity ions away from the substrate, or by postchemical etching of the workpiece to remove any redeposits [32, 37].

8.4 Convergent Polishing

To achieve deterministic control of the *surface figure* during conventional full-aperture polishing, each of the phenomena controlling material removal (discussed in Chapter 2 and summarized in Figure 2.1) ideally needs to be understood, quantified, and controlled. Several methods can be used to achieve the desired *surface figure*. The most common method is to use an iterative method: measure *surface figure*, polish with a process parameter generally directed to improve the *surface figure*, and repeat many times until the desired *surface figure* is achieved. This method relies heavily on the skills of a master optician to make the adjustments necessary to get the desired *surface figure*. Another method is to polish and incorporate all the phenomena in Figure 2.1 quantitatively into a code, such as the SurF code (see Figure 2.43), and calculate the optimum polishing parameters to get the desired *surface figure*. However, this method for deterministic finishing is sometimes challenging, because the large number of process variables leads to difficulty in calculating a unique set of process parameters. A third method for achieving deterministic finishing is to remove all sources of nonuniformities in material removal on the workpiece (through engineering controls and polishing-process parameter controls) except for the workpiece-shape-induced workpiece–lap mismatch (see Section 2.5.6). The surface mismatch at the interface between the workpiece and lap causes a spatial pressure (and hence, removal) differential, which decreases with removal amount (i.e. polishing time), thus allowing the workpiece to converge to the shape of the lap [38]. This technique, called convergent polishing [39–41], allows for polishing workpieces (flat or spheres) in a single iteration using a single set of polishing parameters, regardless of the initial shape of the workpiece and without the need for initial and in-process *surface figure* metrology.

Figure 8.11 illustrates how an initial workpiece *surface figure* may converge to the lap *surface figure* based on the concept of workpiece–lap mismatch during convergent polishing [41]. Consider the flat lap and a hypothetical workpiece of complex shape shown at the top left. During polishing, the workpiece has the highest local pressure in the center (see bottom left of Figure 8.11); hence, this location will observe the highest initial *material removal rate* during polishing. As the material is removed, the pressure differential across the workpiece will decrease due to a decrease in the workpiece–lap mismatch, and the workpiece will converge to the shape of the lap. At convergence, the workpiece pressure distribution, and hence material removal, will be uniform across the workpiece (see right side of Figure 8.11). This example is illustrated for a flat lap, but the same concept applies to either a concave or convex spherical lap. Again, the convergence process works *only* when all other phenomena affecting spatial material nonuniformity have been eliminated. In other words, if even one of these nonuniformities is not controlled, the *surface figure* will not converge to the desired point [41].

The convergent-polishing system and method is a compilation of a number of developed technologies that not only enable *surface figure* convergence but also high *material removal rate*, high *surface quality* (i.e. low SSD), and low *surface roughness*. The key enabling technologies include the following:

1. A novel shaped septum, which compensates for nonuniform pad wear, improves temperature uniformity, improves slurry distribution, and reduces viscoelastic pad edge effects (see Section 7.1.3) [39–42].
2. Bulk acid etching, which allows more rapid removal of SSD and reduces the amount of material needed to be removed from the workpiece during polishing. It also removes grinding stress, which can deform or bend high-aspect-ratio workpieces (see Section 7.1.2) [43–45].
3. Pitch-button blocking, which allows blocking or mounting a high-aspect-ratio workpiece, preventing it from bending during blocking and polishing while reducing the risk of scratching the opposing workpiece surface (see Section 7.1.1) [46].
4. Radial stroke, which results in improved local material removal spatial–time averaging during polishing and prevents high-frequency ripples that can occur on a workpiece surface (see Section 2.5.6.7) [47].
5. Balanced three-body wear of workpiece, septum, and lap, which provides a desired, stable lap shape and improves the stability of the convergent point of the workpiece with polishing time.

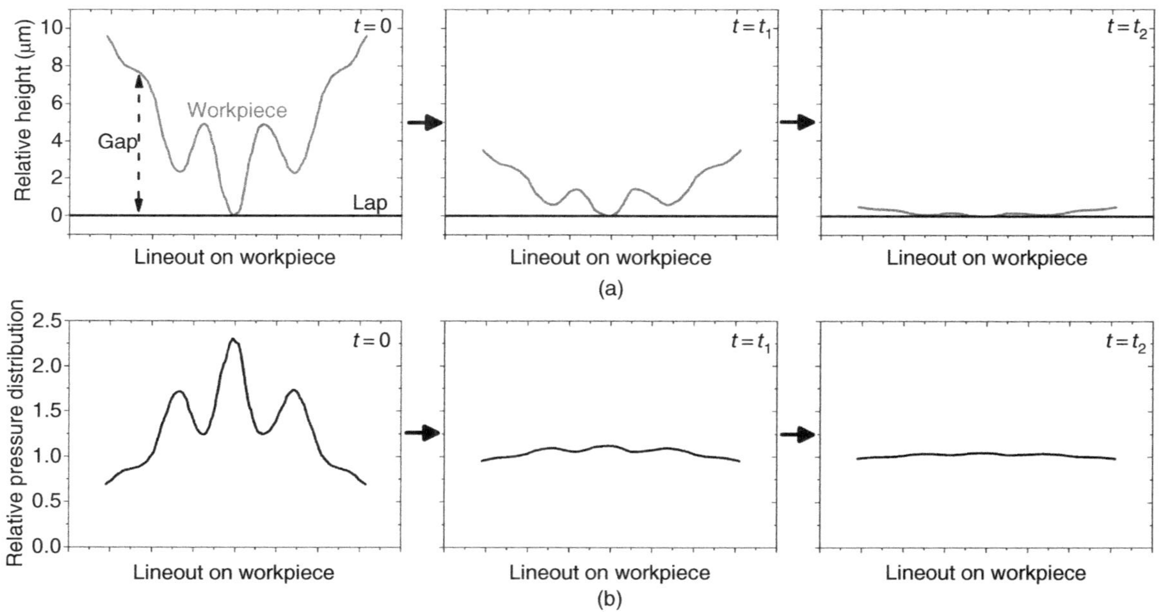

Figure 8.11 An illustration of workpiece shape (a) and interface pressure distribution (b) with polishing time (*t*), shown from left to right, during convergent polishing. Source: Suratwala et al. 2014 [41]. Reproduced with permission from Journal of Visualized Experiments.

6. Hermetically sealed high humidity polishing chamber, which prevents the entry of external rogue particles and formation of dried slurry agglomerates, which are common sources of workpiece scratching; this also reduces a chance of pad drying and permanently deforms the lap shape (see Section 7.1.4) [39, 41, 48].
7. Engineered filtration system, which improves and maintains a desired slurry particle size distribution (PSD) that reduces the overall *surface roughness* of the optic and reduces the probability of scratch formation. This includes the use of fluorinated piping, minimization of dead zones, and controlling flow velocities, preventing slurry settling, agglomeration, and contamination (see Section 7.1.5) [39, 41].
8. Chemical slurry stabilization, which reduces the number and size of agglomerates in the slurry without sacrificing *material removal rate*, using special surfactants that employ the use of a novel charged micelle "halo" chemical stabilization mechanism (see Section 7.1.6) [42, 49].
9. In situ ultrasonic pad treatment, which allows removal of slurry and glass polishing deposits from the surface of the polishing lap, helping maintain high *material removal rates* and minimizing midrange spatial scale length degradation of the workpiece due to preferential material deposit (see Section 7.1.8) [39, 47].
10. Novel diamond conditioning of pad, which reduces the asperities on the polyurethane pad without overall pad wear; this significantly increases the *material removal rate* and lowers the average roughness of the workpiece [50].

In addition, convergent polishing takes advantage of technologies known within the optical fabrication field to eliminate spatial nonuniformities in material removal and improve *surface quality*. These include the following:

- Constant time-averaged velocity kinematics using matched rotation ($R_o = R_L$).
- Edge-driven workpiece to prevent moment force workpiece bending [38] (see Section 2.5.4).
- Uniform applied pressure.
- Applied pressure and relative velocity adjusted to operate in contact mode (see Section 2.3) [38].
- Stiff lap base to prevent bending under load.
- Careful pad selection with relatively high modulus and low wear rate to provide rapid and stable convergence.
- Appropriate pad-groove pattern to provide good slurry transport and replenishment.
- Appropriate Baume (slurry concentration) and maintenance of the pH of polishing slurry.

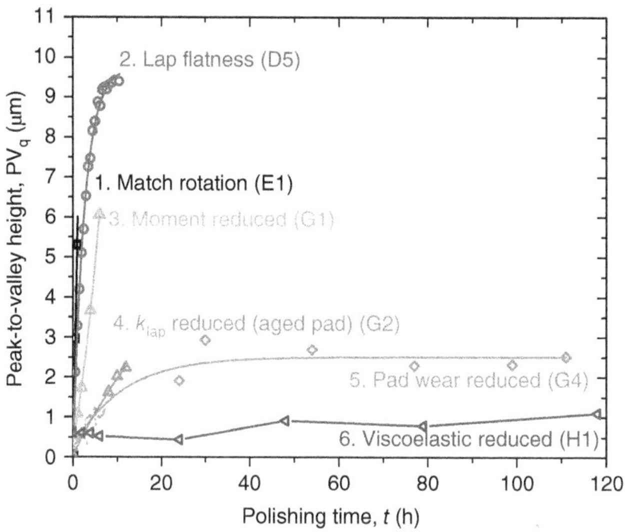

Figure 8.12 A change in polished surface figure PV height of fused-silica workpieces as a function of polishing time for various polishing conditions with progressive removal of spatial material removal nonuniformities. All the samples started nominally flat (PV < 0.5 μm). Source: Suratwala et al. 2012 [40]. Reproduced with permission from John Wiley & Sons.

- Submerged water cleaning processes for the workpiece postpolishing to prevent slurry staining.

Figure 8.12 illustrates how the gradual elimination of phenomena that contribute to spatial material removal nonuniformities reduces the amount of deviation in *surface figure* with polishing time for an initially flat fused-silica workpiece [38, 40]. For the initial baseline polishing conditions (E1), using a spindle drive and a new fibrous polyurethane pad (Suba 550), a large degree of nonuniform removal is present, as indicated by a large PV of 5.3 μm within 1 h of polishing. Improving the lap-base flatness from a PV of 25 μm down to 2 μm (D5), the degree of nonuniformity was slightly reduced. Further reduction in removal nonuniformity was realized by changing to a side-wheel-driven workpiece from a spindle-driven workpiece (reducing the moment force contribution to nonuniform removal) (G1). The uniformity in removal was improved further by using an aged Suba 550 lap that has been used for >100 h of polishing (G2), which is likely due to a reduction in the pad-wear rate. An even higher level of uniform material removal is achieved with a use of a pad-wear-compensating septum (G4). Finally, the best removal uniformity was achieved when the septum was placed at the leading edge of the workpiece, reducing pad viscoelastic effects on removal uniformity (H1). Figure 7.8 shows the lineouts of the *surface figure* after long polishing times (>100 h) under these optimized conditions, illustrating that once all spatial material removal nonuniformities are removed, the converged *surface figure* will not change with continued polishing.

Figure 8.13 shows a 3D *surface figure* evolution as a function of polishing time using the convergent polishing method for round and square 100 mm scale fused-silica workpieces with different initial *surface figures*. Note that all the *surface figures* are plotted on a single scale for a given polishing run. Convergent polishing has also been demonstrated on larger workpieces of up to 430 mm; photos of such polishers are shown in Figure 8.14 [41].

8.5 Tumble Finishing

Another unique polishing method, called tumble finishing, is used for smoothing or locally planarizing the surface of more-complex-shaped workpieces. Historically, tumble finishing has been used for smoothing the rough surfaces of rocks, ceramics, glass, and metals by rotating them in a cylinder in the presence of an abrasive slurry and, sometimes, load-bearing media [51, 52]. In the case of rocks, this method is commonly known by hobbyists as rock tumbling. The small, random contacts between the media and workpiece lead to polishing over the entire surface of the workpiece, usually with little *surface figure* change. Tumble finishing relies on many of the same principles used in ball milling, which is used to prepare ceramic slurries [53].

In recent years, the tumble-finishing method has been extended for the optical polishing of hollow plastic spheres, called ablators, for use in targets for high-peak-power laser systems (see Figure 8.15) [54–57]. These ablator workpieces, typically 2 mm in diameter and ~190 μm thick, are filled with isotopes of hydrogen that are frozen as a layer on the inner surface and compress through laser inertial confinement [58]. The ablators are fabricated using plasma-assisted chemical vapor deposition (PA-CVD), where hydrogen and *trans*-2-butene are broken down to form an amorphous polymer coating on a prefabricated spherical poly-alpha-methylstyrene substrate ("mandrels") produced by microencapsulation [59–61]. The mandrels are later removed through thermal decomposition.

The ablators have stringent *surface roughness* and isolated surface-defect requirements, as these can contribute to hydrodynamic instabilities and thus undesirable nonuniform implosions [62, 63]. Despite the good overall roughness (<10 nm RMS), the ablators may have many isolated surface defects, or domes that are hundreds of nanometers in height and tens of microns wide (Figure 8.16a) [64]. These domes are caused by a small particle or the asperity precursors present on the mandrel that grow into a dome during PA-CVD. Previously, conventional polishing techniques were used to remove domes, usually at the expense of scratching the surface and increasing overall roughness (Figure 8.16b) [65, 66]. Other existing techniques of polishing spheres were insufficient due to the ablators' small size and fragility [67–69].

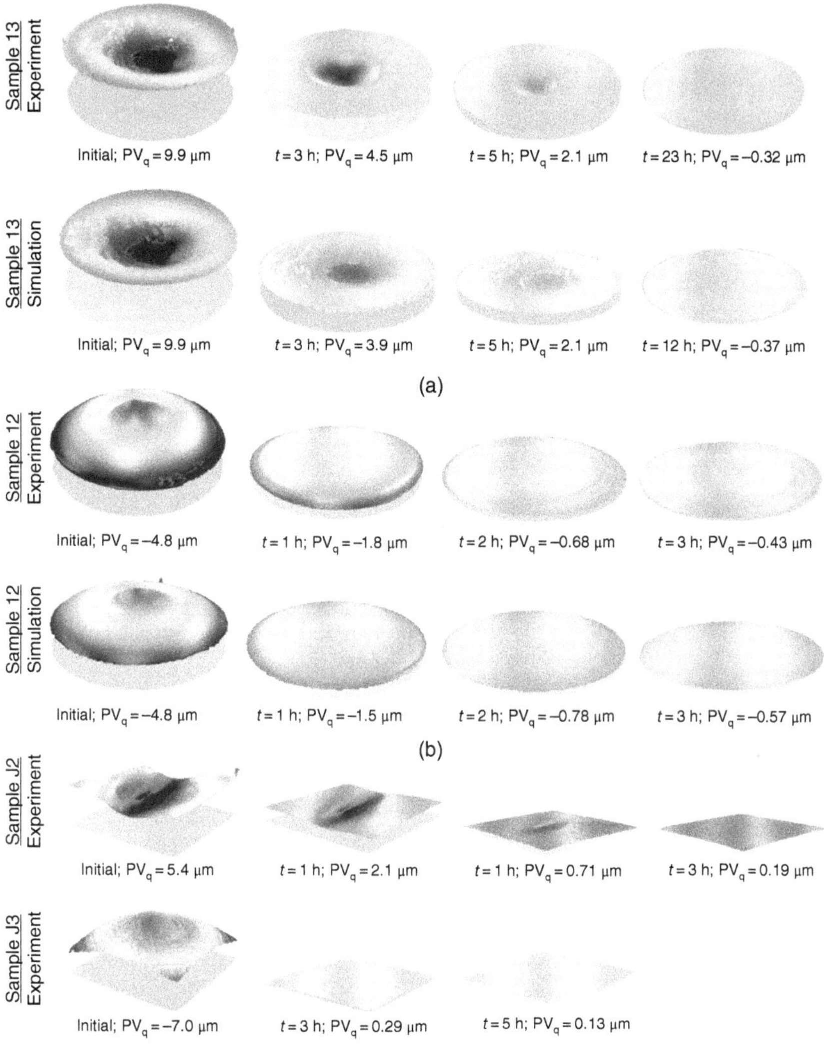

Figure 8.13 Convergent polishing demonstration and corresponding simulation using SurF code on a variety of fused-silica workpiece types and pads: (a) initial concave round workpiece (100 × 10 mm); (b) initial convex round workpiece (100 mm × 10 mm); (c) concave and convex square workpieces (100 mm diagonal × 10 mm) on IC1000. Source: Suratwala et al. 2012 [40]. Reproduced with permission from John Wiley & Sons.

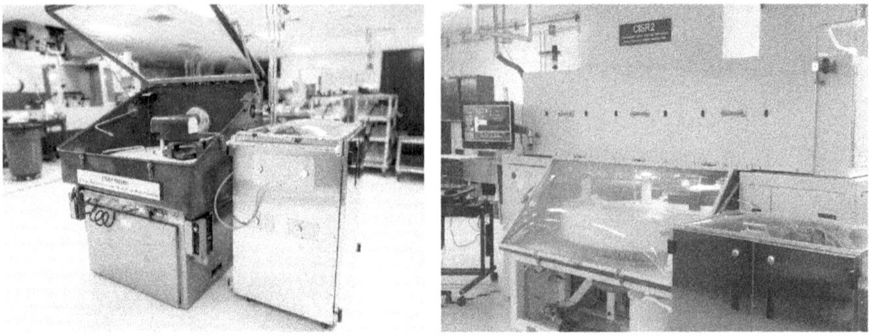

(a) (b)

Figure 8.14 Photos of larger convergent polishing systems capable of finishing (a) ~0.26 m optics (called CISR1) (b) ~0.5 m optics (called CISR2). CISR = convergent, initial-surface independent, rogue-particle free.

Figure 8.15 The photo of a hollow plastic sphere (2 mm diameter) used as an ablator in high peak-power laser targets for fusion-energy research. Source: Suratwala et al. 2012 [54]. Reproduced with permission from Elsevier.

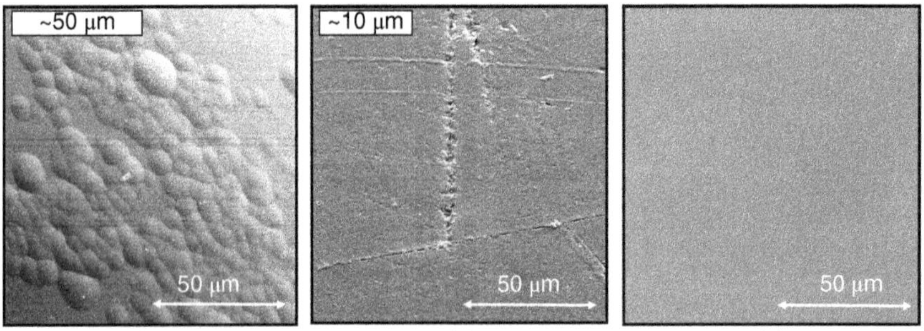

(a) (b) (c)

Figure 8.16 SEM micrographs of the ablator surface (a) as deposited, (b) after conventional polishing, and (c) after tumble finishing. Source: Suratwala et al. 2012 [54]. Reproduced with permission from Elsevier.

Figure 8.17 The photo of acrylic vial cylinder (60 mm long) containing colloidal-silica slurry, glass media, and a single capsule. Source: Suratwala et al. 2012 [54]. Reproduced with permission from Elsevier.

A typical tumble-finishing process for these ablators is conducted by first placing an aqueous colloidal-silica slurry (50 nm in diameter), spherical borosilicate-glass bead media (2.4 mm in diameter), and a single ablator (2 mm in diameter) into a sealed, cylindrical acrylic vial (25 mm diameter × 60 mm long), as shown in Figure 8.17. The vial is rotated about its cylindrical axis (parallel to its length) at 100 rpm for 96 h [54, 57]. During rotation, the glass beads randomly load the ablator workpiece's outer surface via sphere-to-sphere contact with colloidal-silica slurry at the interface, resulting in a small contact zone – hence, this process is analogous to subaperture tool finishing. An optimized tumble-finishing process removes material from the ablator at a slow rate of ~35 nm day^{-1}, which is effective at locally planarizing the preexisting domes (Figure 8.16c).

The tumble-finishing process offers several advantages over conventional polishing techniques for local planarization. First, since the polishing system is hermetically sealed, the ablator is less prone to scratching, as long as the materials used are free of rogue particles and surface asperities. Second, unlike conventional polishing, ablator mounting is not needed, thereby minimizing ablator deformation and mounting interface-induced scratching. Finally, the process is relatively simple and of low cost.

Tumble finishing of ablators is an example of how the polishing principles outlined in Chapters 2–5 can be applied to develop new, novel finishing processes for unique applications. The main principles exploited with tumble finishing of ablators are (i) management of rogue particles and asperities to prevent scratches and digs (see Section 3.1.3); (ii) using the workpiece–lap mismatch concept due to workpiece shape (see Section 2.5.6.1); and (iii) using the conformal versus nonconformal removal concept (see Section 7.1.1). These concepts are extended to random kinematics using a subaperture contact spot. The random kinematics allows for locally changing the roughness or smoothness without changing the overall shape, or long-range *surface figure* of the workpiece. The overall *surface figure* of the as-born ablator is very spherical (out of round <200 nm); hence, it is well suited for tumble finishing. Other conventional techniques can modify the overall shape and improve the sphericity for the workpieces not starting from a spherical shape [67–69].

One challenge to developing an effective tumble-finishing process is the planarizing of isolated surface defects (i.e. domes) without introducing undesirable surface features such as pits and scratches from rogue particles or asperities introduced by external contamination, from corrosion products of the media and vial, and from roughness of media and vial surfaces [48, 54]. Hence, *surface quality* can be improved by using stringent cleaning or filtering processes of the media, slurry, and vial, selecting media materials that do not corrode, such as glass or ceramic, and prepolishing the media and vial surfaces to low *surface roughness*. These controls have led to high *surface quality* surfaces (see Figure 8.16c) [54].

During tumble finishing, an ensemble of random spherical–spherical contacts between the glass media and the ablator occurs, leading to a well-defined contact spot that may be thought of as the tool size. Accounting for the modulus and radius of the glass media and ablator and the load at contact based on impact mechanics, the Hertzian contact zone was determined as ~39 μm, which is in the order of the width of the domes to be removed [54]. Dome removal from the surface of the ablator is caused by the workpiece–lap mismatch, due to workpiece shape applied locally to the contact zone or tool size. To review, when a tool or a lap contacts the workpiece, the physical mismatch at the interface will cause a pressure differential, and, hence, a spatial removal-rate differential. Since the tool typically wears out much slower than the workpiece, the workpiece surface will converge to the shape of tool because material is removed faster where the workpiece–lap mismatch is small. As material is removed, the mismatch at the interface will decrease and the pressure, and hence the removal rate, will become spatially uniform. Examples of isolated domes that converged are shown in Figure 8.18, where surface topologies of the domes before and after tumble finishing were measured using phase-shifting diffractive interferometry (PSDI) [54].

The rate of convergence will depend greatly on the stiffness of the tool and workpiece. For high-stiffness contact, convergence is fast (nonconformal removal). For compliant contact, convergence does not occur, and only uniform material removal occurs (conformal removal). The rate of convergence of a dome can be described by using the following form of the Preston equation, which accounts for the instantaneous spatially dependent removal rate ($\mathrm{d}h_\mathrm{i}/\mathrm{d}t$) due to workpiece–lap mismatch:

$$\frac{\partial h_\mathrm{i}}{\partial t}(x, y, t) \approx (-1 + L\nabla^2 h_\mathrm{i}(x, y, t))\frac{\mathrm{d}h}{\mathrm{d}t}, \tag{8.4}$$

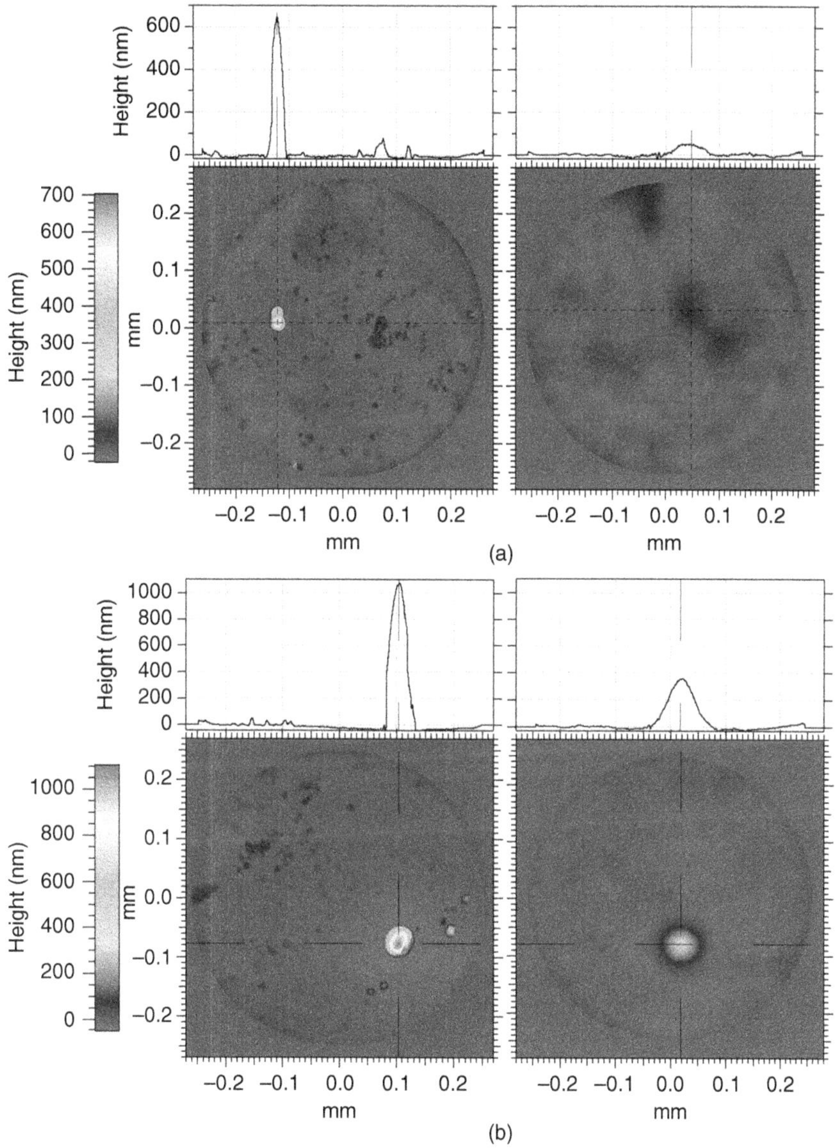

Figure 8.18 The PSDI image of capsule surface pre- (left) and post- (right) tumble finishing of isolated narrow (a) and wide domes (b) on capsule. Crosshairs identify the same location pre- and post-tumble finishing. Source: Suratwala et al. 2012 [54]. Reproduced with permission from Elsevier.

where dh/dt is the spatial and time-averaged thickness removal rate, $h_i(x, y, t)$ is the surface height on a given point on the surface, and L is a characteristic length. The second term in the parentheses describes the workpiece–lap mismatch effect on material removal in terms of local curvature of the surface of the workpiece. In other words, a negative curvature feature on the surface (i.e. a peak) will observe enhanced removal, and a positive curvature feature on the surface (i.e. a valley) will observe reduced removal. The L value incorporates the effect of both the size of the contact spot and relative stiffness of the tool [54].

Figure 8.19 shows the results of solving Equation (8.4) numerically for various isolated defects (a narrow dome, a wide dome, two intersecting domes, and a narrow dimple). The narrow domes converged more quickly than the wider domes (see Figure 8.19a,b), consistent with experimental observation [54]. The intersecting domes have a more complicated behavior; with removal, the domes merge, and the resulting dome has an effective width larger than that of either initial dome, which slows down the convergence rate (Figure 8.19c). Finally, a

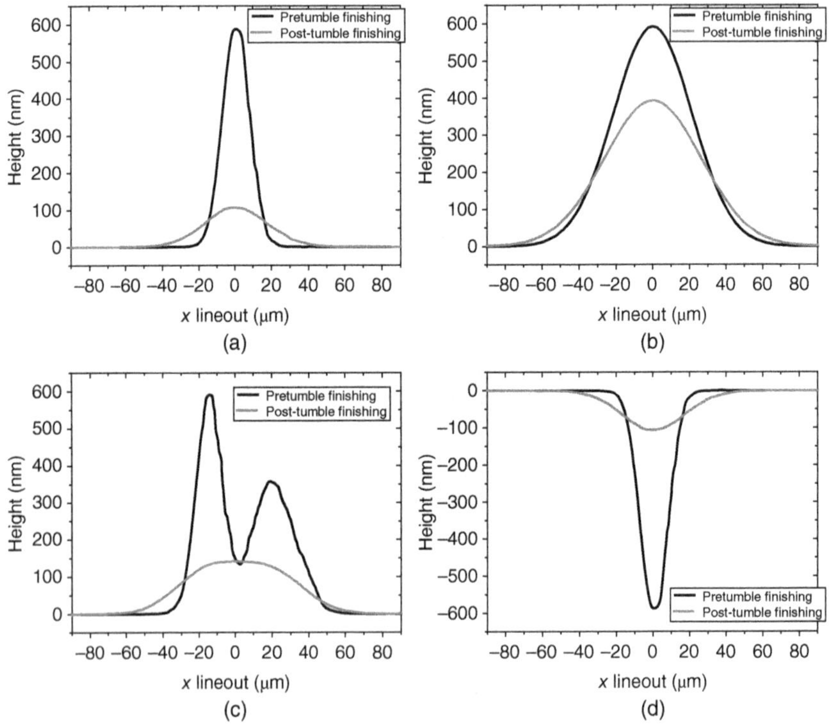

Figure 8.19 Lineouts of 3D simulations of the surface evolution of various dome and dimple defects during tumble finishing (using Equation 8.4), with $L = 820\ \mu m$; $dh/dt = 35\ nm\ day^{-1}$; $t = 96$ h: (a) narrow Gaussian dome (600 nm high; 16.5 μm wide); (b) wide Gaussian dome (600 nm high; 50 μm wide); (c) multiple domes; and (d) narrow dimple (600 nm deep; 16.5 μm wide). Source: Suratwala et al. 2012 [54]. Reproduced with permission from Elsevier.

dimple (i.e. a valley) will also converge by widening and reducing is relative depth (Figure 8.19d). Notice that domes and valleys do not completely go away, especially the wider domes, which indicate that removal is not completely non-conformal. This is because the ablator, being a hollow sphere, has a relatively low stiffness (i.e. effective modulus).

Equation (8.4) can also be solved analytically by approximating the shape of the domes as a Gaussian dome with initial height h_0 and full-width at half-maximum (FWHM) width w_0 [54]. The solution for height (h_i) relative to the baseline surface at time t and distance r from the dome center is given by

$$h_i(r,t) = h_m(t)\, e^{-\ln 2\left(\frac{2r}{w(t)}\right)^2}, \tag{8.5}$$

where

$$w(t) = w_0\sqrt{1 + \frac{16\ \ln 2L\frac{dh}{dt}t}{w_0^2}}, \tag{8.6}$$

and where $h_m(t)$ is the maximum dome height as a function of time given by

$$h_m(t) = \frac{h_0}{1 + \frac{16\ \ln 2L\frac{dh}{dt}t}{w_0^2}}. \tag{8.7}$$

Equation (8.7) also shows that the convergence rate will decrease as the width of the dome increases. Figure 8.20 compares the effectiveness of the

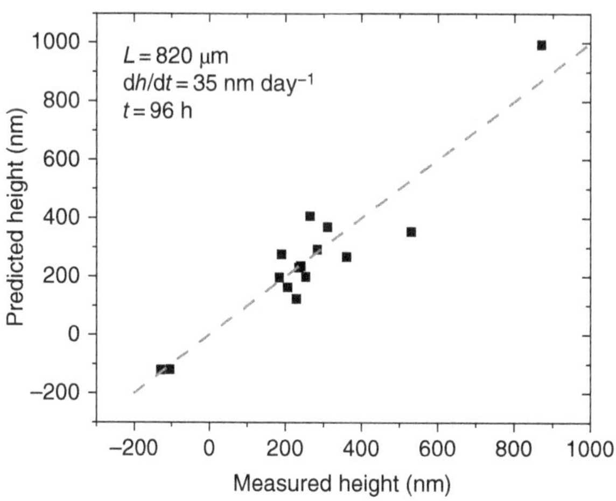

Figure 8.20 Calculated post-tumble finishing heights of various measured domes and dimples, using Equations (8.5)–(8.7) and initial measured heights (h_o) and width (w_o), cross-plotted with the measured final height of the same corresponding domes and dimples. The dashed line corresponds to a one-to-one correlation between calculated and measured values. Source: Suratwala et al. 2012 [54]. Reproduced with permission from Elsevier.

analytical solution to the convergence model in predicting the final height of isolated domes after polishing, using a single value for the characteristic length ($L = 820\,\mu m$). The measured final height of the dome after polishing is cross-plotted against the predicted final height of the dome after polishing. The dashed line represents a one-to-one correspondence between measured and calculated heights. The model shows a reasonably good agreement with experimental results over a wide range of isolated dome heights, and even some dimples.

8.6 Other Subaperture Polishing Methods

Subaperture polishing methods have greatly expanded the ability to finish asymmetric optics, such as aspheres, and other complex-shaped optics (referred to as freeform optics). In addition to MRF and IBF, there are numerous methods such as computer-controlled optical surfacing (CCOS), computer-controlled polishing (CCP), stress-lap polishing, fluid-jet polishing, bonnet polishing; elastic emission machining (EEM), and plasma chemical-vapor machining (PCVM). All these methods have a tool with a defined or controllable removal function in which kinematics and dwell time are used to alter the overall *surface figure* of the workpiece. The differences between these methods stem from the shape and size of the tool, the lap properties, the contact regime, and the removal mechanism.

One unique subaperture polishing method is stress-lap polishing, where the workpiece–lap mismatch (Section 2.5.6) is directly controlled by a mechanically active tool that can adjust its shape in real time using a set of electromechanical actuators [70–72]. This leads to a more conformal removal at the contact spot. A stress-lap tool is particularly advantageous in polishing a relatively large contact area on parabolic and aspheric surfaces. This technology has been used on a number of large (1.8–8.4 m) telescope optics [70–72]. The complex control system that adjusts the tool shape during its rotation and translation is "transparent to the optician, so that for practical purposes the optician can polish a highly aspheric surface as if he or she were polishing a sphere." [70]

An alternative method to achieve a more conformal removal using a subaperture tool on complex-shaped workpieces is to use a more compliant lap tool. Bonnet polishing is an example, where a curved-shape compliant material rotates against the surface of the workpiece (see Figure 8.21). This polishing method has been used for fabricating X-ray optics with *surface roughness* requirements <0.3 nm RMS and deviation from the desired aspheric shape of <50-nm PV on molding dies 15 cm in size with Ni surfaces [73]. An example set of process conditions uses 0.2 μm alumina slurry with a napped, porometric cloth bonnet-polishing tool of 41 mm radius, 0.05 MPa applied pressure, and a feed rate of 250–1000 mm min^{-1}. Because of the compliant

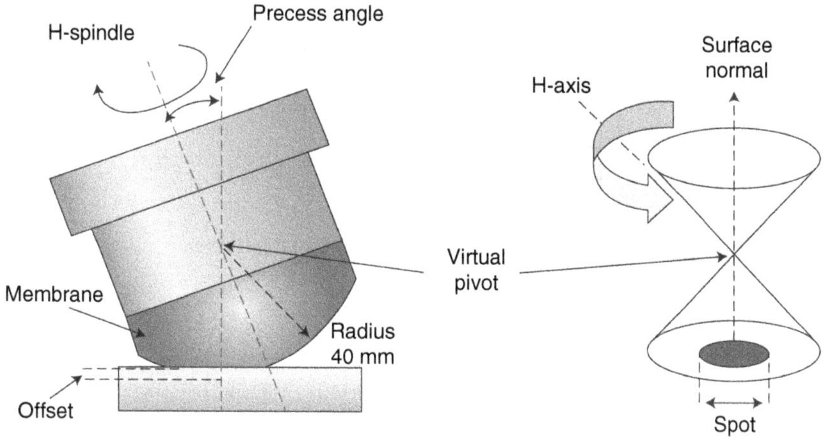

Figure 8.21 A schematic of bonnet polishing illustrating local kinematics of tool with workpiece. Source: Beaucamp and Namba 2013 [73]. Reproduced with permission from Elsevier.

tool, the load per particle at the interface is significantly reduced, and, as in the ensemble Hertzian multi-gap (EHMG) model (see Section 4.7.1) leads to relatively low-roughness surfaces. This also leads to relatively low *material removal rates* – hence, this technique is largely used for smoothing. In one study, the combined methods of diamond turning, fluid-jet polishing for *surface figure*, and bonnet polishing for smoothing was demonstrated for X-ray optical applications [73]. Some level of *surface figure* control has also been demonstrated using bonnet polishing on 50 × 100 mm cylinder borosilicate glass using a ceria slurry [73]. Various removal-spot shapes can be created, depending on the kinematics and orientation of the bonnet; several process studies have attempted to understand this [74–77]. The smoothness of the edge is particularly important, since it affects the magnitude of the tool spatial frequency left behind on the workpiece.

Fluid-jet polishing is a subaperture method that uses a slurry stream (water plus polishing particles) from a nozzle contacting the workpiece at a defined angle and standoff [78]. The particles are typically abrasive particles such as SiC or Al_2O_3, with a size from 5 to 15 μm. A typical slurry velocity is 20 m s^{-1}, and the local removal rate is around 1 μm min^{-1}. This technique is analogous to sand blasting, but in a fluid environment. In contrast to sandblasting, however, the resulting load per particle is below fracture initiation, and hence, removal is dominated by nanoplastic removal. The volume *material removal rate* was found to scale as $E_s^{0.5}/H_s^{1.8}$ of the workpiece and linearly with fluid pressure [79]. Tests of *surface figure* improvement using fluid-jet polishing have been demonstrated on flat borosilicate glass (BK7) samples (20 mm diameter) from ~1.5λ to ~0.3λ PV.

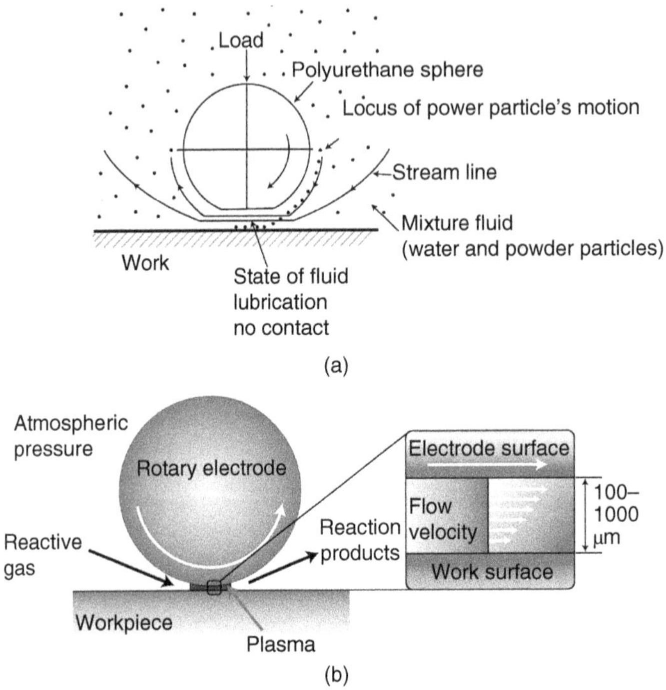

Figure 8.22 A schematic of (a) the EEM process and (b) the PCVM process. Source: Adapted from Yamamura et al. 2002 [81] and Mori et al. 1990 [82].

EEM is a subaperture method that operates in hydrodynamic mode, leading to very low load per particle and hence very low roughnesses and fine figure control [80]. A typical EEM system uses a polyurethane sphere rotated under a load against a workpiece in the presence of a slurry, typically containing silica particles 2 μm in diameter (Figure 8.22a). The fluid velocity and viscosity are such that the contact interface is in hydrodynamic mode (see Section 2.3). The interface gap is typically 1 μm, the slurry shear rate is a very high 1.0 m s^{-1} μm $^{-1}$, and the typical contact zone size ranges from 100 to 5 mm [83]. EEM has been largely used for metal and semiconductor polishing but has also been demonstrated on optics. For example, using EEM on Zerodur and ULE has led to 0.1 nm RMS roughnesses [84]. In this process, the volumetric *material removal rate* is quiet low (0.001–0.05 mm^3 h^{-1}) and hence, it is largely been used on smooth process or prefabricated optics to reduce roughness. The technique is particularly used in X-ray optics [80].

A complimentary method called PCVM uses neutral atomic and molecular radicals generated in a radio-frequency plasma to react with the surface [81, 83], similar to plasma etching, but not performed in a vacuum. The atmospheric pressure allows for high radical density and relative high removal rates

(0.1–50 mm^3 h^{-1}). By using a high-speed rotary electrode with a narrow gap between the workpiece and the electrode, local enhancement of the *material removal rate* is achieved (see Figure 8.22b). Impressive *surface figures* have been achieved on X-ray mirrors down to 2 nm PV.

References

1 Kordonsky, W.I. ed. (1993). Adaptive structures based on magnetorheological fluids. 3rd International Conference on Adaptive Structures.

2 Prokhorov, I., Kordonsky, W., Gleb, L., and Gorodkin, G. (1992). WB8 New high-precision magnetorheological instrument-based method of polishing optics. Technical Digest Series-Optical Society of America, Volume 24, p. 134.

3 Jacobs, S.D., Kordonski, W., and Prokhorov, I.V. (2000). Deterministic magnetorheological finishing. Google Patents.

4 Harris, D.C. ed. (2011). History of magnetorheological finishing. SPIE Defense, Security, and Sensing. International Society for Optics and Photonics.

5 Shorey, A.B., Kordonski, W., and Tricard, M. ed. (2004). Magnetorheological finishing of large and lightweight optics. Optical Science and Technology, The SPIE 49th Annual Meeting. International Society for Optics and Photonics.

6 Golini, D., Kordonski, W.I., Dumas, P., and Hogan, S.J. ed. (1999). Magnetorheological finishing (MRF) in commercial precision optics manufacturing. SPIE's International Symposium on Optical Science, Engineering, and Instrumentation. International Society for Optics and Photonics.

7 Kordonski, W. and Gorodkin, S. (2011). Material removal in magnetorheological finishing of optics. *Appl. Opt.* 50 (14): 1984–1994.

8 Golini, D., Jacobs, S.D., Kordonski, V., and Dumas, P. ed. (1997). Precision optics fabrication using magnetorheological finishing. Advanced Materials for Optics and Precision Structures: A Critical Review.

9 Menapace, J.A., Ehrmann, P.E., Bayramian, A.J. et al. (2016). Imprinting high-gradient topographical structures onto optical surfaces using magnetorheological finishing: manufacturing corrective optical elements for high-power laser applications. *Appl. Opt.* 55 (19): 5240–5248.

10 Menapace, J.A., Dixit, S.N., Génin, F.Y., and Brocious, W.F. ed. 2004). Magnetorheological finishing for imprinting continuous-phase plate structures onto optical surfaces. XXXV Annual Symposium on Optical Materials for High Power Lasers: Boulder Damage Symposium. International Society for Optics and Photonics.

11 Menapace, J.A., Davis, P.J., Steele, W.A. et al. ed. 2006). MRF applications: on the road to making large-aperture ultraviolet laser resistant continuous phase plates for high-power lasers. Boulder Damage Symposium XXXVIII: Annual Symposium on Optical Materials for High Power Lasers. International Society for Optics and Photonics.

12 Shorey, A.B. (2000). *Mechanisms of Material Removal in Magnetorheological Finishing (MRF) of Glass*. University of Rochester.

13 Shorey, A.B., Jacobs, S.D., Kordonski, W.I., and Gans, R.F. (2001). Experiments and observations regarding the mechanisms of glass removal in magnetorheological finishing. *Appl. Opt.* 40 (1): 20–33.

14 Menapace, J.A., Peterson, J.E., Penetrante, B.M. et al. (2005). Combined advanced finishing and UV laser conditioning process for producing damage resistant optics. Google Patents.

15 Kordonski, W. and Golini, D. (1999). Fundamentals of magnetorheological fluid utilization in high precision finishing. *J. Intell. Mater. Syst. Struct.* 10 (9): 683–689.

16 DeGroote, J.E., Marino, A.E., Wilson, J.P. et al. (2007). Removal rate model for magnetorheological finishing of glass. *Appl. Opt.* 46 (32): 7927–7941.

17 Cook, L. (1990). Chemical processes in glass polishing. *J. Non-Cryst. Solids* 120 (1–3): 152–171.

18 Miao, C., Shafrir, S.N., Lambropoulos, J.C. et al. (2009). Shear stress in magnetorheological finishing for glasses. *Appl. Opt.* 48 (13): 2585–2594.

19 Miao, C., Lambropoulos, J.C., and Jacobs, S.D. (2010). Process parameter effects on material removal in magnetorheological finishing of borosilicate glass. *Appl. Opt.* 49 (10): 1951–1963.

20 Lambropoulos, J.C., Miao, C., and Jacobs, S.D. (2010). Magnetic field effects on shear and normal stresses in magnetorheological finishing. *Opt. Express* 18 (19): 19713–19723.

21 Miao, C. (2009). *Frictional Forces in Material Removal for Glasses and Ceramics Using Magnetorheological Finishing*. Citeseer.

22 Lambropoulos, J.C., Xu, S., and Fang, T. (1997). Loose abrasive lapping hardness of optical glasses and its interpretation. *Appl. Opt.* 36 (7): 1501–1516.

23 Kordonski, W. and Jacobs, S. (1996). Model of magnetorheological finishing. *J. Intell. Mater. Syst. Struct.* 7 (2): 131–137.

24 Namba, Y. and Tsuwa, H. (1978). Mechanism and some applications of ultra-fine finishing. *Ann. CIRP* 27 (1): 511–516.

25 Namba, Y. and Tsuwa, H. (1977). Ultra-fine finishing of sapphire single crystal. *Ann. CIRP* 26 (1): 325.

26 Namba, Y., Tsuwa, H., Wada, R., and Ikawa, N. (1987). Ultra-precision float polishing machine. *CIRP Ann. Manuf. Technol.* 36 (1): 211–214.

27 Namba, Y., Ohnishi, N., Yoshida, S. et al. (2004). Ultra-precision float polishing of calcium fluoride single crystals for deep ultra violet applications. *CIRP Ann. Manuf. Technol.* 53 (1): 459–462.

28 Soares, S., Baselt, D., Black, J.P. et al. (1994). Float-polishing process and analysis of float-polished quartz. *Appl. Opt.* 33 (1): 89–95.

29 Arnold, T., Böhm, G., Fechner, R. et al. (2010). Ultra-precision surface finishing by ion beam and plasma jet techniques – status and outlook. *Nucl. Instrum. Methods Phys. Res., Sect. A* 616 (2): 147–156.

30 Haensel, T., Nickel, A., and Schindler, A. ed. (2008). Ion beam figuring of strongly curved surfaces with a (*x*, *y*, *z*) linear three-axes system. Optical Fabrication and Testing. Optical Society of America.

31 Drueding, T.W., Fawcett, S.C., Wilson, S.R., and Bifano, T.G. (1995). Ion beam figuring of small optical components. *Opt. Eng.* 34 (12): 3565–3571.

32 Ghigo, M., Canestrari, R., Spiga, D., and Novi, A. ed. (2007). Correction of high spatial frequency errors on optical surfaces by means of Ion Beam Figuring. Optical Engineering + Applications. International Society for Optics and Photonics.

33 Gailly, P., Collette, J.-P., Renson, L.F., and Tock, J.P. ed. (1999). Ion beam figuring of small BK7 and Zerodur optics: thermal effects. Optical Systems Design and Production. International Society for Optics and Photonics.

34 Mori, Y., Yamamura, K., Endo, K. et al. (2005). Creation of perfect surfaces. *J. Cryst. Growth* 275 (1): 39–50.

35 Weiser, M. (2009). Ion beam figuring for lithography optics. *Nucl. Instrum. Methods Phys. Res., Sect. B* 267 (8): 1390–1393.

36 Schindler, A., Haensel, T., Flamm, D. et al. ed. (2001). Ion beam and plasma jet etching for optical component fabrication. International Symposium on Optical Science and Technology. International Society for Optics and Photonics.

37 Suratwala, T.I., Miller, P.E., Bude, J.D. et al. (2011). HF-based etching processes for improving laser damage resistance of fused silica optical surfaces. *J. Am. Ceram. Soc.* 94 (2): 416–428.

38 Suratwala, T.I., Feit, M.D., and Steele, W.A. (2010). Toward deterministic material removal and surface figure during fused silica pad polishing. *J. Am. Ceram. Soc.* 93 (5): 1326–1340.

39 Suratwala, T., Steele, R., Feit, M. et al. (2012). Method and system for convergent polishing. WO 2012129244 A1.

40 Suratwala, T., Steele, R., Feit, M. et al. (2012). Convergent pad polishing of amorphous silica. *Int. J. Appl. Glass Sci.* 3 (1): 14–28.

41 Suratwala, T., Steele, R., Feit, M. et al. (2014). Convergent polishing: a simple, rapid, full aperture polishing process of high quality optical flats & spheres. *J. Vis. Exp.* 94: doi: 10.3791/51965.

42 Dylla-spears, R., Feit, M., Miller, P.E. et al. (2015). Method for preventing agglormeration of charged colloids without loss of surface activity. US Patent 20,150,275,048.
43 Wong, L., Suratwala, T., Feit, M.D. et al. (2009). The effect of HF/NH_4F etching on the morphology of surface fractures on fused silica. *J. Non-Cryst. Solids* 355 (13): 797–810.
44 Suratwala, T., Wong, L., Miller, P. et al. (2006). Sub-surface mechanical damage distributions during grinding of fused silica. *J. Non-Cryst. Solids* 352 (52–54): 5601–5617.
45 Miller, P.E., Suratwala, T.I., Wong, L.L. et al. ed. (2005). The distribution of subsurface damage in fused silica. Proceedings of SPIE, Volume 5991, 1–25.
46 Feit, M.D., DesJardin, R.P., Steele, W.A., and Suratwala, T.I. (2012). Optimized pitch button blocking for polishing high-aspect-ratio optics. *Appl. Opt.* 51 (35): 8350–8359.
47 Suratwala, T., Feit, M.D., Steele, W.A., and Wong, L.L. (2014). Influence of temperature and material deposit on material removal uniformity during optical pad polishing. *J. Am. Ceram. Soc.* 97 (6): 1720–1727.
48 Suratwala, T., Steele, R., Feit, M.D. et al. (2008). Effect of rogue particles on the sub-surface damage of fused silica during grinding/polishing. *J. Non-Cryst. Solids* 354 (18): 2023–2037.
49 Dylla-Spears, R., Wong, L., Miller, P.E. et al. (2014). Charged micelle halo mechanism for agglomeration reduction in metal oxide particle based polishing slurries. *Colloids Surf., A* 447: 32–43.
50 Suratwala, T., Steele, W., Feit, M. et al. (2016). Mechanism and simulation of removal rate and surface roughness during optical polishing of glasses. *J. Am. Ceram. Soc.* 99 (6): 1974–1984.
51 Degarmo, E., Black, J., and Kohser, R. (2012). *Material and Processes in Manufacturing*. Wiley.
52 Madigan, G. and Middleton, W. (1960). Process for tumble finishing. Google Patents.
53 Reed, J.S. (1988). *Introduction to the Principles of Ceramic Processing*. Wiley.
54 Suratwala, T.I., Steele, W.A., Feit, M.D. et al. (2012). Polishing and local planarization of plastic spherical capsules using tumble finishing. *Appl. Surf. Sci.* 261: 679–689.
55 Moses, E.I. and Wuest, C.R. (2005). The National Ignition Facility: laser performance and first experiments. *Fusion Sci. Technol.* 47 (3): 314–322.
56 Lindl, J.D., Amendt, P., Berger, R.L. et al. (2004). The physics basis for ignition using indirect-drive targets on the National Ignition Facility. *Phys. Plasmas* (1994–present) 11 (2): 339–491.
57 Suratwala, T., Steele, W., Feit, M. et al. (2013). Method and system for polishing solid hollow spheres. Provisional US Patent Application. 61.

58 Lindl, J.D., Landen, O.L., Edwards, J. et al. (2014). Review of the National Ignition Campaign 2009–2012. *Phys. Plasmas* 21 (12): 020501.
59 Chen, K., Cook, R., Huang, H. et al. (2006). Fabrication of graded germanium-doped CH shells. *Fusion Sci. Technol.* 49 (4): 750–756.
60 Letts, S., Myers, D., and Witt, L. (1981). Ultrasmooth plasma polymerized coatings for laser fusion targets. *J. Vac. Sci. Technol.* 19 (3): 739–742.
61 Nikroo, A., Bousquet, J., Cook, R. et al. (2004). Progress in 2 mm glow discharge polymer mandrel development for NIF. *Fusion Sci. Technol.* 45 (2): 165–170.
62 Haan, S., Salmonson, J., Clark, D. et al. (2011). NIF ignition target requirements, margins, and uncertainties: status February 2010. *Fusion Sci. Technol.* 59 (1): 1–7.
63 Clark, D.S., Haan, S.W., Hammel, B.A. et al. (2010). Plastic ablator ignition capsule design for the National Ignition Facility. *Phys. Plasmas (1994–present)* 17 (5): 052703.
64 Chen, K., Nguyen, A., Huang, H. et al. (2009). Update on germanium-doped CH capsule production for NIF: scale-up issues and current yields. *Fusion Sci. Technol.* 55 (4): 429–437.
65 Chen, K., Lee, Y., Huang, H. et al. (2007). Reduction of isolated defects on Ge doped CH capsules to below ignition specifications. *Fusion Sci. Technol.* 51 (4): 593–599.
66 Czechowicz, D., Chen, C., Dorman, J., and Steinman, D. (2007). Investigations to remove domes from plastic shells by polishing. *Fusion Sci. Technol.* 51 (4): 600–605.
67 Vickers, B.L. and Thill, R.E. (1969). A new technique for preparing rock spheres. *J. Phys. E: Sci. Instrum.* 2 (10): 901.
68 Folks, L., Street, R., Warburton, G., and Woodward, R. (1994). A sphere forming and polishing machine. *Meas. Sci. Technol.* 5 (7): 779.
69 Gray, W. and Spedding, F. (1969). A technique for cutting and polishing small metal spheres. *Rev. Sci. Instrum.* 40 (11): 1427–1428.
70 West, S., Martin, H., Nagel, R. et al. (1994). Practical design and performance of the stressed-lap polishing tool. *Appl. Opt.* 33 (34): 8094–8100.
71 Johns, M., Angel, J.R.P., Shectman, S. et al. ed. (2004). Status of the giant magellan telescope (GMT) project. SPIE Astronomical Telescopes+ Instrumentation. International Society for Optics and Photonics.
72 Martin, H.M., Anderson, D.S., Angel, J.R.P. et al. ed. (1990). Progress in the stressed-lap polishing of a 1.8-mf/1 mirror. Astronomy'90, Tucson AZ (11–16 February 1990). International Society for Optics and Photonics.
73 Beaucamp, A. and Namba, Y. (2013). Super-smooth finishing of diamond turned hard X-ray molding dies by combined fluid jet and bonnet polishing. *CIRP Ann. Manuf. Technol.* 62 (1): 315–318.

74 Cheung, C., Kong, L., Ho, L., and To, S. (2011). Modelling and simulation of structure surface generation using computer controlled ultra-precision polishing. *Precis. Eng.* 35 (4): 574–590.

75 Wang, W., Xu, M., Yu, G. et al. ed. (2010). Research on edge control in the process of polishing using ultra precise bonnet on optical elements. 5th International Symposium on Advanced Optical Manufacturing and Testing Technologies. International Society for Optics and Photonics.

76 Gong, J., Xie, D., and Song, J. (2008). Study on influences of processing parameters on polishing spot for curved optical work-piece in bonnet polishing. *J. Yanshan University* 32 (3): 197–200.

77 Li, H., Yu, G., Walker, D., and Evans, R. (2011). Modelling and measurement of polishing tool influence functions for edge control. *J. Eur. Opt. Soc. Rapid Publ.* 6: 1104801–1104806.

78 Booij, S.M., Van Brug, H., Braat, J.J., and Fa, O.W. (2002). Nanometer deep shaping with fluid jet polishing. *Opt. Eng.* 41 (8): 1926–1931.

79 Fang, H., Guo, P., and Yu, J. (2006). Surface roughness and material removal in fluid jet polishing. *Appl. Opt.* 45 (17): 4012–4019.

80 Mori, Y., Yamauchi, K., and Endo, K. (1987). Elastic emission machining. *Precis. Eng.* 9 (3): 123–128.

81 Yamamura, K., Mimura, H., Yamauchi, K. et al. ed. (2002). Aspheric surface fabrication in nm-level accuracy by numerically controlled plasma chemical vaporization machining (CVM) and elastic emission machining (EEM). International Symposium on Optical Science and Technology. International Society for Optics and Photonics.

82 Mori, Y., Yamauchi, K., Endo, K. et al. (1990). Evaluation of elastic emission machined surfaces by scanning tunneling microscopy. *J. Vac. Sci. Technol., A* 8 (1): 621–624.

83 Mori, Y., Yamauchi, Y., Yamamura, K. et al. ed. (2001). Development of plasma chemical vaporization machining and elastic emission machining systems for coherent X-ray optics. International Symposium on Optical Science and Technology. International Society for Optics and Photonics.

84 Kanaoka, M., Liu, C., Nomura, K. et al. (2007). Figuring and smoothing capabilities of elastic emission machining for low-thermal-expansion glass optics. *J. Vac. Sci. Technol., B* 25 (6): 2110–2113.

9

Laser Damage Resistant Optics

The fundamental concepts outlined in this book have thus far been applied generally for improving *surface quality, surface figure, surface roughness, material removal rate,* and fracture yield for various types of optical workpieces (Chapters 2–6), and these concepts have been applied to specific characterizations, processes, and polishing methods (Chapters 7 and 8). These fundamentals can be further extended for end-use applications such as improving the optical- and laser-damage performance of laser optics, improving the resolution of X-ray optics, improving the optical performance of large-telescope optics, improving the cost-effectiveness of manufacturing conventional imaging optics, and reducing the feature size printable with UV lithography optics.

In this final chapter, one of these end-use applications – high-fluence laser optics – is covered. Advances over the past several decades have significantly increased the energy, power, and performance [1] of fusion-class lasers. Many of these advances can be linked to an improved fundamental understanding of the materials science behind optical fabrication and the physics of laser–matter interactions.

High-energy, high-power laser systems are often limited by the optical material's ability to survive high fluence and power. This is ultimately linked to laser-light absorption in the workpiece material, where high absorption leads to material change and, typically, thermal and mechanical damage. There is a variety of ways a material can absorb, including the following:

- Direct linear absorption, when interacting light is greater than the bandgap of the workpiece.
- Bulk defects induced by linear absorption due to sub-bandgap energy states present in the workpiece.
- Multiphoton or excited-state absorption occurring at very high intensities.
- Local extrinsic precursors present in the bulk or on the surface of the workpiece.

Materials Science and Technology of Optical Fabrication, First Edition. Tayyab I. Suratwala.

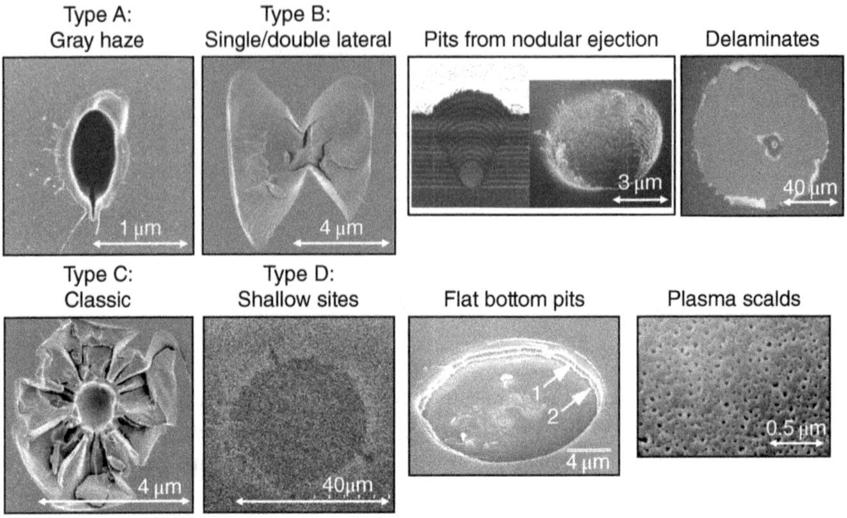

Figure 9.1 Various types of microscopic laser damage observed on high-fluence glass optics. Source: Adapted from Carr et al. 2007 [3] and Genin and Stolz 1996 [4]. Reproduced with permission from authors.

Laser damage is also a function of the many characteristics of the light involved, such as wavelength, energy, pulse characteristics (length, shape, and rate), and polarization (see for example [2]). Some examples of the different morphologies of isolated microscopic laser damage that occurs on glass surfaces with and without coatings are shown in Figure 9.1. Small micron-scale damage alone is generally not problematic; the issue arises with subsequent laser shots, where damage sites can grow, often exponentially, and ultimately shorten the lifetime of the optics (Figure 9.2). The physics of these laser–matter interactions has been a rich field of research and development.

Often the absorption source is not intrinsic to the material itself. Hence, how the workpiece material is manufactured or processed strongly influences its laser-damage resistance. For materials limited by surface damage, optical fabrication and cleaning methods can strongly influence its performance. This is the case for fused-silica glass optics used in fusion-class laser systems, such as National Ignition Facility (NIF), where the glass's surface-damage resistance is significantly lower than its bulk laser-damage resistance. The NIF laser is a flashlamp-pumped Nd: glass laser system that is frequency tripled, delivering 1.8 MJ of 351 nm, 3 ns light to a small target. There are 192 laser beamlines with a ~37 cm square-beam aperture, resulting in a fluence of 8–9 J cm^{-2} (351 nm, 3 ns) through the fused-silica optics [5]. At the time NIF was built, the optics needed to survive an order-of-magnitude increase in laser fluence, relative to predecessor lasers, over an order-of-magnitude-greater optical area

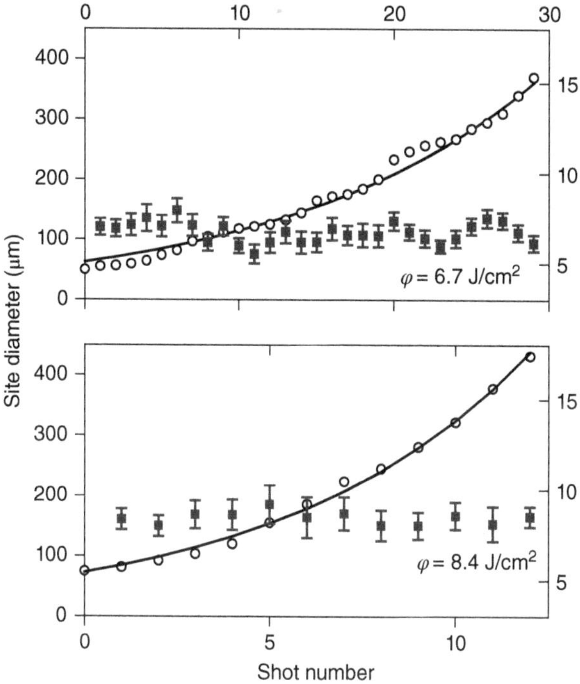

Figure 9.2 Exponential growth observed in damage sites on fused-silica glass as a function of subsequent laser shots. Source: After Spaeth et al. 2016 [1]. Reproduced with permission from Taylor & Francis.

(see Figure 9.3). The strategy to resolve this requirement involved improving the damage resistance of the critical optics and developing an optics-recycle loop to repair optics with a small amount of laser damage at minimal expense (see review by Spaeth et al. [1]).

Figure 9.4 shows the progression of the *surface quality* of fused-silica optics, leading to improvements in laser damage in terms of the number density of the isolated damage sites as a function of laser fluence. Each curve, labeled by year, represents a quality level of the optic workpiece; the further the curve shifts rightward on the plot, the greater the laser-damage resistance. Also overlaid on the plot is the fluence distribution for a 1.8 MJ laser shot. A typical fused-silica optic finished in 1997 would have ~17 000 damage sites per optic. Improvements from 1997 to 2007 dropped the damage to ~90 sites per optic, largely due to dramatic improvements in optical fabrication that reduced tiny surface microfractures using the sub-surface mechanical damage (SSD) and rogue-particle management principles outlined in Section 3.1. Improvements from 2010 to 2014 reduced the damage even further, through the development of a novel, engineered chemical-etch process called the advanced mitigation

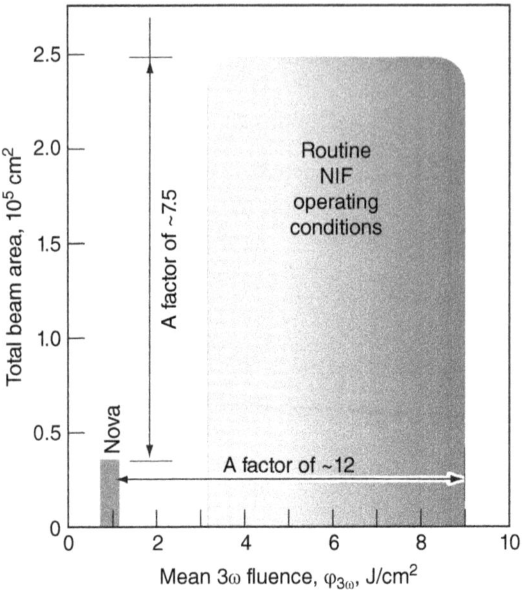

Figure 9.3 A comparison of NIF laser with predecessor laser, with respect to laser fluence and optical area. Source: After Spaeth et al. 2016 [1]. Reproduced with permission from Taylor & Francis.

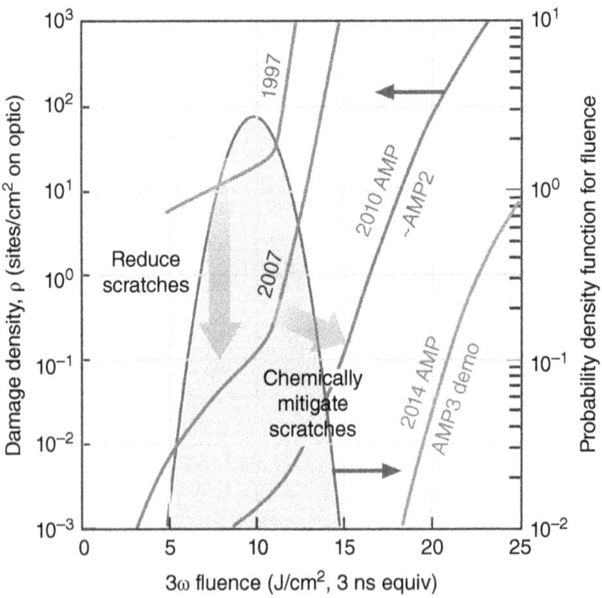

Figure 9.4 Laser damage density on a fused-silica optic surface as a function of 3ω laser fluence. The lines represent the quality level of the optics at various times and the shaded region indicates the relative fluence distribution expected on a 1.8-MJ NIF laser shot. Source: After Spaeth et al. 2016 [1]. Reproduced with permission from Taylor & Francis.

process (AMP), which mitigated the few remaining surface fractures and mitigated the higher-fluence precursors leading to damage [6–8]. These higher-fluence precursors are nanometer-level impurities found in the Beilby layer (Section 4.2) and cleaning residue (Section 3.2.2). The precursors, SSD reduction, and the AMP process are briefly described below.

9.1 Laser Damage Precursors

Through a sustained research program, the surface precursors that lead to laser damage were isolated and identified [1, 7–12]. The absorbing precursors can be summed in four basic categories (see the schematic in Figure 9.5):

(1) *Fracture surface.* Intrinsic (nonimpurity related) silica defects present on a fracture surface. This is strongly linked to SSD creation during the optical fabrication described in Section 3.1.
(2) *Precipitates.* Redeposition of by-products or foreign species on the surface. This is directly related to surface cleaning and was covered in Section 3.2.
(3) *Debris particles.* Settling or impacting debris particles from the external environment, which can occur during manufacture, transport or in use. This is also partly related to surface cleaning, covered in Section 3.2.
(4) *Chemical impurities.* Foreign chemical impurities, typically from polishing compounds such as Ce, present in the polished Beilby layer and fracture crevices during polishing. The characteristics of the Beilby layer and its control are covered in Section 3.2.3.

The challenge in identifying some of these precursors was that often they could not be detected until after laser damage had occurred. At that point, evidence of the precursor is often destroyed. A powerful nondestructive technique called fast photoluminescence (PL) was developed to identify and characterize

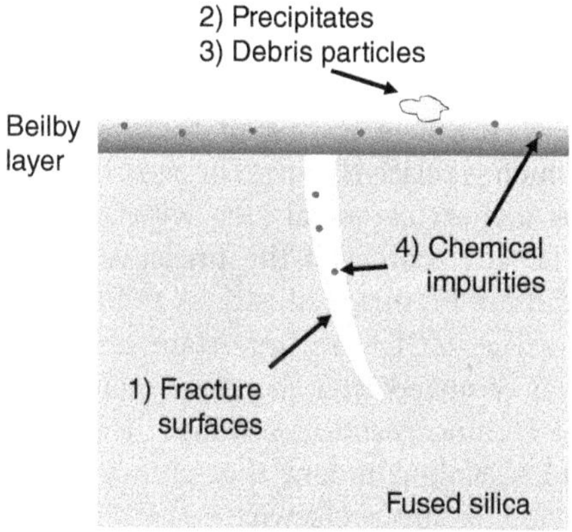

Figure 9.5 Identified surface precursors leading to surface laser damage on fused silica.

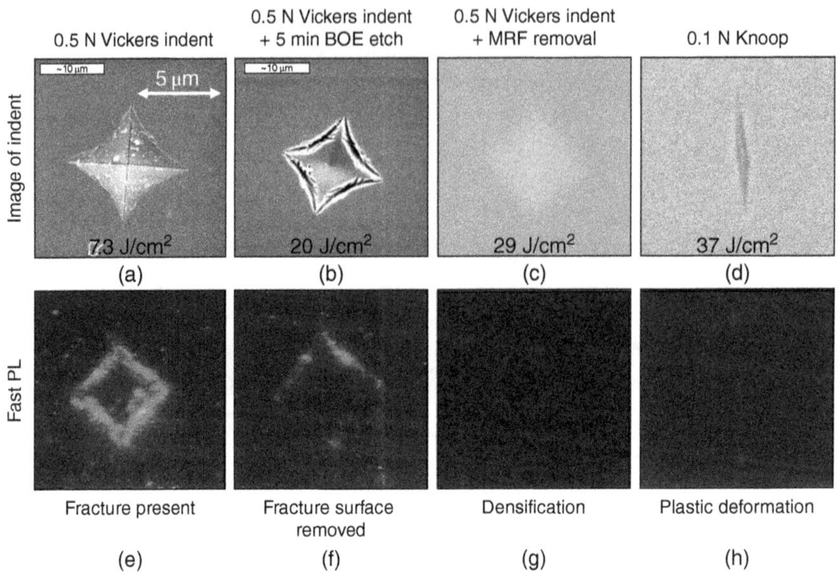

Figure 9.6 (a–d) The images of various surface indents and post treatment and (e–h) corresponding fast photoluminescence image. The laser damage resistance of each feature is listed in the middle. Source: After Laurence et al. 2009 [10] and Miller et al. 2010 [13]. Reproduced with permission from Optical Society of America.

the presence of absorbing defects related to laser damage before the damage event [10]. This is illustrated in Figures 9.6 and 9.7.

Figure 9.6a–d shows microscopic images of a series of isolated indents and subsequent treatments on a polished fused-silica workpiece; Figure 9.6e–h shows the corresponding PL image. The first is a 0.5 N Vickers indent, resulting in both plastic deformation and densification on the surface. The load is sufficiently low so as not to initiate radial cracks (see Figure 3.2); however, some shallow shear-band fractures are present at the boundary of the diamond-shaped deformed region. The corresponding PL images suggest that an absorber is also present around the boundary of the diamond-shape indent. Laser-damage testing of these sites shows damage initiation at 7.3 J cm^{-2}, which is relatively low. The next indent is also a 0.5 N Vickers indent, followed by a short chemical etch with 20 : 1 buffered oxide etch (BOE) (NH_4F:HF). The etch confirms the presence of shear-band fractures on the boundary, and the PL image illustrates reduced absorption; this argues that the fracture surface itself is an important absorbing precursor. Note that not all the PL was removed after the static etch, which was found to be owing to redeposit of etching reaction products (see discussion below). The third indent is a 0.1 N Knoop indent that shows plastic deformation and densification (see evidence given elsewhere [14]), but not fracture. Here the PL image shows

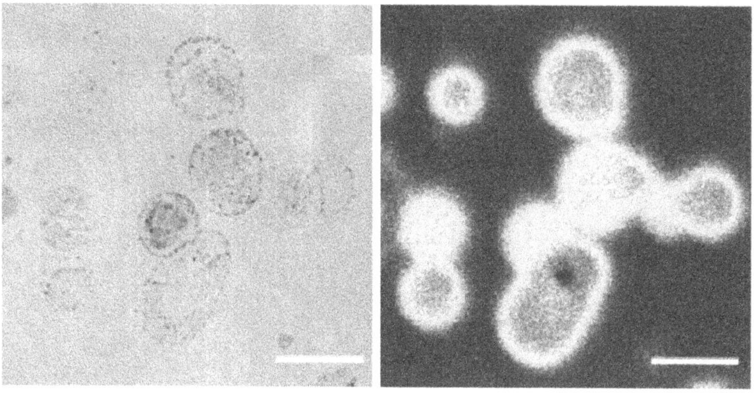

Figure 9.7 (a) An optical micrograph of NaCl crystals left by aerosol deposition. The precipitates congregate in a pattern reflecting the impinging droplets due to the lack of bulk flow at these length scales. (b) Fast photoluminescence (PL) of the same region. Scale bar is 100 μm. Source: After Bude et al. 2014 [8]. Reproduced with permission from Optical Society of America.

no absorption, and laser-damage resistance was significantly increased to 37 J cm^{-2}. The final indent is a 0.5 N Vickers followed by magnetorheological finishing (MRF) polishing to remove the plastic-deformed region, leaving the densified region behind. This is evidenced by a small change in reflectivity from the surface due to a change in index or density and the fact that the image surface was physically flat, as measured by profilometry. Here there is also no observable absorption, and the laser-damage resistance was as high as 29 J cm^{-2}.

The major conclusion drawn is that a very shallow-fracture surface layer contains defects that are absorbing, leading to laser damage, and plastic deformation and densification do not contribute to the damage resistance of fused silica in the fluence range explored. Identifying surface defects that do not contribute to laser damage is also important.

Figure 9.7 shows another microscopic image of an engineered precipitate of NaCl crystals after aerosol deposition [8]. Despite NaCl having a large bandgap (i.e. it should not absorb in the visible), such precipitation still leads to significant PL and hence absorption – and, ultimately, laser damage. This data supports the observation that the presence of dried precipitates tends to laser damage, regardless of their bulk absorption characteristics. A semiquantitative correlation between the integrated PL signal of various isolated surface features and its laser-damage resistance is illustrated in Figure 9.8.

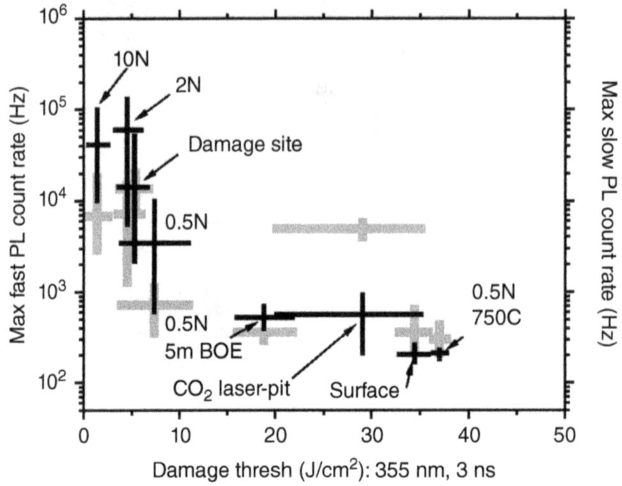

Figure 9.8 A correlation between PL signal versus laser damage resistance. Source: After Laurence et al. 2009 [10]. Reproduced with permission from AIP Publishing LLC.

The chemical, molecular-level impurities present in the Beilby layer and within fracture surfaces are another important precursor leading to laser damage. A common impurity for glass is Ce (as discussed in Section 3.3) from using CeO_2 polishing slurry. To illustrate this, a series of fused-silica glass workpieces was first polished using CeO_2 slurry. Then the workpieces were chemically treated with BOE at room temperature or HNO_3:H_2O_2 (nitric peroxide) at 50 °C. Nitric peroxide is a known strong oxidizer that can preferentially remove impurities without removal of the base SiO_2 glass. Secondary ion mass spectroscopy (SIMS) measurements of these samples showed that Ce surface concentration was significantly reduced with the chemical treatment – in this case, most effectively with nitric peroxide (Figure 9.9a) [13]. Laser testing of the samples showed that the damage number density decreased with a decrease in the Ce remaining on the surface (see Figure 9.9b) [13].

With precursors for laser damage now identified, a critical physics question arose. In the case of chemical impurities such as nanometer-scale Ce, how can this precursor lead to a damage site at the micrometer scale? This ~1000× change in size was difficult to account for with normal thermal transport alone. The physics of this novel laser–matter interaction was explained using the absorption-front model [15], illustrated in Figure 9.10. This model starts with nanometer-scale impurity at the near surface of a fused-silica glass workpiece. When a high-fluence laser pulse contacts the absorber, it heats up to very high temperatures (10^4 K) very rapidly. As the surrounding local bulk-glass material heats up, the bulk material itself also becomes absorbing (a phenomenon known as temperature-activated bulk absorption). Hence, the continued laser

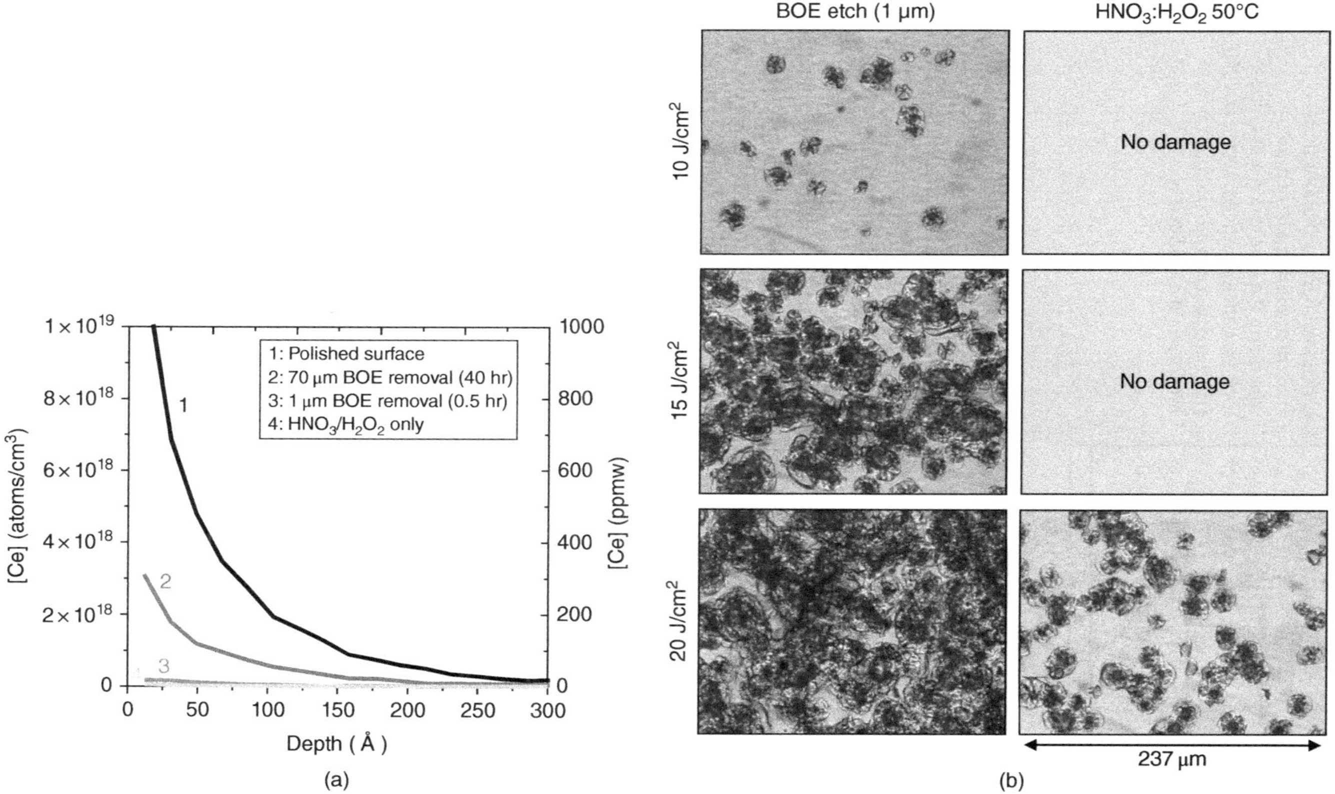

Figure 9.9 (a) Ce surface concentration in the polishing Beilby layer after various posttreatments as measured by SIMS; (b) Optical micrographs of surface after laser-damage testing of various fluences for the BOE static etch treatment and HNO_3:H_2O_2 treatment at 50 °C. Source: Miller et al. 2009 [14]. Reproduced with permission from SPIE.

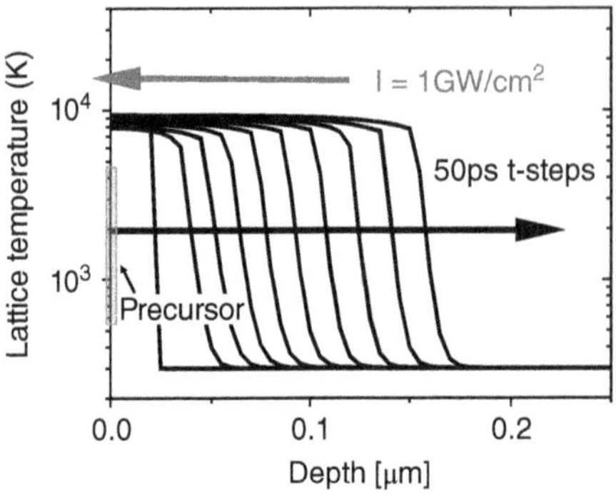

Figure 9.10 Lattice temperature as a function of depth at various time increments during a laser pulse, illustrating the absorption-front model for laser damage. Source: Carr et al. 2010 [15]. Reproduced with permission from American Physical Society.

pulse starts to further heat the bulk to very high temperatures, leading to thermal runaway. This temperature-activated bulk absorption is complimented by temperature-activated thermal conduction, transporting heat farther into the bulk and repeating to cause more temperature-activated bulk absorption. Note that these effects are happening at the picosecond time scale, while the pulse is much longer, in the nanosecond time scale. This repeated domino effect manifests as an absorption front that migrates further into the material than would be effected by normal thermal transport, leading to a significantly larger region of damage. This model also predicts the experimentally observed phenomena that an increase in length of the laser pulse results in a larger resulting damage site due to greater time for the absorption front to propagate [15].

9.2 Reduction of SSD in Laser Optics

As discussed above, advances in the laser-damage resistance of fused-silica optics from 1997 to 2007 were largely a result of improvements in optical fabrication (grinding and polishing) leading to a significant reduction in SSD (i.e. fracture-surface precursors, see Figure 9.4). Developing a fracture mechanics and tribological understanding of SSD creation, as outlined in Section 3.1, and implementing the strategies outlined in Section 3.1.5 contributed to these improvements. Some of the key principles included (i) statistically measuring SSD depth distributions and optimizing removal amounts at each step, (ii) controlling slurry-particle size distribution during grinding and polishing, and (iii) preventing rogue particles from contacting optical surfaces during polishing, handling, and cleaning of the workpiece.

Several optical fabricators developed processes, referred to as the 3ω finishing process, which use these principles, coupled with proprietary techniques, to achieve significant reductions in SSD. A final 3ω-finished fused-silica optic (430 mm × 430 mm) often had extremely tight scratch specifications,

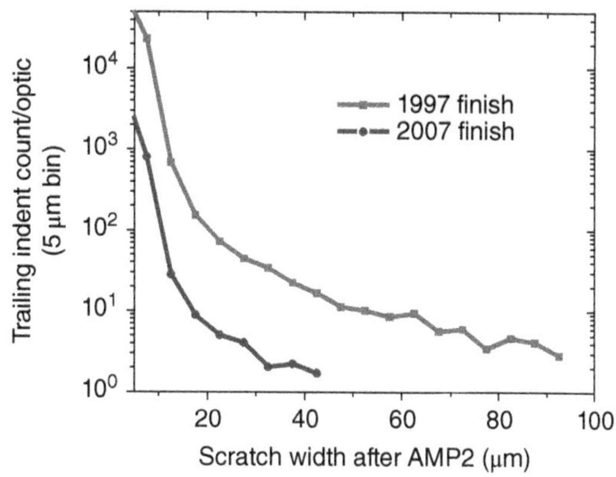

Figure 9.11 Scratch-width distributions on a typical fused-silica optic fabricated using 1997 and 2007 finishing procedures, measured after etching with the AMP2 process. Note the trailing indent count per optic was calculated over 5 μm scratch-width bins. Source: After Spaeth et al. 2016 [1]. Reproduced with permission from Taylor & Francis.

such as no scratches >15 μm in width and <12 scratches with scratch widths between 8 and 15 μm after a 1 μm strip etch (see Table 1.1). A comparison of the scratch distributions of the 1997 and 2007 finishing processes on production fused-silica optics for NIF is shown in Figure 9.11. A 50× reduction in surface-fracture number density for smaller-width scratches and complete elimination of the larger scratch widths were achieved. Note that the scratch width on the x-axis of the figure is after the employment of AMP2 (a 26 μm etch discussed below), which significantly increases the scratch width originally present.

9.3 Advanced Mitigation Process

Despite the dramatic improvements in SSD reduction, the number of damage sites on NIF-sizes fused silica optics was still ~50–100 after an 8 J cm^{-2} fluence laser shot, and additional mitigations were required. This leads to an additional effort to reduce remaining laser-damaging precursors with a postoptical fabrication treatment. At the time of its development, static chemical etching of the surface to remove fracture surfaces led to mixed results, sometimes improving laser-damage resistance and sometimes making it worse. Additional investigation showed that new laser damage precursors were depositing on the surface during etching, consisting of the precipitation of etch reaction products or foreign atomic or molecular impurities. Figure 9.12a shows a scanning electron microscopy (SEM) image of an nonoptimized etched scratch on fused silica, illustrating the presence of precipitation, largely collecting on the edge of trailing-indent etch cusps. Figure 9.12b,c shows that laser damage occurs where precipitates are observed, a key evidence that precipitation is the likely absorbing precursor. Thus if the etch process could be chemically engineered

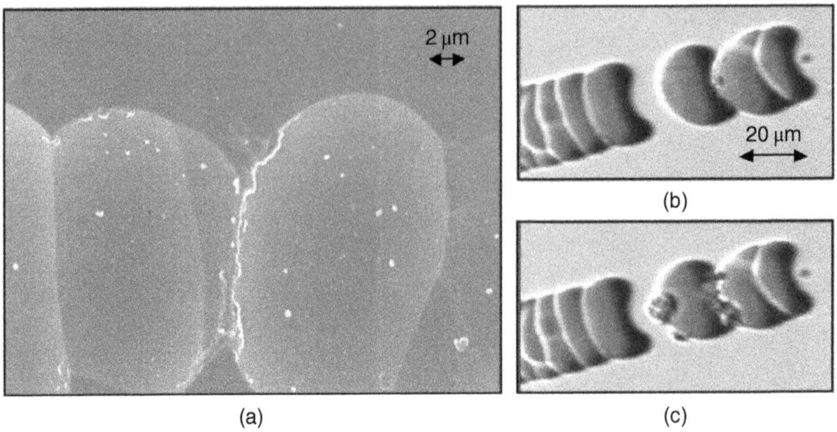

Figure 9.12 (a) SEM image of precipitate that is sometimes observed at the edges of etched open scratches on fused silica. An optical micrograph of a similar etched scratch (b) before and (c) after laser damage initiation (initiation fluence = 19 J cm^{-2}). Source: Suratwala et al. 2011 [7]. Reproduced with permission from John Wiley & Sons.

to prevent redeposition or precipitation, a high laser damage surface might be achieved.

Figure 9.13 shows the progressive improvements of the laser-damage resistance of surface fractures as a function of critical process changes (labeled as "series") during etching and rinsing [7]. The plot shows the fluences at which an individual scratch of a given scratch width will cause damage. The control is shown for Series A, which defines when untreated surface fractures will yield laser damage. Generally, narrow scratches have higher laser-damage thresholds than wider ones, likely because of the amount of absorption present on the fracture surface. Performing a static 20 : 1 BOE etch (Series B) may improve the laser damage resistance of narrow scratches, but generally reduces the damage resistance of wider scratches, often significantly.

Adding ultrasonic agitation during the BOE etch resulted in significantly increased damage resistance. This yielded greater transport of the etch reaction products, namely hexafluorosilicate anion (SiF_6^{2-}), away from the surface, effectively reducing boundary-layer thickness and reducing precipitation. The precipitation products are likely salts formed from the hexafluorosilicate anion, such as ammonium hexafluorosilicate or $(NH_4)_2SiF_6$. Further reduction in precipitation, and hence increased laser damage resistance, was realized by changing the composition of the etch, effectively reducing the concentration of NH_4^+ (Series D and F).

An even further increase in laser damage resistance was realized, particularly for wider scratches, upon etching >20 μm, as in Series H (Figure 9.13). The improvement in laser damage resistance as a function of etched amount

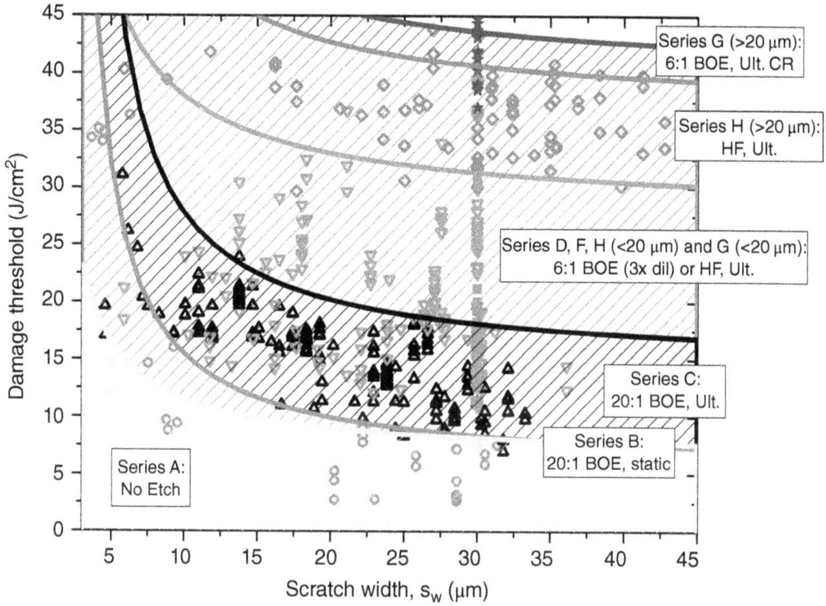

Figure 9.13 Results from a small-beam laser damage test, where damage is measured as a function of scratch width for various etch processes. Ratios indicate relative ammonium-fluoride to hydrofluoric-acid concentration. The solid lines represent the upper bound of the laser damage threshold for each series. Ult. = ultrasonic or megasonic rinsed; CR = processed in a cleanroom. Source: Suratwala et al. 2011 [7]. Reproduced with permission from John Wiley & Sons.

is shown in Figure 9.14a. Etching such large amounts is not needed to remove surface-fracture-absorbing precursors, that are likely <100 nm. Rather, the reason for the long etch times is to alter the geometry of the trailing-indent fracture to improve the mass transport of the etch reaction products. In other words, it is more difficult to remove etch reaction products from narrow fractures at short etch times than wider fractures at long etch times. The morphology changes to a scratch after etching different amounts are shown in Figure 9.14b [7].

A 2D advective and diffusional mass-transport model during etching and rinsing aided in understanding and developing the optimized etch and rinse times for the process. In this model, a parabolic-shaped crack of initial width and depth is isotropically etched, and its geometry will change as outlined in Figures 3.40 and 3.41. The transport of reaction products beyond the boundary layer will be governed by rapid fluid motion incurred by ultrasonic agitation. Within the boundary layer and the etching fracture, the removal of reaction products will be governed largely by diffusion (see Ref. [7] for details of the model).

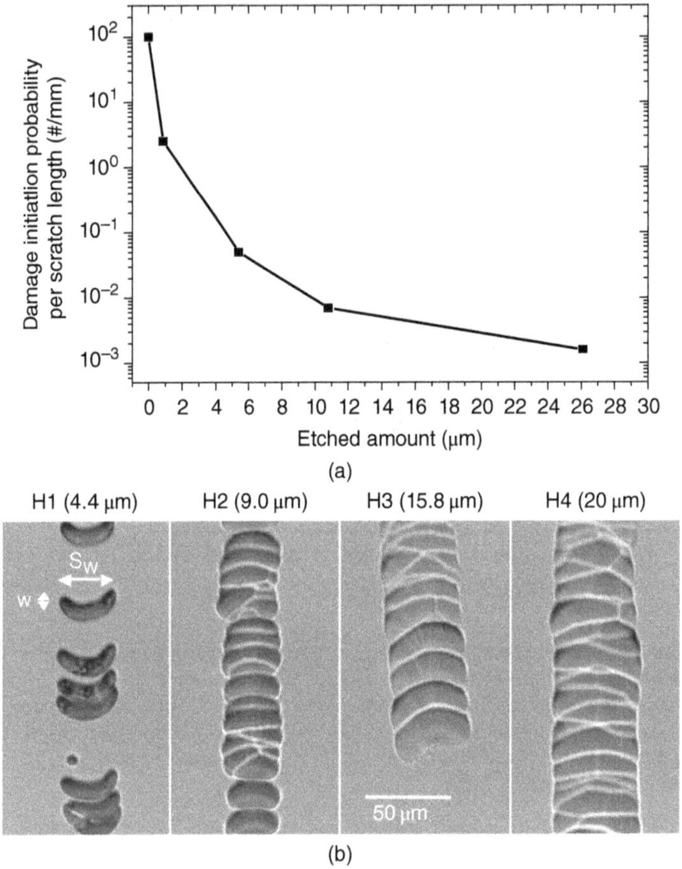

Figure 9.14 (a) The probability of large-beam laser damage initiation (351 nm, 12 J cm^{-2}) as a function of etched amount for variations in AMP treatment; (b) morphology of a scratch after etching different amounts from the surface. Source: Suratwala et al. 2011 [7]. Reproduced with permission from John Wiley & Sons.

Figure 9.15 summarizes the model results for an optimized etch-and-rinse process. The simulation was conducted on 30 μm deep scratch, showing the absolute and relative distribution of SiF_6^{2-} reaction product concentration at various times during the etch, rinse, and spray-rinse processes. Initially, the crack is very narrow at time t_1. During etching, a concentration gradient of reaction products in the boundary layer forms as the crack opening width increases. After etching is completed at time t_3, it is transferred into a submerged water-rinse tank, taking a finite thickness of fluid with reaction products with it. During the submerged rinse, the crack shape does not change (i.e. there is no moving boundary), and the reaction products have sufficient

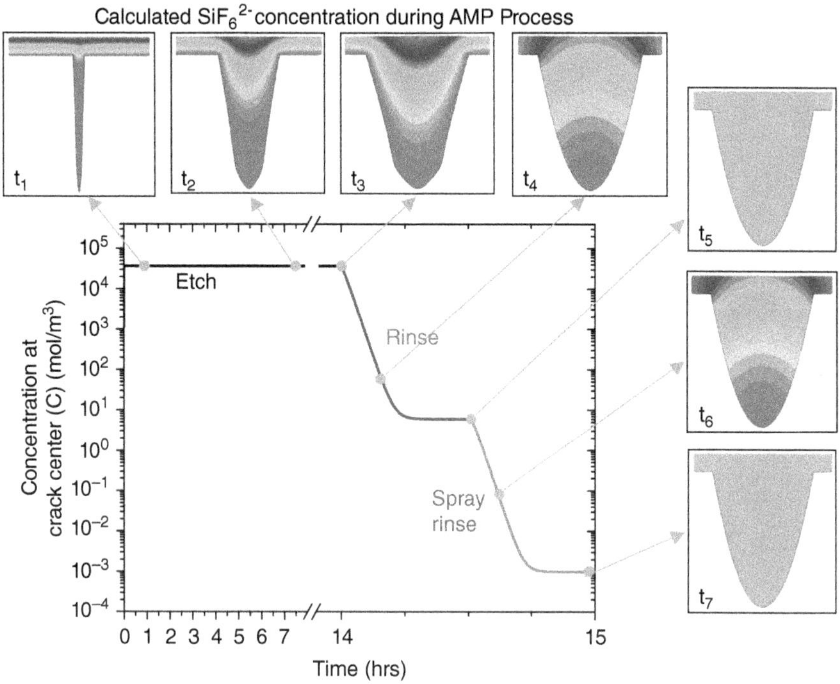

Figure 9.15 Calculated SiF_6^{2-} concentration as a function of process time (during etching, submerged rinsing, and spray rinsing) using a 2D mass-transport model. For selected times, the concentration profiles within the cracks are shown, with the highest concentration at the bottom of the crack and the lowest concentration at the top. The crack is 30 μm deep and the full horizontal scale is 60 μm. Source: Suratwala et al. 2011 [7]. Reproduced with permission from John Wiley & Sons.

time to be removed from the boundary layer. However, because the tank size is finite, all the reaction products will evenly distribute into the rinse tank, setting a finite concentration of reaction products at time t_5. Next the workpiece is transferred to a water spray-rinsing process, where diffusion of the reaction products from the boundary layer continues to match the purity level of the rinse water at time t_7.

A later study showed that other improvements in supply water purity and rinsing methods may additionally reduce the concentration of reaction products and impurities, further reducing nanometer-level precipitation products and resulting in extremely damage-resistant surfaces (as indicated by the 2014 curve in Figure 9.4) [8]. A photo of the production-level AMP facility (using a process called AMP3) is shown in Figure 9.16. An example of the impact of AMP process on scratched fused-silica samples is illustrated in Figure 9.17. Without AMP treatment, a high density of damage sites is observed, while

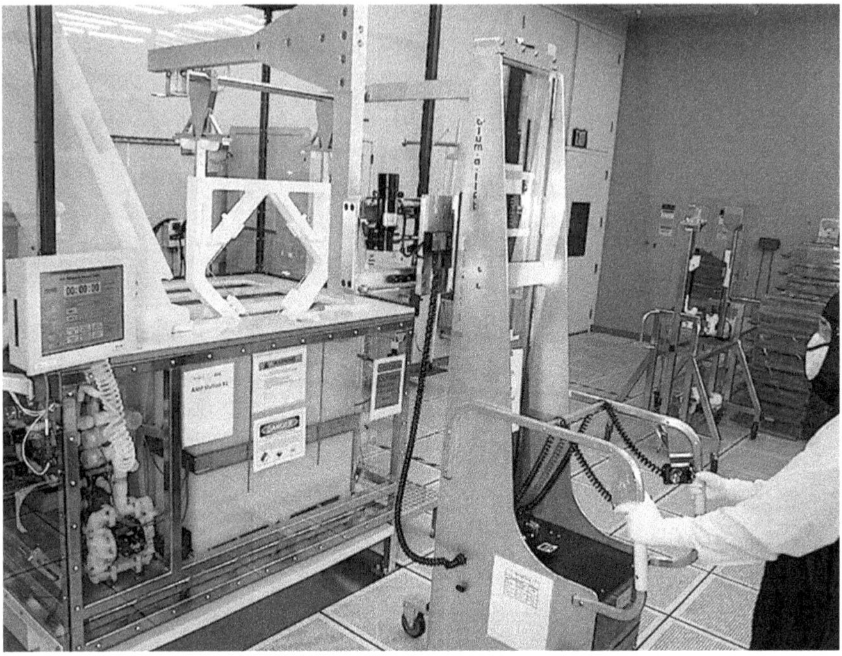

Figure 9.16 Production-scale AMP2/AMP3 process station for chemical treatment of 3ω fused-silica optics to increase their laser damage resistance.

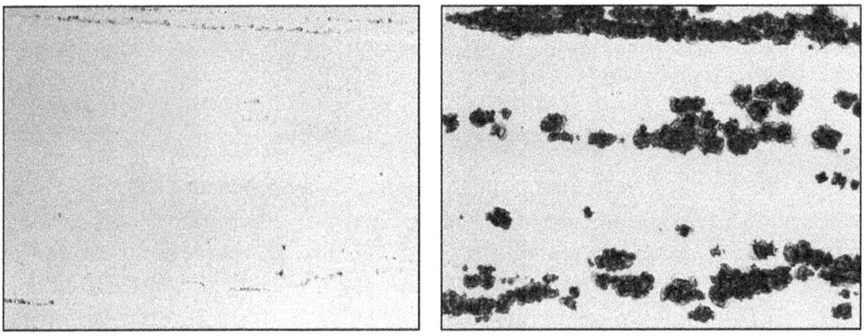

(a) (b)

Figure 9.17 Optical micrographs of scratches generated on fused silica before and after laser shots at 8, 10, and 12 J cm^{-2} (3-cm beam, 3 ns, 351 nm) for (a, b) an untreated sample and (c, d) a scratched sample after the AMP2 etch treatment. Source: Suratwala et al. 2011 [7]. Reproduced with permission from John Wiley & Sons.

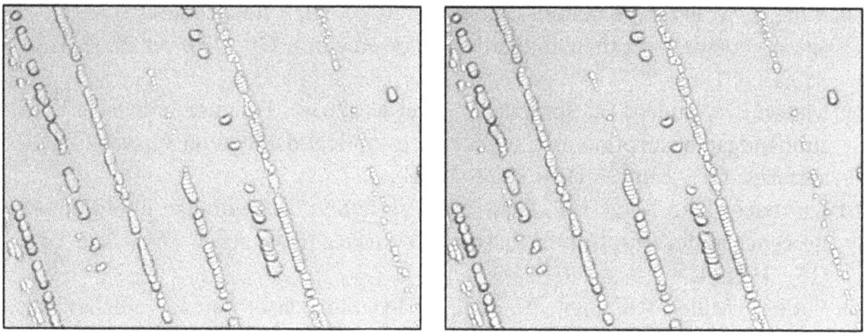

(c) (d)

Figure 9.17 (*Continued*)

with the treatment, no laser damage on scratches was observed at these fluences. The AMP process has been implemented on all 3ω fused-silica optics at NIF, reducing damage to a level that can be effectively managed through the optics-recycling loop [1].

References

1 Spaeth, M.L., Wegner, P.J., Suratwala, T.I. et al. (2016). Optics recycle loop strategy for NIF operations above UV laser-induced damage threshold. *Fusion Sci. Technol.* 69 (1): 265–294.

2 Negres, R.A., Norton, M.A., Cross, D.A., and Carr, C.W. (2010). Growth behavior of laser-induced damage on fused silica optics under UV, ns laser irradiation. *Opt. Express* 18 (19): 19966–19976.

3 Carr, C.W., Matthews, M.J., Bude, J.D., and Spaeth, M.L. ed. (2007). The effect of laser pulse duration on laser-induced damage in KDP and SiO^2. *Boulder Damage Symposium XXXVIII: Annual Symposium on Optical Materials for High Power Lasers*. SPIE.

4 Genin, F.Y., and Stolz, C.J. ed. (1996). Morphologies of laser-induced damage in hafnia-silica multilayer mirror and polarizer coatings. *3rd International Workshop on Laser Beam and Optics Characterization*. SPIE.

5 Spaeth, M.L., Manes, K.R., Kalantar, D.H. et al. (2016). Description of the NIF laser. *Fusion Sci. Technol.* 69 (1): 25–145.

6 Miller, P.E., Suratwala, T.I., Bude, J.D. et al. (2012). Methods for globally treating silica optics to reduce optical damage. Google Patents.

7 Suratwala, T.I., Miller, P.E., Bude, J.D. et al. (2011). HF-based etching processes for improving laser damage resistance of fused silica optical surfaces. *J. Am. Ceram. Soc.* 94 (2): 416–428.

8 Bude, J., Miller, P., Baxamusa, S. et al. (2014). High fluence laser damage precursors and their mitigation in fused silica. *Opt. Express* 22 (5): 5839–5851.
9 Miller, P.E., Bude, J.D., Suratwala, T.I. et al. (2010). Fracture-induced subbandgap absorption as a precursor to optical damage on fused silica surfaces. *Opt. Lett.* 35 (16): 2702–2704.
10 Laurence, T.A., Bude, J.D., Shen, N. et al. (2009). Metallic-like photoluminescence and absorption in fused silica surface flaws. *Appl. Phys. Lett.* 94 (15): 151114.
11 Bude, J., Miller, P.E., Shen, N. et al. (2014). Silica laser damage mechanisms, precursors and their mitigation. In: *Laser-Induced Damage in Optical Materials: 2014. Proceedings of SPIE*, vol. 9237 (ed. G.J. Exarhos, V.E. Gruzdev, J.A. Menapace, et al.).
12 Baxamusa, S., Miller, P., Wong, L. et al. (2014). Mitigation of organic laser damage precursors from chemical processing of fused silica. *Opt. Express* 22 (24): 29568–29577.
13 Miller, P., Suratwala, T., Bude, J. et al. ed. (2009). *Laser Damage Precursors in Fused Silica. Laser-Induced Damage in Optical Materials: 2009.* International Society for Optics and Photonics.
14 Menapace, J.A., Davis, P.J., Steele, W.A. ed. (2005). Utilization of magnetorheological finishing as a diagnostic tool for investigating the three-dimensional structure of fractures in fused silica. *Boulder Damage Symposium XXXVII: Annual Symposium on Optical Materials for High Power Lasers.* International Society for Optics and Photonics.
15 Carr, C., Bude, J., and DeMange, P. (2010). Laser-supported solid-state absorption fronts in silica. *Phys. Rev. B.* 82 (18): 184304.

Index

d

e

f

q

r

t

www.ingramcontent.com/pod-product-compliance
Lightning Source LLC
Chambersburg PA
CBHW072315070426
R18095600001BA/R180956PG41189CBX00002BA/3
* 9 7 8 1 1 1 9 4 2 3 6 8 3 *